Organic Pollutants in Wastewater II
Methods of Analysis, Removal and Treatment

Edited by
Inamuddin
Mohd Imran Ahamed
Shadi Wajih Hasan

Wastewater represents an alternative to freshwater if it can be treated successfully for re-use applications. Promising techniques involve photocatalysis, photodegradation, adsorption, bioreactors, nanocomposites, nanofiltration and membranes. The book focuses on the following topics:

Biological Wastewater Treatment Technologies. Pesticides and their Treatment in Wastewater. Adsorption Removal of Organic Pollutants using Graphene-based Nanocomposites. Reverse Osmosis for the Removal of Organic Compounds from Wastewater. Treatment of Refractory Organic Pollutants using Ionic Liquids. Biohydrogen and Bioethanol Production from Agro-Industrial Wastewater. Methods for the Treatment of Dairy Wastewater. Membrane Bioreactors for the Removal Pesticides and Hormones in Municipal Wastewater. Carbon Nanotubes and their Composites for Treating Industrial and Municipal Wastewater. Low-cost Adsorbents for the Removal of Malachite Green from Water and Wastewater.

Keywords: Wastewater Treatment, Biohydrogen Production, Bioethanol Production, Biological Wastewater, Carbon Nanotubes, Dairy Wastewater, Graphene-based Nanocomposites, Hormones in Wastewater, Malachite Green Removal, Membrane Bioreactors, Nanocomposites, Nanofiltration, Nanomembranes, Nanotubes, Organic Pollutants, Pesticides Removal, Photocatalysis, Photodegradation, Reversed Osmosis, Textile Wastewater

Organic Pollutants in Wastewater II
Methods of Analysis, Removal and Treatment

Edited by

Inamuddin[1,2], Mohd Imran Ahamed[3], Shadi Wajih Hasan[4]

[1] Chemistry Department, Faculty of Science, King Abdulaziz University, Jeddah 21589, Saudi Arabia

[2] Department of Applied Chemistry, Faculty of Engineering and Technology, Aligarh Muslim University, Aligarh-202 002, India

[3] Department of Chemistry, Faculty of Science, Aligarh Muslim University, Aligarh-202 002, India

[4] Department of Chemical Engineering, Khalifa University of Science and Technology – Masdar City Campus, PO Box 54224, Abu Dhabi, United Arab Emirates

Published by **Materials Research Forum LLC**
Millersville, PA 17551, USA

Published as part of the book series
Materials Research Foundations
Volume 32 (2018)
ISSN 2471-8890 (Print)
ISSN 2471-8904 (Online)

Print ISBN 978-1-945291-70-8
ePDF ISBN 978-1-945291-71-5

Distributed worldwide by

Materials Research Forum LLC
105 Springdale Lane
Millersville, PA 17551
USA
http://www.mrforum.com

Manufactured in the United States of America
10 9 8 7 6 5 4 3 2 1

Table of Contents

Preface

Wastewater refers to used water discharged from homes, industries, farms and commercial establishments. It contains various types of contaminants such as dissolved or suspended solids, pathogens, organic and inorganic particles, nutrients, pharmaceuticals, heavy metals and hazardous toxins. Wastewater can be classified into grey water, black water, and yellow water, and can originate from domestic, industrial, commercial, recreational, residential, and institutional activities. It can also originate from surface runoff, stormwater, produced water from oil and gas exploration, industrial cooling towers, and sewer infiltration. Wastewater treatment is the process of removing contaminants from water and generate treated effluents which can be safely discharged to water bodies or can be re-used for other applications such as irrigation, cleaning, and washing. The principal requirement for the treatment of wastewater is the protection of public health and the environment. A successful treatment process must meet the requirement of community concerns, as enforced by governments through the enactment of stringent water quality standards. Several treatment processes have been developed over the past years. A typical wastewater treatment plant consists of several unit operations as each unit is dedicated to the removal of specific contaminants. For example, preliminary treatment processes include screens, grit removal, pre-chlorination and fat and oil removal through which large and floating solids are removed to avoid clogging downstream channels, pumps, and tanks. Primary treatment processes such as primary sedimentation tanks are used to remove suspended solids and some organics while secondary treatment processes such as activated sludge processes, membrane bioreactors, trickling filters, and extended aeration rely on microorganisms to remove colloidal and soluble organic contaminants, nutrients and heavy metals from wastewater. Enhanced treatment systems enable some wastewater plants to produce discharges that contain less nitrogen than plants using conventional treatment methods. Advanced tertiary treatment processes such as advanced oxidation processes, photocatalysis, nitrification-denitrification, adsorption, and sand filters are integrated for further quality polishing. Advanced treatment plants offer higher performance and operational efficiencies with reliabilities that have set new standards in wastewater treatment.

In this volume, topics such as graphene and carbon nanotube-based composites, reverse osmosis, ionic liquids, biofuel production, hormones and pesticides, biological processes, and adsorption are highlighted. For instance, the first chapter discusses the potential use of graphene-based nanocomposites as efficient adsorbents of organics from wastewater. Inorganic particles modified graphene oxide (GO) and reduced GO (rGO), organic components modified GO and rGO, three-dimensional (3D) GO and rGO-based composites are discussed. In the second chapter, a state-of-the-art on distributed models and associated performances of the most recent wastewater treatment methods based on the reverse osmosis (RO) process for the removal of high toxicological organic compounds from wastewater is presented. An interesting topic on the removal of refractory organic pollutants using ionic liquids is highlighted in the third chapter. The solubility of halogenated hydrocarbons (HHCs) in ionic liquids, extraction of HHCs from water using ionic liquids, and different methods used for the degradation of the extracted HHCs are discussed. The fourth chapter focuses on the production of biofuels from agro-industrial wastewater. More specifically, the co-digestion of cassava wastewater with swine residues was evaluated for hydrogen and methane production. The fifth chapter presents an overview of the dairy wastewater and treatment methods. This includes screening, flow and pH balancing and control, separation of fat, oil, and grease, and biological processes. Also, the sixth and the seventh chapters set the fundamentals both positive and negative aspects for the presence of hormones and pesticides in wastewater. They also discuss the available treatment methods employed for the detoxification of pesticides from the contaminated wastewater such as photodegradation, chemical degradation, membrane bioreactors, nanofiltration, and others. The eight chapter addresses the importance of biological processes in the treatment of textile wastewater. It critically evaluates the potential of hybrid wastewater treatment systems such as anaerobic-aerobic membrane bioreactor for the treatment of textile wastewater and various biological processes with respect to their pros and cons. The ninth chapter focuses on the use of multi-walled carbon nanotubes alone and in combination with TiO_2 for the removal of organic contaminants from treated water containing pharmaceuticals, and oil produced wastewater containing BTEX (Benzene, Toluene, Ethylbenzene, and Xylene) and other organic contaminants. The last chapter, chapter ten, investigates the use of low-cost adsorbents for the removal of malachite green from water and wastewater. This volume will be of a great interest to a broad audience of environmental engineers, water professionals, graduate students, and water management operators and technologists working in the field of wastewater treatment.

Finally, our great appreciation goes to all authors for their time and efforts in the preparation of this comprehensive collection of research articles. We extend our great appreciation to the contributing authors for their countless efforts towards the preparation of this book. We may like a need to thank all distributors, authors, and other peoples who conceded consent to utilize their figures, tables, and schemes. However, every effort has been made to get the copyright approvals from the individual proprietors to join reference to the reproduced materials; we might need to offer our sincere articulations of disappointment to any copyright holder if unintentionally their benefit is being infringed.

Inamuddin[1,2], Mohd Imran Ahamed[3], Shadi Wajih Hasan[4]

[1] Chemistry Department, Faculty of Science, King Abdulaziz University, Jeddah 21589, Saudi Arabia

[2] Department of Applied Chemistry, Faculty of Engineering and Technology, Aligarh Muslim University, Aligarh-202 002, India

[3] Department of Chemistry, Faculty of Science, Aligarh Muslim University, Aligarh-202 002, India

[4] Department of Chemical Engineering, Khalifa University of Science and Technology – Masdar City Campus, PO Box 54224, Abu Dhabi, United Arab Emirates

Chapter 1

Adsorption Removal of Organic Pollutants using Graphene-based Nanocomposites: Methodologies and Protocols

Jin-Gang Yu*

College of Chemistry and Chemical Engineering, Hunan Provincial Key Laboratory of Efficient and Clean Utilization of Manganese Resources, Central South University, Changsha, Hunan 410083, P. R. China

Email: yujg@csu.edu.cn

Abstract

The fascinating two-dimensional (2D) carbon-based materials, including both graphene and its oxide, have attracted a great deal of attention as efficient adsorbents for the removal of organic pollutants. In order to improve their removal properties such as efficiency, selectivity and so on, it is important to modify their physio-chemical and chemical properties. Various graphene-based composites developed in recent years have revealed a great potential in the treatment of environmental pollutants. Based on their excellent performance for the removal of organic contaminants, undoubtedly graphene-based nanocomposites will find wider practical applications in environmental pollution control.

Keywords

Graphene, Composite, Removal, Organic Pollutants, Methodology, Protocol

Contents

1.   Introduction

Due to the chemical reagent leaking, industrial wastewater discharging, antibiotics used in aquaculture and medical devices, and leakage of crude oil, water pollution has attracted global concerns in recent years. Organic pollutants including pesticides, dyes, phenols, pharmaceutical antibiotics and so on are among the most important pollution sources in aquatic systems [1]. The magnitudes of organic pollutants in the environment determine the individual limiting loads not only in water but also in soil as well as in the atmosphere. Because the recalcitrant organic pollutants continue to increase in air and wastewater streams all the time, more and more strict environmental laws and regulations have been made to control the situation. Due to their impacts on human health, a number of efforts have been made to remove organic pollutants, especially the key persistent organic pollutants (POPs) which are ubiquitous in the environment as a result of modern industrial processes [2]. The development of various novel methods to destroy these pollutants has been an imperative task.

Recently, research activities have focused on adsorption for the removal of synthetic organic species besides other conventional methods such as osmosis and reverse osmosis [3], ion exchange [4], pervaporation [5], biological method [6], Fenton and photo-Fenton reaction [7], photodegradation [8,9], electrodeposition [10], advanced oxidation [11], ozone treatment [12], ultrasonic treatment [13] and so on. The most attractive feature of adsorption is that this method allows the removal of a wide range of organic chemical substrates with high efficiencies. And adsorbent materials with high-performance for the effective treatment of wastewater are always anticipated.

Owing to its layered structure and atomic thickness, graphene, also named as reduced graphene oxide (rGO), has a huge theoretical specific surface area of around 2630 m^2/g

[14]. Therefore, graphene could be used as a novel and efficient adsorbent for the removal of environmental pollutants including both organic and inorganic substances due to its large specific surface area and the possible multiple interactions such as π-π stacking and electrostatic interaction with the adsorbates. Furthermore, graphene oxide (GO), has also been used as more effective adsorbent for pesticides, antibiotics, organic dyes, metal ions and so on due to the abundant oxygen-containing functional groups such as epoxy (C-O-C), hydroxyl (-OH) and carboxyl (-COOH) groups on its basal planes and edges [15,16]. To improve their adsorption efficiency and selectivity, graphene and GO-based composites have been further functionalized to fabricate various composites which contain versatile parts [17-19]. The graphene-based nanocomposites developed in recent years will find their potential applications in the removal of organic contaminants from the environment in the future.

2.   Reduced graphene oxide-based adsorbents

Development of rGO-based composites is important for a wide range of practical applications such as wastewater treatment, separation and purification, desalination and electrochemical recognition. However, the substantial aggregation of pristine rGO nanosheets would greatly decrease their powerful adsorption capacity [20,21]. To overcome this shortcoming, the modification of rGO is necessary. For example, the facilely thermal treatment of rGO with KOH could realize an alteration in its morphology, and the adsorption of the activated rGO for p-nitrotoluene, naphthalene and phenanthrene was found to be greatly enhanced in comparison to that of as-prepared rGO, indicating that more available adsorption sites such as wrinkles, functional groups and folds would be created to enhance adsorption capacities [22]. The KOH-activated graphene (G-KOH) also showed an excellent adsorption capacity of 194.6 mg/g for ciprofloxacin (CIP) [23]. Graphene-based composites have opened up some new opportunities as effective adsorbents for the removal of various organic contaminants.

2.1   Inorganic particles modified rGO

The elimination of organic matters from the contaminated water remains a longstanding challenge in wastewater treatment. Enhanced adsorption of p-nitrophenol from aqueous solution by aluminum metal-organic framework MIL-68(Al)/rGO composite was found due to the significant changes of the morphologies of the MIL-68(Al) and the increases of its surface area [24]. The removal of methylene blue (MB) by an economical and environmentally friendly graphene/palygorskite/TiO_2 (Gil/Pal/TiO_2) composite showed an adsorption efficiency of approximately 100% [25]. Microwave-assisted TiO_2 reduced

GO exhibited a higher adsorption capacity of 845.6 mg/g than that (467.6 mg/g) of directly TiO_2 reduced GO 467.6 mg/g [26].

Incorporation of magnetic nanoparticle materials into graphene-based composites is interesting. The adsorption of 1-naphthol and 1-naphthylamine onto iron oxide modified rGO was better than iron oxide modified GO, indicating the oxygen-containing groups on the surface of GO might prevent the organic substances from adsorbing [27]. Additionally, the iron oxide modified adsorbents could be easily separated by magnetic separation. By immobilizing the $MnFe_2O_4$ microspheres on rGO, the $MnFe_2O_4$-rGO composite could be used for highly efficient removal of glyphosate from contaminated water [28]. Water-soluble magnetic-graphene (Fe_3O_4@GNs-SO$_3$H) nanocomposite showed high removal efficiency for malachite green [29].

There have been extensive investigations of other carbon nanomaterials in the field of adsorption, and the combination of these carbon nanomaterials has shown a synergistic effect on adsorption. Graphene quantum dots (GQDs) modified graphene exhibited an excellent removal capacity (497 mg/g) and record-breaking adsorption rate (475 mg/g.min) for Rhodamine B (RhB), which were nearly three-fold compared to that of graphene due to the increased surface area [30]. Due to good toughness and hydrophobicity, carbon nanotube (CNT) has been widely used in the removal of pollutants from wastewater [31]. The CNT embedded graphene aerogel (GA) network (GA-CNT) showed an enhanced adsorption capacity and improved mechanical property in comparison to as-prepared GA, and higher adsorption capacities for methylene blue (MB) and methyl orange (MO) could be 100-270 times of its own weight. Especially, the GA-CNT possessed both excellent reusability and mechanical strength [32].

Silica is widely used in the field of adsorption and separation, and the fabrication of silica-based composites has always attracted a great deal of attention from researchers. The stacked interlamination of graphene would be pillared by SiO_2, and the powerful sites in the interlayers would be opened for more efficient adsorption. The silica/rGO (SiO_2/rGO) composite showed rapid and stable adsorption performance for MB and thionine (TH) due to its larger surface area and good aqueous disperse ability [33]. Effective dynamic adsorption of polybrominated diphenyl ethers (PBDEs) and 5-chloro-2-(2,4-dichlorophenoxy) phenol (triclosan) at the surface of rGO-coated silica (SiO_2)-Pt Janus magnetic micromotors could be achieved in a very short time. The rGO-coated SiO_2 functional material-based micromotors also showed outstanding capabilities for both the disposition of persistent organic pollutants (POPs) and the adsorption of phenanthrene, opening up some new opportunities for the efficient environmental remediation [34,35]. Besides the high adsorption capability for MB, Fe_3O_4-graphene@mesoporous SiO_2 nanocomposite can be easily separated from the solution by an external magnet, for use as a novel sorbent for the removal of organic pollutants from large volumes of aqueous solutions [36].

To improve removal efficiencies, metal oxides are often anchored to graphene-based materials. SnO_2 quantum dots (QDs) decorated rGO exhibited excellent removal capacity and fast adsorption rate for MB, good separation efficiency of 94% for the binary solution of MB/MO and 76% for MB/RhB mixture [37]. Specifically, Mn_3O_4 nanoparticle anchored rGO also showed enhanced efficiency for 1-naphthylamine [38], and magnetic rGO showed an excellent reusability, photodegradation ability, fast adsorption rates besides extraordinary adsorption capacity [39-45]. The metal oxide-rGO hybrids have opened up an opportunity for fabrication of highly efficient graphene-based adsorbents [46-48].

2.2   Organic components modified rGO

Due to the conjugate aromatic structure of rGO, the π-π interactions between rGO and organic dyes play a key role in the adsorption of organic contaminants. The maximum adsorption capacities of L-cysteine modified rGO (rGO-Cys) for anionic indigo carmine (IC) and cationic neutral red (NR) were 1005.7 and 1301.8 mg/g, respectively [49]. Besides normal organic molecules, smart RNA could be used as an aptamer and covalently immobilized on rGO for the removal of contamination of peptide toxins in drinking water [50].

2.3   Three-dimensional (3D) rGO-based composites

As one of the most promising methodologies for fabrication of bottom-up nanomaterials, three-dimensional (3D) graphene-based composites have been recognized as one of the most active research fields in recent years. In general, the basic structural features of 3D graphene-based composites exhibit large surface area and well-defined porous structure, qualifying them to be ideal adsorbents for pollutant removal with excellent recyclability. Undoubtedly, porous 3D graphene-based materials with hierarchical structures have generated great interest among researchers [21].

Three-dimensional graphene foam (3D-GF) possesses high porosity, good hydrophobicity, and excellent thermal stability. Compared to the as-prepared GF, 3D-GF showed better adsorption performance for oil in separating oil-water mixtures [51]. Due to the interconnected 3D porous network and the synergistic effects of the assembled graphene sheets and Fe_3O_4 nanoparticles, the graphene and Fe_3O_4 (G/Fe_3O_4) composite exhibited excellent adsorption capacity for Rhodamine B (RhB) with a maximum removal ratio of 99.6% [52]. Graphene/polyester staple composite (GPSC) could be used for effectively removing oils and organic solvents, especially crude oil [53]. The ultrasonic-microwave assisted synthesis of rGO modified melamine foam (RGMF) would shorten the reduction time and enhance the firmness of anchoring rGO. In addition, the excellent selective adsorption ability, good utilization and high adsorption capacity for oils and organic solvents from aqueous solutions would improve its wider application in practice [54,55].

Another 3D porous structure, graphene sponges (GS), has also attracted extensive interest as efficient adsorbents. However, the preparation GS using toxic reducing reagents might cause additional environmental pollution [53]. The hydrothermal reduction of GO by glucose to form GS could avoid potential contamination, and the obtained GS could be easily regenerated by evaporating or burning. It showed excellent adsorption properties for oils and various organic solvents [56]. In addition, the hydrophobic nature of graphene would hinder the aqueous dispersion of GS. The attachment of a super-hydrophilic melamine skeleton onto rGO has been achieved, and the rGO-melamine-sponge (rGOMS) possessed excellent adsorption capacities of 286.5 and 80.51 mg/g for MB and orange G (OG), respectively [57]. Especially, the rGOMS could be easily regenerated, and 95% adsorption capacity is possible even after 10 cycles. The rGOMS composite sponge also exhibited an enormous adsorption capacity of 99.0 g/g for diesel oil [58].

Slightly reduced GO (srGO) with large meso-and macroporosity exhibited high oils and organic solvents adsorption capacities (up to 319 times its own weight) on its hydrophilic surface due to its porous nature [59]. Especially, a porous and green 3D polydimethylsiloxane (PDMS)-rGO sponge exhibited high adsorption selectivity for petroleum products, organic solvents and emulsified oil-water mixtures due to its hydrophobic and oleophilic properties [60]. Highly macroporous graphene-based 3D structures with good adsorption selectivity might be the next-generation adsorbents for removal of organic pollutants.

More facile and greener approaches are always expected. With the assistance of glutathione, the straightforward assembly of 3D nitrogen and sulfur co-doped graphene hydrogels (N/S-GHs) was carried out. The as-obtained N/S-GHs possessed superior adsorption properties for several organic dyes including MB, malachite green (MG) and crystal violet (CV) [61]. A facile, environment-friendly and mild *in situ* reduction-assembly approach was developed for constructing rGO hydrogel, the prepared rGO hydrogel also showed excellent adsorption performance for organic dyes [62]. A newly developed facile one-pot hydrothermal process proved to be feasible for fabricating 3D porous nitrogen-doped rGO hydrogels [63]. The spinach assisted the reduction of GO was green and facile [64]. With polystyrene (PS) particle as a sacrificial template, the 3D hierarchical porous rGO aerogel (HrGOA) prepared by a "fishing" process exhibited highly efficient adsorption performance for removal of oils and organic solvents due to its tunable porous architecture [65].

With the development of the research on rGO-based composites, various rGO-based adsorbents are constantly emerging (Table 1). The successful use of rGO-based adsorbents in water treatment is only a matter of time.

Table 1: Adsorption of organic pollutant by rGO-based adsorbents.

rGO-based adsorbents	Adsorption properties	Ref.
rGO	Adsorption capacity of 323 mg/g for Ciprofloxacin (CIP)	[66]
rGO	Adsorption capacity of 13.52 mg/g for malachite green (MG)	[67]
rGO	Removal efficiencies of 98-100% for acid orange 7	[68]
rGO	Adsorption capacity of 44.2 mg/g for bisphenol A (BPA)	[69]
rGO modified Nickel chromium layered double hydroxides	Adsorption capacity of 312.5 mg/g for MO	[70]
Ni@rGO	A removal efficiency of 99.5% for Rhodamine-B (RhB) even after 16 times recycling	[71]
Magnetic rGO/Zeolitic imidazole framework	Adsorption capacity of similar to 300 mg/g for benzotriazole (BTA)	[72]
rGO/zeolitic imidazolate framework hybrid composite	An ultra-high adsorption capacity for MG (3000 mg/g)	[73]
Magnetite-rGO aerogels	Excellent adsorption capacity for oils, organic solvents, and dyes	[74]
rGO-silver nanocomposite (rGO@Ag)	Unusual adsorption capacity as high as 1534 mg/g for organohalides	[75]
Cerium oxide modified rGO	Adsorption capacities of 62.33 and 41.31 mg-As/g at pH 4.0, respectively for arsenate	[46]
Tb$_4$O$_7$ complexed rGO	Almost 100% removal of RhB	[76]
Aluminum and iron-doped rGO	Selectively removal of trivalent and pentavalent methylated arsenic pollutants at various pH	[77]
rGO/wheat straw composite	Higher removal efficiency of phenanthrene	[78]
Poly(4-styrenesulfonic acid-co-maleic acid) -sodium-modified magnetic rGO	Excellent adsorption capacity for basic fuchsin (BF), crystal violet (CV) and MB	[79]
Exfoliated graphite nanoplatelets	Adsorption capacity of 850 mg/g for BPA	[80]

Nanoporous graphene	Adsorption capacities of 118.83,123.45 and 125.36 mg/g for benzene, toluene and xylenes, respectively	[81]
Magnetite/rGO nanocomposite	Adsorption capacity of 39.0 mg/g for MB	[82]
Magnetite/rGO nanocomposite	Adsorption capacity of 144.9 1/mg for MB	[83]
Few layered graphene sheets	Highly efficient adsorption selectivity and capacity for petroleum products and organic solvents such as ethanol, cyclohexane, and chloroform	[84]
Ce-Fe oxides dispersed on graphene	Adsorption capacity of 179.5 mg/g for congo red (CR)	[85]
PVC@graphene-polyaniline fiber bundles	The adsorption capacity of 40.0 mg/g for CR.	[86]
3D rGO macrostructure	Adsorption capacities of 300.60 and 50.61 mg/g⁻ for MB and acid red 1 (AC1), respectively	[87]
Magnetic rGO	Adsorption capacities of 63.96 and 48.74 mg/g for 4-n-nonylphenol (4-n-NP) and BPA	[88]
3D Ni-carbon-rGO hybrid	Adsorption capacity of 21.1 mg/g for RhB	[89]
Magnetic polypyrrole composite (Fe3O4@PPy/RGO)	Adsorption capacity of 270.3 mg/g for MB	[90]
Few-layered rGO	Adsorption capacities in the range 11.40~ 18.60 mg/g for humic acid (HA)	[91]
Magnetic rGO	Adsorption capacity of 186.40 mg/g for RhB	[92]
Graphene–CNT aerogels	High recycling performance and maintaining of adsorption capacity (~80% of the first maximum) after 8 cycles for gasoline	[93]
Sulfonated graphene nanosheets	Adsorption capacity of 6.4 mmol/g for 1-naphthol	[94]
Graphene-FA composite	Removal efficiency of 72.4% for methylmercury	[95]

3.　GO-based adsorbents

Because of their outstanding properties like tremendous adsorption capacity, high removal efficiency, facile process, convenient operation and low cost, GO-based composites have found wider potential application in adsorption and removal of organic contaminants from wastewater than that of rGO-based adsorbents [96-98]. The aqueous GO suspension produced directly from chemical exfoliation of flake graphite by oxidation exhibited enormous removal capabilities of 1020.848, 947.294, 1049.643, and 1036.739 mg/g for p-nitrophenol, o-nitrophenol, m-nitrophenol, and 1-naphthol, respectively [99]. Electrochemical exfoliation was a newly developed technology, and the maximum adsorption capacity of the electrochemically exfoliated GO for MB was about 511.7 mg/g [100]. Few-layered GO also showed stronger adsorptive interactions for 17 β-Estradiol (E2), and the adsorption capacity of 149.4 mg/g was the highest value for E2 adsorption of the reported adsorbents [101]. Even for colored wastewater, GO showed high adsorption capacities of 1429, 1250, and 476 mg/g for Basic blue 41 (BB41), basic red 18 (BR18), and basic red 46 (BR46), respectively [102].

Usually, a high adsorption capacity for the removal of various kinds of organic pollutants from aqueous solutions could be obtained due to the high density of oxygen-containing functional groups on GO. However, the lack of selective removal ability for target organic pollutant and lower removal efficiencies need to be addressed. To solve these obvious problems, various methodologies have been developed.

3.1   Inorganic particles modified GO

Although monodisperse GO microspheres showed rapid and high removal efficiency for perfluorooctane sulfonate (PFOS), the Mg^{2+} modified GO composite exhibited significantly improved PFOS removal efficiency through the introduced bridging and interaction between Mg^{2+} and PFOS [103]. In comparison with as-prepared GO, the amine functionalized GO (AF-GO) also exhibited good adsorption capacities for acid Blue 92 (AB92), Acid Red 14 (AR14), and direct green 6 (DG6) from their multicomponent (binary) systems [104]. A silica-crosslinked GO membrane with unique adsorption capability for neutral organic molecules was prepared by soaking a layer-stacked GO film in saturated silica solution. The negatively charged GO membrane was found to more efficiently remove glucose and sucrose (84 to 90%) than the tested negatively charged ionic species (trisodium citrate/TSC) [105].

SnO_2 functionalized GO (SOG) nanocomposite exhibited fast adsorption rates for several dyes such as RhB, MB, methyl green (MG) and methyl red (MR) with the maximum adsorption capabilities of 115.4, 72.2, 76.5 and 108.3 mg/g, respectively. Especially, the generation of the adsorbent could be easily carried out by ethanol washing [106].

Magnetic GO (MGO) was used to simultaneously remove humic acid/fulvic acid (HA/FA), and the regenerated MGO still showed high efficiency, providing the possibility of application of MGO in real wastewaters treatment [107]. A magnetic calcium silicate GO (MGSi) composite adsorbent, Fe_3O_4 nanoparticles coated with calcium silicate and then firmly immobilized on the surface of the GO, exhibited highly selective adsorption ability toward alkaline dye and acridine orange (AO) [108]. Magnetic porous silica-GO ($Fe_3O_4@mSiO_2/GO$) hybrid composite showed a maximum adsorption capacity of 1548.78 mg/g for p-nitrophenol at around room temperature, and the property of easy separation by an external magnetic force made it very attractive [109]. By tuning the dimensions and geometries of Fe_3O_4, the sulfonated GO-Fe_3O_4 hybrids could easily regenerate and exhibited impressively high removal rates (nearly 100%) for MB and Rhodamine B (RhB) [110]. It is possible for the practical application of magnetic GO hybrids in environmental management due to the newly developed ultrasound-assisted co-precipitation method [111] and click reaction [112].

The removal of volatile organic compounds (VOCs) has been an interesting research field. In contrast with pristine activated carbon nanofiber and conventional activated carbon fiber, GO/carbon nanofibers with well-developed mesoporous structure due to the embedded GO exhibited improved benzene and butanone adsorption capacities [113].

3.2   Organic molecules modified GO

Taking advantage of the strong adhesive abilities of mussel adhesive proteins (MAPs) and dopamine, the linkage of amino-terminated poly(sodium styrene sulphonate) (NH_2-PSPSH) with GO- poly(amide acid) (GO-PDA) *via* Michael addition reaction was carried out [114]. The GO-PDA-PSPSH showed increased removal efficiency and faster adsorption speed for organic dyes methylene blue (MB) than pristine GO. The polyaniline (PANI)-coated GO was further doped with $SrTiO_3$, which could be used for efficient removal of both cationic dyes such as MB and an anionic dye such as MO [115].

Besides mussels, natural polysaccharides composed of bonded monosaccharide units in long chains possess good ability to bind with other compounds and exhibit great potentials in adsorption. The polysaccharides (such as inulin, xylan or κ-carrageenan) functionalized GO composites showed higher adsorption abilities for MB, rhodamine 6G (Rh6G), acid fuchsin (AF) and orange II (OII) in comparison with as-prepared GO [116].

A novel β-cyclodextrin (β-CD) functionalized GO-isophorone diisocyanate (IPDI) composite (GO-IPDI-CDs) exhibited an adsorption capacity of 83.40 mg/g for MO [117]. And the magnetic β-CD functionalized GO acid showed several times, even hundreds of times of adsorption capacities higher than that of the reported carbonaceous materials for fuchsin and rhodamine 6G [118]. A versatile adsorbent, chitosan modified GO (CS-GO), exhibited pH-tunable surface charge and morphology, resulting in a flexible and tunable adsorption performance for MB [119].

Multifunctional hyperbranched polyamine modified GO (HPA-GO) showed an equilibrium adsorption capacity of 740.7 mg/g for MB due to the π-π stacking interactions and electrostatic attraction [120]. Besides its excellent adsorption performance for MB, GO-perylene bisimides-containing poly(N-isopropyl acrylamide) hybrid (TGO) (GO-PBI-PNIPAM) also showed reversible temperature-dependent self-assembly and disassembly properties [121].

3.3   3D GO-based composites

Due to the high porosity, increased surface area, remarkable adsorption capability, diverse surface modified functional groups and environmental compatibility, graphene aerogel has been further applied as a promising adsorbent [122]. N-doped GO aerogel showed excellent adsorption capacity because of its high specific surface area and superhydrophobic nature. The good adsorption recyclability has added the advantage of this material making it more practically useful [123]. The fabricated 3D GO aerogels-mesoporous silica (GOAS-MS) framework exhibited high removal efficiencies of 68.6, 86.6, 91.1 and 94.7% for phenol, catechol, resorcinol, and hydroquinone, respectively due to its narrow mesopore size distribution, high surface area of 1000.80 m^2/g, and hierarchical macro- and mesoporous structures [124]. The as-prepared GO/polyethyleneimine (PEI) hydrogels exhibited high removal rates and adsorption capacities for both MB and RhB [125]. And the konjac glucomannan (KGM)-GO hydrogel showed much enhanced adsorbing performance for both MO and MB [126]. Besides the improved adsorption capacity for organic contaminants, the GO/polyacrylamide hydrogel showed superior mechanical strength, high toughness, prominent self-healing ability and fatigue resistance [127]. This new class of strong and fully physically cross-linked hydrogels might be promising as efficient recyclable adsorbents in the future.

Multifunctional 3D materials are always anticipated. Magnetic GO/poly(vinyl alcohol) (PVA) composite gels were developed as a convenient and remarkable adsorbent for MB and methyl violet (MV). Simultaneously, *in situ* growth of noble metal catalyst such as Pt, Au and so on onto the magnetic GO/PVA provides additional catalytic performance [128].

As previously mentioned, inorganic nanomaterials could act as pillared components for preventing the aggregation of graphene. The montmorillonite-pillared GO exhibited higher removal efficiencies toward MO and MB [129]. As time goes on, GO-based composites might be one of the most attractive adsorbents in wastewater treatment (Table 2).

Table 2: The adsorption of organic pollutants using GO-based composites.

GO-based adsorbents	Adsorption properties	Ref.
GO	Adsorption capacity of 27.16 mg/g for MG	[67]
GO	Adsorption capacities of 16.83 and 63.69 mg/g for MO and basic red 12 (BR 12), respectively	[130]
GO	Adsorption capacity of 49.26 mg/g for BPA	[131]
GO	Adsorption capacity of 148.36 mg/g for RhB	[132]
GO suspensions	Removed efficiencies of 75 and 30% for diclofenac (DCF) and sulfamethoxazole (SMX), respectively	[133]
GO	Adsorption capacities of 67 and 116 mg/g for propranolol (PRO) and atenolol (ATL), respectively	[134]
Hybrid copper oxide nanoneedles on GO (GO-CuONNs)	Strong electrostatic interaction is necessary for the adsorption	[135]
Polyacrylonitrile yarn waste/GO composite	Removal efficiency was still nearly 95% for MB	[136]
GO/chitin nanofibrils (GO-CNF) composite foam	Removal efficiency was still nearly 90% after 3 cycles for MB	[137]
GO-loaded agarose hydrogel	Adsorption capacity of 355.9 mg/g for Nile red	[138]
Macroporous polystyrene/ GO composite monolith	Adsorption capacity of 197.9 mg/g at pH 6 for tetracycline	[139]
4-Aminothiophenol functionalized GO	Adsorption capacity of 763.30 mg/g for MB	[140]
3-Aminopropyltriethoxysilane functionalized GO	Adsorption capacity of 676.22 mg/g for MB	[140]
Poly(acrylic acid) functionalized magnetic GO	Adsorption capacity of 291 $mg \cdot g^{-1}$ for MB	[141]
Anionic polyacrylamide/GO aerogels	Adsorption capacity of 1034.3 mg/g for basic fuchsin	[142]
GO decorated $Fe_3O_4@SiO_2$	Adsorption capacity of 44.05 mg/g for MB	[143]
β-CD functionalized GO	Better excellent adsorption capacity for fuchsin acid than MB and MO	[144]
β-CD functionalized magnetic GO	Avery high adsorption capacity of 1102.58 $mg \cdot g^{-1}$ for p-phenylenediamine	[145]
Magnetic GO	Removal efficiencies of 7-19% to 78-98% for disinfection by-product (DBP) precursors	[146]
Tannic acid functionalized GO	Adsorption capacity of 201 mg/g for RhB	[147]
Magnesium hydroxide $(Mg(OH)_2)$-GO composite	Adsorption capacity of 779.4 mg/g for MB	[148]
GO/polypyrrole composite	Adsorption capacities of 2.14 and 6.74 mmol/g for phenol and aniline, respectively	[149]
Few-layered GO	Adsorption capacities of 73.36~ 82.91 mg/g for humic acid (HA)	[91]
GO-PBI-PNIPAM	Adsorption capacity of 568 mg/g for MB	[121]
GO and β-FeOOH bundles	Removal efficiency of 60% for congo red in 10 min	[150]

Magnetic citric acid functionalized GO	Adsorption capacity of 112 mg/g for MB	[151]
Polydopamine (PDA)-GO	Adsorption capacity of up to 2.1 mg/g for MB	[152]
Amino-functionalized Fe_3O_4 (NH_2-Fe_3O_4) modified GO	Adsorption capacities of 167.2 and 171.3 mg/g for MB and Neutral Red (NR), respectively	[153]

4. Adsorption mechanisms

Nanosized graphene-based composites are being considered as a class of highly efficient adsorbents for wastewater treatment. However, the adsorption affinity of graphene-based composites for the existing contaminants should be improved to decrease the costs. Therefore, it's necessary to investigate the graphene adsorption behaviors and uncover the adsorption mechanisms between graphene and the adsorbates [154].

A comparative study revealed that the adsorption of benzene and toluene onto rGO was more favorable than CNT. It indicates that the specific surface area and pore's volume of rGO has a significant contribution to the adsorption [155]. The stability of the pollutant after adsorption also plays a key role, silicon doped graphene (SiG) showed adsorption selectivity for the removal of organic arsenic pollutants in their trivalent and pentavalent oxidation states [156].

The presence of oxygenated functional groups such as hydroxyl and epoxy groups in GO and nitrogen-containing functionalities such as imine groups and amine groups are beneficial to its adsorportion properties [157, 158]. A double-oxidized GO (DGO) with high BET specific surface area of 51.2 m^2/g and a high density of oxygen-containing functional groups was developed, and a maximum adsorption capacity of 704 mg·g^{-1} for acetaminophen (ACT) was achieved [159].

The weak interactions between graphene-based adsorbents and the organic contaminants such as van der Waals force, hydrogen bonding and π-π interaction all contribute to the adsorption. In addition, GO-CuONNs exhibited a higher adsorption rate for cationic dyes such as Coomassie brilliant blue (CBB) and MB than anionic dyes such as congo red (CR) and amido black 10B (AB), indicating a strong electrostatic interaction is necessary for the adsorption [135]. Besides the π-π electron donor-acceptor and van der Waals interactions, hydrophobic interactions would also dominate the adsorption of aliphatic synthetic organic compounds (SOCs) by graphene-based adsorbents [160]. The adsorption of three aromatic organic compounds (AOCs) including aniline, nitrobenzene, and chlorobenzene by GO was investigated at molecular levels, where both preferential hydrophobic interactions including π-π stacking and hydrophobic effect in the unoxidized region and the hydrogen bonding in the oxidized region of GO were observed [161].

The increased porosity of a composite is also beneficial to the adsorption. The addition of GO in a metal-organic framework (MOF) of MIL-101 (Cr-benzene dicarboxylate) was carried out, which showed excellent adsorption capacity for nitrogen-containing compounds (NCCs) such as indole and quinoline [162]. Granular activated carbon (GAC) and single-walled CNT (SWCNT) both showed higher adsorption capacities for the selected aliphatic semivolatile organic compounds (SVOCs) than graphene due to their microporous volumes [160]. The inorganic nanoparticles could be acted as spacers, aiding in the segregation of GO or rGO sheets, and therefore increases the effective surface area of the nanocomposite and enhance its adsorption capacity [115].

Besides the physisorption procedure, a possible chemisorption procedure would be beneficial to the removal of organic contaminants from aqueous solutions. The adsorption of m-xylene onto tetramer clusters of platinum (Pt), palladium (Pd), gold (Au) and silver (Ag) deposited pristine and defective graphene were investigated. It was found that the adsorption of m-xylene onto Pt_4^- and Pd_4-DG was a chemisorption procedure, while the adsorption of m-xylene onto Au_4^- and Ag_4-DG was a physisorption, indicating the chemisorption counterparts of a physisorption-driven material is suitable as catalysts for the degradation of organic contaminants [163]. GO-titania hybrid Nafion membranes showed both the most promising adsorption and photocatalytic degradation ability for MO [164], so did TiO_2-graphene composite [165], MIL-53(Fe)-rGO hybrid [166], Fe_3O_4-rGO composite [167], iron(III)-based metal-organic framework/GO composite [168] and GO@ZnO composite [169] . Especially, PDA-GO composites with high surface area, a series of sub-nano thick PDA layer could selectively adsorb the dyes containing an Eschenmoser structure due to the possible 1,4-Michael addition reaction between the ortho position of the catechol phenolic hydroxyl group of PDA and Eschenmoser groups of the dyes [152].

5.   Conclusions and future directions

Fabrication of recyclable, stable, efficient, environmentally benign and cost-effective graphene-based composite is expected. Therefore, GO/graphene could be encapsulated into or grafted by environmentally benign, biocompatible, non-toxic and cost-effective components such as alginate (SA), cellulose, CDs, chitosan and so on. The functionalization of graphene-based composites should be more rational. The environmental-friendly synthesis technologies undoubtedly raised the graphene-based composites' profile in the adsorption field.

The fabrication of graphene-based hydrogels and aerogels also provides a new opportunity for developing novel carbon-based adsorbent materials. In addition, high-strength hydrogels in the fields of wastewater treatment should be expanded. The addition of several nanoparticles such as zirconia proved to improve the thermal and mechanical stability of graphene-based composites.

Adsorption selectivity is an interesting problem being concerned, and some researcher have already focused on it so far. Responsive components including a thermo-responsive polymer such as poly(2-hydroxyethyl methacrylate) (PHEMA), pH-sensitive polymers such as polyaniline and carboxymethyl chitosan have been grafted to GO, which showed high selectivity towards organic pollutants. Graphene-based composites with sensitive adsorption abilities are always anticipated.

To establish "structure-property-application" relationships for the adsorption removal of organic pollutants using graphene-based nanocomposites, various approaches such as the tunable synthesis of graphene-based adsorbents, the reasonable decoration of graphene, and the suitable regulation of the micro-, meso-, and macro-structure of 3D graphene-based frameworks, and the possibility of providing more active sites are expected.

Acknowledgments

Financial support from National Natural Science Foundation of China (Nos. 21471163 and 51674292), Provincial Natural Science Foundation of Hunan (No. 2016JJ1023), Project of Innovation-driven Plan in Central South University (No. 2016CX007) and the Hunan Provincial Science and Technology Plan Project, China (No. 2016TP1007) are gratefully acknowledged.

References

[1]   L.A. Al-Khateeb, S. Almotiry, M.A. Salam, Adsorption of pharmaceutical pollutants onto graphene nanoplatelets, Chem. Eng. J. 248 (2014) 191-199. https://doi.org/10.1016/j.cej.2014.03.023

[2]   J.G. Yu, L.Y. Yu, H. Yang, Q. Liu, X.H. Chen, X.Y. Jiang, X.Q. Chen, F.P. Jiao, Graphene nanosheets as novel adsorbents in adsorption, preconcentration and removal of gases, organic compounds and metal ions, Sci. Total Environ. 502 (2015) 70-79. https://doi.org/10.1016/j.scitotenv.2014.08.077

[3]   M. Xie, L.D. Nghiem, W.E. Price, M. Elimelech, Comparison of the removal of hydrophobic trace organic contaminants by forward osmosis and reverse osmosis, Water Res. 46 (2012) 2683-2692. https://doi.org/10.1016/j.watres.2012.02.023

[4]   J. Choi, J.-O. Kim, J. Chung, Removal of isopropyl alcohol and methanol in ultrapure water production system using a 185 nm ultraviolet and ion exchange

system, Chemosphere 156 (2016) 341-346.
https://doi.org/10.1016/j.chemosphere.2016.04.085

[5]   C. Lipski, P. Côté, The use of pervaporation for the removal of organic
        contaminants from water, Environl. Prog. 9 (1990) 254-261.
        https://doi.org/10.1002/ep.670090420

[6]   D. Mohan, A. Sarswat, Y.S. Ok, C.U. Pittman, Organic and inorganic
        contaminants removal from water with biochar, a renewable, low cost and
        sustainable adsorbent–a critical review, Bioresource Technol. 160 (2014) 191-202.
        https://doi.org/10.1016/j.biortech.2014.01.120

[7]   M. Perez, F. Torrades, J.A. García-Hortal, X. Domènech, J. Peral, Removal of
        organic contaminants in paper pulp treatment effluents under Fenton and photo-
        Fenton conditions, App. Catal B-Environ. 36 (2002) 63-74.
        https://doi.org/10.1016/S0926-3373(01)00281-8

[8]   J. Tang, Z. Zou, J. Ye, Efficient photocatalytic decomposition of organic
        contaminants over $CaBi_2O_4$ under visible-light irradiation, Angew Chem. Int. Ed.
        43 (2004) 4463-4466. https://doi.org/10.1002/anie.200353594

[9]   I. Alkorta, C. Garbisu, Phytoremediation of organic contaminants in soils,
        Bioresource Technol. 79 (2001) 273-276. https://doi.org/10.1016/S0960-
        8524(01)00016-5

[10]   A. Ribeiro, J. Rodrıguez-Maroto, E. Mateus, H. Gomes, Removal of organic
        contaminants from soils by an electrokinetic process: The case of atrazine:
        Experimental and modeling, Chemosphere 59 (2005) 1229-1239.
        https://doi.org/10.1016/j.chemosphere.2004.11.054

[11]   Y. Deng, C.M. Ezyske, Sulfate radical-advanced oxidation process (SR-AOP) for
        simultaneous removal of refractory organic contaminants and ammonia in landfill
        leachate, Water Res. 45 (2011) 6189-6194.
        https://doi.org/10.1016/j.watres.2011.09.015

[12]   A.F. Bollmann, W. Seitz, C. Prasse, T. Lucke, W. Schulz, T. Ternes, Occurrence
        and fate of amisulpride, sulpiride, and lamotrigine in municipal wastewater
        treatment plants with biological treatment and ozonation, J. Hazard Mater. 320
        (2016) 204-215. https://doi.org/10.1016/j.jhazmat.2016.08.022

[13]   R.A. Al-Juboori, T. Yusaf, V. Aravinthan, L. Bowtell, Investigating natural
        organic carbon removal and structural alteration induced by pulsed ultrasound, Sci.

Total Environ. 541 (2016) 1019-1030.
https://doi.org/10.1016/j.scitotenv.2015.09.143

[14] K.S. Novoselov, V. Fal, L. Colombo, P. Gellert, M. Schwab, K. Kim, A roadmap for graphene, Nature 490 (2012) 192-200. https://doi.org/10.1038/nature11458

[15] G.Z. Kyzas, E.A. Deliyanni, K.A. Matis, Graphene oxide and its application as an adsorbent for wastewater treatment, J. Chem. Technol. Biot. 89 (2014) 196-205. https://doi.org/10.1002/jctb.4220

[16] L. Ji, W. Chen, Z. Xu, S. Zheng, D. Zhu, Graphene nanosheets and graphite oxide as promising adsorbents for removal of organic contaminants from aqueous solution, J. Environ. Qual. 42 (2013) 191-198. https://doi.org/10.2134/jeq2012.0172

[17] S. Samiee, E.K. Goharshadi, Graphene nanosheets as efficient adsorbent for an azo dye removal: kinetic and thermodynamic studies, J. Nanopart. Res. 16 (2014). https://doi.org/10.1007/s11051-014-2542-8

[18] M.I. Ahamed, Inamuddin, Lutfullah, G. Sharma, A. Khan, A.M. Asiri, Turmeric/polyvinyl alcohol Th(IV) phosphate electrospun fibers: Synthesis, characterization and antimicrobial studies, J. Taiwan Inst. Chem. Eng. 68 (2016) 407–414. https://doi.org/10.1016/j.jtice.2016.08.024

[19] N.K. Rotte, S. Yerramala, J. Boniface, V.V.S.S. Srikanth, Equilibrium and kinetics of Safranin O dye adsorption on MgO decked multi-layered graphene, Chem. Eng. J. 258 (2014) 412-419. https://doi.org/10.1016/j.cej.2014.07.065

[20] T. Liu, Y. Li, Q. Du, J. Sun, Y. Jiao, G. Yang, Z. Wang, Y. Xia, W. Zhang, K. Wang, H. Zhu, D. Wu, Adsorption of methylene blue from aqueous solution by graphene, Colloid Surface B 90 (2012) 197-203. https://doi.org/10.1016/j.colsurfb.2011.10.019

[21] F. Liu, S. Chung, G. Oh, T.S. Seo, Three-dimensional graphene oxide nanostructure for fast and efficient water-soluble dye removal, ACS Appl. Mater. Inter. 4 (2012) 922-927. https://doi.org/10.1021/am201590z

[22] J. Wang, B. Chen, B. Xing, Wrinkles and folds of activated graphene nanosheets as fast and efficient adsorptive sites for hydrophobic organic contaminants, Environ. Sci. Technol. 50 (2016) 3798-3808. https://doi.org/10.1021/acs.est.5b04865

[23] F. Yu, J. Ma, D. Bi, Enhanced adsorptive removal of selected pharmaceutical antibiotics from aqueous solution by activated graphene, Environ. Sci. Pollut. Res. 22 (2015) 4715-4724. https://doi.org/10.1007/s11356-014-3723-9

[24] Z. Wu, X. Yuan, H. Zhong, H. Wang, G. Zeng, X. Chen, H. Wang, L. Zhang, J. Shao, Enhanced adsorptive removal of p-nitrophenol from water by aluminum metal-organic framework/reduced graphene oxide composite, Sci. Reports 6 (2016) 25638. https://doi.org/10.1038/srep25638

[25] H. Zhao, F. Qiu, J. Yan, J. Wang, X. Li, D. Yang, Preparation of economical and environmentally friendly graphene/palygorskite/TiO_2 composites and its application for the removal of methylene blue, Appl. Clay Sci. 121 (2016) 137-145. https://doi.org/10.1016/j.clay.2015.12.009

[26] H. Wang, H. Gao, M. Chen, X. Xu, X. Wang, C. Pan, J. Gao, Microwave-assisted synthesis of reduced graphene oxide/titania nanocomposites as an adsorbent for methylene blue adsorption, Appl. Clay Sci. 360 (2016) 840-848. https://doi.org/10.1016/j.apsusc.2015.11.075

[27] X. Yang, C. Chen, J. Li, G. Zhao, X. Ren, X. Wang, Graphene oxide-iron oxide and reduced graphene oxide-iron oxide hybrid materials for the removal of organic and inorganic pollutants, RSC Adv. 2 (2012) 8821-8826. https://doi.org/10.1039/c2ra20885g

[28] N.U. Yamaguchi, R. Bergamasco, S. Hamoudi, Magnetic $MnFe_2O_4$-graphene hybrid composite for efficient removal of glyphosate from water, Chem. Eng. J. 295 (2016) 391-402. https://doi.org/10.1016/j.cej.2016.03.051

[29] L. Zhang, S. Wu, Y. Tai, C. Lv, X. Zhang, Water-soluble magnetic-graphene nanocomposites: Use as high-performance adsorbent for removal of dye pollutants, Fuller Nanotube Carb N 24 (2016) 116-122.

[30] Y. Ying, P. He, G. Ding, X. Peng, Ultrafast adsorption and selective desorption of aqueous aromatic dyes by graphene sheets modified by graphene quantum dots, Nanotechnology 27 (2016) 245703. https://doi.org/10.1088/0957-4484/27/24/245703

[31] S.C. Smith, F. Ahmed, K.M. Gutierrez, D.F. Rodrigues, A comparative study of lysozyme adsorption with graphene, graphene oxide, and single-walled carbon nanotubes: Potential environmental applications, Chem. Eng. J. 240 (2014) 147-154. https://doi.org/10.1016/j.cej.2013.11.030

[32] W. Wan, R. Zhang, W. Li, H. Liu, Y. Lin, L. Li, Y. Zhou, Graphene-carbon nanotube aerogel as an ultra-light, compressible and recyclable highly efficient absorbent for oil and dyes, Environ. Sci-Nano 3 (2016) 107-113. https://doi.org/10.1039/C5EN00125K

[33] W. Li, W. Liu, H. Wang, W. Lu, Preparation of silica/reduced graphene oxide nanosheet composites for removal of organic contaminants from water, J. Nanosci. Nanotechnol. 16 (2016) 5734-5739. https://doi.org/10.1166/jnn.2016.11729

[34] J. Orozco, L.A. Mercante, R. Pol, A. Merkoci, Graphene-based Janus micromotors for the dynamic removal of pollutants, J. Mater. Chem. A 4 (2016) 3371-3378. https://doi.org/10.1039/C5TA09850E

[35] K. Yang, B. Chen, L. Zhu, Graphene-coated materials using silica particles as a framework for highly efficient removal of aromatic pollutants in water, Sci. Reports 5 (2015) 11641. https://doi.org/10.1038/srep11641

[36] X.L. Wu, Y. Shi, S. Zhong, H. Lin, J.R. Chen, Facile synthesis of Fe_3O_4-graphene@mesoporous SiO_2 nanocomposites for efficient removal of methylene blue, Appl. Surf. Sci. 378 (2016) 80-86. https://doi.org/10.1016/j.apsusc.2016.03.226

[37] D. Dutta, S. Thiyagarajan, D. Bahadur, SnO_2 quantum dots decorated reduced graphene oxide nanocomposites for efficient water remediation, Chem. Eng. J. 297 (2016) 55-65. https://doi.org/10.1016/j.cej.2016.03.130

[38] L. Duan, Z. Wang, Y. Hou, Z. Wang, G. Gao, W. Chen, P.J.J. Alvarez, The oxidation capacity of Mn3O4 nanoparticles is significantly enhanced by anchoring them onto reduced graphene oxide to facilitate regeneration of surface-associated Mn(III), Water Res. 103 (2016) 101-108. https://doi.org/10.1016/j.watres.2016.07.023

[39] S. Thakur, N. Karak, One-step approach to prepare magnetic iron oxide/reduced graphene oxide nanohybrid for efficient organic and inorganic pollutants removal, Mater. Chem. Phys. 144 (2014) 425-432. https://doi.org/10.1016/j.matchemphys.2014.01.015

[40] J.Z. Sun, Z.H. Liao, R.W. Si, G.P. Kingori, F.X. Chang, L. Gao, Y. Shen, X. Xiao, X.Y. Wu, Y.C. Yong, Adsorption and removal of triphenylmethane dyes from water by magnetic reduced graphene oxide, Water Sci. Technol. 70 (2014) 1663-1669. https://doi.org/10.2166/wst.2014.427

[41] C. Santhosh, P. Kollu, S. Doshi, M. Sharma, D. Bahadur, M.T. Vanchinathan, P. Saravanan, B.-S. Kim, A.N. Grace, Adsorption, photodegradation and antibacterial study of graphene-Fe_3O_4 nanocomposite for multipurpose water purification application, RSC Adv. 4 (2014) 28300-28308. https://doi.org/10.1039/C4RA02913E

[42] Y. Yao, S. Miao, S. Liu, L.P. Ma, H. Sun, S. Wang, Synthesis, characterization, and adsorption properties of magnetic Fe_3O_4@graphene nanocomposite, Chem. Eng. J. 184 (2012) 326-332. https://doi.org/10.1016/j.cej.2011.12.017

[43] C. Wang, C. Feng, Y. Gao, X. Ma, Q. Wu, Z. Wang, Preparation of a graphene-based magnetic nanocomposite for the removal of an organic dye from aqueous solution, Chem. Eng. J. 173 (2011) 92-97. https://doi.org/10.1016/j.cej.2011.07.041

[44] H. Sun, L. Cao, L. Lu, Magnetite/reduced graphene oxide nanocomposites: One step solvothermal synthesis and use as a novel platform for removal of dye pollutants, Nano Res. 4 (2011) 550-562. https://doi.org/10.1007/s12274-011-0111-3

[45] L. Ai, C. Zhang, Z. Chen, Removal of methylene blue from aqueous solution by a solvothermal-synthesized graphene/magnetite composite, J. Hazard. Mater. 192 (2011) 1515-1524. https://doi.org/10.1016/j.jhazmat.2011.06.068

[46] L. Yu, Y. Ma, C.N. Ong, J. Xie, Y. Liu, Rapid adsorption removal of arsenate by hydrous cerium oxide-graphene composite, RSC Adv. 5 (2015) 64983-64990. https://doi.org/10.1039/C5RA08922K

[47] G.Q. Liu, M.X. Wan, Z.H. Huang, F.Y. Kang, Preparation of graphene/metal-organic composites and their adsorption performance for benzene and ethanol, New Carbon Mater. 30 (2015) 566-571. https://doi.org/10.1016/S1872-5805(15)60205-0

[48] K.R. Parmar, I. Patel, S. Basha, Z.V.P. Murthy, Synthesis of acetone reduced graphene oxide/Fe_3O_4 composite through simple and efficient chemical reduction of exfoliated graphene oxide for removal of dye from aqueous solution, J. Mater. Sci. 49 (2014) 6772-6783. https://doi.org/10.1007/s10853-014-8378-x

[49] J. Xiao, W. Lv, Z. Xie, Y. Tan, Y. Song, Q. Zheng, Environmentally friendly reduced graphene oxide as a broad-spectrum adsorbent for anionic and cationic dyes via π-π interactions, J Mater. Chem. A 4 (2016) 12126-12135. https://doi.org/10.1039/C6TA04119A

[50]   X. Hu, L. Mu, J. Wen, Q. Zhou, Immobilized smart RNA on graphene oxide nanosheets to specifically recognize and adsorb trace peptide toxins in drinking water, J. Hazard. Mater. 213 (2012) 387-392. https://doi.org/10.1016/j.jhazmat.2012.02.012

[51]   S. Yang, L. Chen, L. Mu, B. Hao, J. Chen, P.C. Ma, Graphene foam with hierarchical structures for the removal of organic pollutants from water, RSC Adv. 6 (2016) 4889-4898. https://doi.org/10.1039/C5RA23820J

[52]   Y. Yang, Y. Zhao, S. Sun, X. Zhang, L. Duan, X. Ge, W. Lu, Self-assembled three-dimensional graphene/Fe$_3$O$_4$ hydrogel for efficient pollutant adsorption and electromagnetic wave absorption, Mater. Res. Bull. 73 (2016) 401-408. https://doi.org/10.1016/j.materresbull.2015.09.032

[53]   R. Wu, B. Yu, X. Liu, H. Li, Y. Bai, Z. Ming, L. Chen, S.-T. Yang, X.-L. Chang, Graphene/polyester staple composite for the removal of oils and organic solvents, Mater. Res. Express 3 (2016) 065601. https://doi.org/10.1088/2053-1591/3/6/065601

[54]   S. Song, H. Yang, C. Su, Z. Jiang, Z. Lu, Ultrasonic-microwave assisted synthesis of stable reduced graphene oxide modified melamine foam with superhydrophobicity and high oil adsorption capacities, Chem. Eng. J. 306 (2016) 504-511. https://doi.org/10.1016/j.cej.2016.07.086

[55]   H. Zhu, D. Chen, W. An, N. Li, Q. Xu, H. Li, J. He, J. Lu, A robust and cost-effective superhydrophobic graphene foam for efficient oil and organic solvent recovery, Small 11 (2015) 5222-5229. https://doi.org/10.1002/smll.201501004

[56]   R. Wu, B. Yu, X. Liu, H. Li, W. Wang, L. Chen, Y. Bai, Z. Ming, S.T. Yang, One-pot hydrothermal preparation of graphene sponge for the removal of oils and organic solvents, Appl. Surf. Sci. 362 (2016) 56-62. https://doi.org/10.1016/j.apsusc.2015.11.215

[57]   Q. Du, Y. Zhou, X. Pan, J. Zhang, Q. Zhuo, S. Chen, G. Chen, T. Liu, F. Xu, C. Yan, A graphene-melamine-sponge for efficient and recyclable dye adsorption, RSC Adv. 6 (2016) 54589-54596. https://doi.org/10.1039/C6RA08412E

[58]   T. Liu, G. Zhao, W. Zhang, H. Chi, C. Hou, Y. Sun, The preparation of superhydrophobic graphene/melamine composite sponge applied in treatment of oil pollution, J. Porous Mater. 22 (2015) 1573-1580. https://doi.org/10.1007/s10934-015-0040-8

[59] H.S. Park, S.O. Kang, Sorption behavior of slightly reduced, three-dimensionally macroporous graphene oxides for physical loading of oils and organic solvents, Carbon Lett. 18 (2016) 24-29. https://doi.org/10.5714/CL.2016.18.024

[60] D.N.H. Tran, S. Kabiri, T.R. Sim, D. Losic, Selective adsorption of oil-water mixtures using polydimethylsiloxane (PDMS)-graphene sponges, Environ. Sci-Water Res. Technol. 1 (2015) 298-305. https://doi.org/10.1039/C5EW00035A

[61] Y.C. Shi, A.J. Wang, X.L. Wu, J.R. Chen, J.J. Feng, Green-assembly of three-dimensional porous graphene hydrogels for efficient removal of organic dyes, J. Colloid Interf. Sci. 484 (2016) 254-262. https://doi.org/10.1016/j.jcis.2016.09.008

[62] S. Yang, L. Zhang, Q. Yang, Z. Zhang, B. Chen, P. Lv, W. Zhu, G. Wang, Graphene aerogel prepared by thermal evaporation of graphene oxide suspension containing sodium bicarbonate, J. Mater. Chem. A 3 (2015) 7950-7958. https://doi.org/10.1039/C5TA01222H

[63] Z. Tian, C. Xu, J. Li, G. Zhu, P. Li, J. Dai, Z. Shi, One-pot hydrothermal synthesis of nitrogen-doped reduced graphene oxide hydrogel, Sci. Adv. Mater. 7 (2015) 1415-1423. https://doi.org/10.1166/sam.2015.2060

[64] D. Suresh, P.C. Nethravathi, Udayabhanu, H. Nagabhushana, S.C. Sharma, Spinach assisted green reduction of graphene oxide and its antioxidant and dye absorption properties, Ceram. Int. 41 (2015) 4810-4813. https://doi.org/10.1016/j.ceramint.2014.12.036

[65] C. Chi, H. Xu, K. Zhang, Y. Wang, S. Zhang, X. Liu, X. Liu, J. Zhao, Y. Li, 3D hierarchical porous graphene aerogels for highly improved adsorption and recycled capacity, Mater. Sci. Eng. B-Adv. 194 (2015) 62-67. https://doi.org/10.1016/j.mseb.2014.12.026

[66] X. Zhu, D.C.W. Tsang, F. Chen, S. Li, X. Yang, Ciprofloxacin adsorption on graphene and granular activated carbon: Kinetics, isotherms, and effects of solution chemistry, Environ. Technol. 36 (2015) 3094-3102. https://doi.org/10.1080/09593330.2015.1054316

[67] D. Robati, M. Rajabi, O. Moradi, F. Najafi, I. Tyagi, S. Agarwal, V.K. Gupta, Kinetics and thermodynamics of malachite green dye adsorption from aqueous solutions on graphene oxide and reduced graphene oxide, J. Mol. Liq. 214 (2016) 259-263. https://doi.org/10.1016/j.molliq.2015.12.073

[68] M.R. Samarghandi, A. Poormohammadi, N. Fatemeh, M. Ahmadian, Removal of acid orange 7 from aqueous solution using activated carbon and graphene as adsorbents, Fresen. Environ. Bull. 24 (2015) 1841-1851.

[69] J. Kwon, B. Lee, Bisphenol A adsorption using reduced graphene oxide prepared by physical and chemical reduction methods, Chem. Eng. Res. Des. 104 (2015) 519-529. https://doi.org/10.1016/j.cherd.2015.09.007

[70] X. Ruan, Y. Chen, H. Chen, G. Qian, R.L. Frost, Sorption behavior of methyl orange from aqueous solution on organic matter and reduced graphene oxides modified Ni-Cr layered double hydroxides, Chem. Eng. J. 297 (2016) 295-303. https://doi.org/10.1016/j.cej.2016.01.041

[71] C. Zhao, J. Guo, Q. Yang, L. Tong, J. Zhang, J. Zhang, C. Gong, J. Zhou, Z. Zhang, Preparation of magnetic Ni@graphene nanocomposites and efficient removal organic dye under assistance of ultrasound, Appl. Surf. Sci. 357 (2015) 22-30. https://doi.org/10.1016/j.apsusc.2015.08.031

[72] K.Y.A. Lin, W.D. Lee, Self-assembled magnetic graphene supported ZIF-67 as a recoverable and efficient adsorbent for benzotriazole, Chem. Eng. J. 284 (2016) 1017-1027. https://doi.org/10.1016/j.cej.2015.09.063

[73] K.Y.A. Lin, W.D. Lee, Highly efficient removal of Malachite green from water by a magnetic reduced graphene oxide/zeolitic imidazolate framework self assembled nanocomposite, Appl. Surf. Sci. 361 (2016) 114-121. https://doi.org/10.1016/j.apsusc.2015.11.108

[74] Y. Li, R. Zhang, X. Tian, C. Yang, Z. Zhou, Facile synthesis of Fe_3O_4 nanoparticles decorated on 3D graphene aerogels as broad-spectrum sorbents for water treatment, Appl Surf Sci, 369 (2016) 11-18. https://doi.org/10.1016/j.apsusc.2016.02.019

[75] D. Koushik, S. Sen Gupta, S.M. Maliyekkal, T. Pradeep, Rapid dehalogenation of pesticides and organics at the interface of reduced graphene oxide-silver nanocomposite, J. Hazard Mater. 308 (2016) 192-198. https://doi.org/10.1016/j.jhazmat.2016.01.004

[76] H. Gao, Y. Zhou, K. Chen, X. Li, Synthesis of Tb_4O_7 complexed with reduced graphene oxide for Rhodamine-B absorption, Mater. Res. Bull. 77 (2016) 111-114. https://doi.org/10.1016/j.materresbull.2016.01.016

[77]   D. Cortes-Arriagada, A. Toro-Labbe, Aluminum and iron doped graphene for adsorption of methylated arsenic pollutants, Appl. Surf. Sci. 386 (2016) 84-95. https://doi.org/10.1016/j.apsusc.2016.05.154

[78]   J. Tang, H. Lv, Y. Gong, Y. Huang, Preparation and characterization of a novel graphene/biochar composite for aqueous phenanthrene and mercury removal, Bioresource Technol. 196 (2015) 355-363. https://doi.org/10.1016/j.biortech.2015.07.047

[79]   Y.-B. Song, X.-D. Song, C.-J. Cheng, Z.-G. Zhao, Poly(4-styrenesulfonic acid-co-maleic acid)-sodium-modified magnetic reduced graphene oxide for enhanced adsorption performance toward cationic dyes, RSC Adv. 5 (2015) 87030-87042. https://doi.org/10.1039/C5RA18255G

[80]   E. Radu, A.C. Ion, F. Sirbu, I. Ion, Adsorption of endocrine disruptors on exfoliated graphene nanoplatelets, Environmental Engineering and Management Journal 14 (2015) 551-558.

[81]   S. Pourmand, M. Abdouss, A.M. Rashidi, Preparation of nanoporous graphene via nanoporous zinc oxide and its application as a nanoadsorbent for benzene, toluene and xylenes removal, International Journal of Environmental Research 9 (2015) 1269-1276.

[82]   K. Luo, Y. Mu, P. Wang, X. Liu, Effect of oxidation degree on the synthesis and adsorption property of magnetite/graphene nanocomposites, Appl. Surf. Sci. 359 (2015) 188-195. https://doi.org/10.1016/j.apsusc.2015.10.083

[83]   G. Liu, N. Wang, J. Zhou, A. Wang, J. Wang, R. Jin, H. Lv, Microbial preparation of magnetite/reduced graphene oxide nanocomposites for the removal of organic dyes from aqueous solutions, RSC Adv. 5 (2015) 95857-95865. https://doi.org/10.1039/C5RA18136D

[84]   D. Liu, W. Lei, Y. Chen, Scalable production of wrinkled and few-layered graphene sheets and their use for oil and organic solvent absorption, Phys. Chem. Chem. Phys. 17 (2015) 6913-6918. https://doi.org/10.1039/C4CP05864J

[85]   Q. Ling, M. Yang, C. Li, A. Zhang, Preparation of monolayered Ce-Fe oxides dispersed on graphene and their superior adsorptive behavior, Fuller. Nanotube. Carb. N 23 (2015) 158-164. https://doi.org/10.1080/1536383X.2013.863759

[86]   R. Kumar, M.O. Ansari, N. Parveen, M.A. Barakat, M.H. Cho, Simple route for the generation of differently functionalized PVC@graphene-polyaniline fiber

bundles for the removal of Congo red from wastewater, RSC Adv. 5 (2015) 61486-61494. https://doi.org/10.1039/C5RA10378A

[87]   H. Kim, S.O. Kang, S. Park, H.S. Park, Adsorption isotherms and kinetics of cationic and anionic dyes on three-dimensional reduced graphene oxide macrostructure, J. Ind. Eng. Chem. 21 (2015) 1191-1196. https://doi.org/10.1016/j.jiec.2014.05.033

[88]   Z. Jin, X. Wang, Y. Sun, Y. Ai, X. Wang, Adsorption of 4-n-nonylphenol and bisphenol-A on magnetic reduced graphene oxides: A Combined experimental and theoretical studies, Environ. Sci. Technol. 49 (2015) 9168-9175. https://doi.org/10.1021/acs.est.5b02022

[89]   H. Du, Z. Wang, Y. Chen, Y. Liu, Y. Liu, B. Li, X. Wang, H. Cao, Anchoring superparamagnetic core-shells onto reduced graphene oxide: Fabrication of Ni-carbon-rGO nanocomposite for effective adsorption and separation, RSC Adv. 5 (2015) 10033-10039. https://doi.org/10.1039/C4RA14651D

[90]   L. Bai, Z. Li, Y. Zhang, T. Wang, R. Lu, W. Zhou, H. Gao, S. Zhang, Synthesis of water-dispersible graphene-modified magnetic polypyrrole nanocomposite and its ability to efficiently adsorb methylene blue from aqueous solution, Chem. Eng. J. 279 (2015) 757-766. https://doi.org/10.1016/j.cej.2015.05.068

[91]   S. Yang, L. Li, Z. Pei, C. Li, X.Q. Shan, B. Wen, S. Zhang, L. Zheng, J. Zhang, Y. Xie, R. Huang, Effects of humic acid on copper adsorption onto few-layer reduced graphene oxide and few-layer graphene oxide, Carbon 75 (2014) 227-235. https://doi.org/10.1016/j.carbon.2014.03.057

[92]   W. Lu, Y. Wu, J. Chen, Y. Yang, Facile preparation of graphene-Fe_3O_4 nanocomposites for extraction of dye from aqueous solution, Cryst. Eng. Comm. 16 (2014) 609-615. https://doi.org/10.1039/C3CE41833B

[93]   S. Kabiri, D.N.H. Tran, T. Altalhi, D. Losic, Outstanding adsorption performance of graphene-carbon nanotube aerogels for continuous oil removal, Carbon 80 (2014) 523-533. https://doi.org/10.1016/j.carbon.2014.08.092

[94]   G. Zhao, J. Li, X. Wang, Kinetic and thermodynamic study of 1-naphthol adsorption from aqueous solution to sulfonated graphene nanosheets, Chem. Eng. J. 173 (2011) 185-190. https://doi.org/10.1016/j.cej.2011.07.072

[95]   X. Hu, L. Mu, K. Lu, J. Kang, Q. Zhou, Green synthesis of low-toxicity graphene-fulvic acid with an open band gap enhances demethylation of methylmercury, ACS Appl. Mater. Inter. 6 (2014) 9220-9227. https://doi.org/10.1021/am501334j

[96]   P. Sharma, B.K. Saikia, M.R. Das, Removal of methyl green dye molecule from aqueous system using reduced graphene oxide as an efficient adsorbent: Kinetics, isotherm and thermodynamic parameters, Colloid Surface A 457 (2014) 125-133. https://doi.org/10.1016/j.colsurfa.2014.05.054

[97]   W. Zhang, C. Zhou, W. Zhou, A. Lei, Q. Zhang, Q. Wan, B. Zou, Fast and considerable adsorption of methylene blue dye onto graphene oxide, Bulletin of Environmental Contamination and Toxicology, 87 (2011) 86-90. https://doi.org/10.1007/s00128-011-0304-1

[98]   S.-T. Yang, S. Chen, Y. Chang, A. Cao, Y. Liu, H. Wang, Removal of methylene blue from aqueous solution by graphene oxide, J. Colloid Interf. Sci. 359 (2011) 24-29. https://doi.org/10.1016/j.jcis.2011.02.064

[99]   L. Yu, X. Wu, Q. Liu, L. Liu, X. Jiang, J. Yu, C. Feng, M. Zhong, Removal of phenols from aqueous solutions by graphene oxide nanosheet suspensions, J. Nanosci. Nanotechnol. 16 (2016) 12426-12432. https://doi.org/10.1166/jnn.2016.12974

[100]   Z. Xue, S. Zhao, Z. Zhao, P. Li, J. Gao, Thermodynamics of dye adsorption on electrochemically exfoliated graphene, J. Mater. Sci. 51 (2016) 4928-4941. https://doi.org/10.1007/s10853-016-9798-6

[101]   L.H. Jiang, Y.G. Liu, G.M. Zeng, F.Y. Xiao, X.J. Hu, X. Hu, H. Wang, T.T. Li, L. Zhou, X.F. Tan, Removal of 17 β-estradiol by few-layered graphene oxide nanosheets from aqueous solutions: External influence and adsorption mechanism, Chem. Eng. J. 284 (2016) 93-102. https://doi.org/10.1016/j.cej.2015.08.139

[102]   Z. Hosseinabadi-Farahani, H. Hosseini-Monfared, N.M. Mahmoodi, Graphene oxide nanosheet: Preparation and dye removal from binary system colored wastewater, Desal. Water Treat. 56 (2015) 2382-2394. https://doi.org/10.1080/19443994.2014.960462

[103]   C. Zhao, J. Fan, D. Chen, Y. Xu, T. Wang, Microfluidics-generated graphene oxide microspheres and their application to removal of perfluorooctane sulfonate from polluted water, Nano Research 9 (2016) 866-875. https://doi.org/10.1007/s12274-015-0968-7

[104]   Z. Hosseinabadi-Farahani, N.M. Mahmoodi, H. Hosseini-Monfared, Preparation of surface functionalized graphene oxide nanosheet and its multicomponent dye removal ability from wastewater, Fiber Polym. 16 (2015) 1035-1047. https://doi.org/10.1007/s12221-015-1035-4

[105] S. Zheng, B. Mi, Emerging investigators series: silica-crosslinked graphene oxide membrane and its unique capability in removing neutral organic molecules from water, Environ. Sci-Water Res. 2 (2016) 717-725. https://doi.org/10.1039/C6EW00070C

[106] W. Yin, H. Cao, One-step synthesis of SnO_2-reduced graphene oxide (SOG) composites for efficient removal of organic dyes from wastewater, RSC Adv. 6 (2016) 100636-100642. https://doi.org/10.1039/C6RA21856C

[107] J. Zhang, J.L. Gong, G.M. Zenga, X.M. Ou, Y. Jiang, Y.N. Chang, M. Guo, C. Zhang, H.-Y. Liu, Simultaneous removal of humic acid/fulvic acid and lead from landfill leachate using magnetic graphene oxide, Appl. Surf. Sci. 370 (2016) 335-350. https://doi.org/10.1016/j.apsusc.2016.02.181

[108] H. Wang, Y. Chen, Y. Wei, A novel magnetic calcium silicate/graphene oxide composite material for selective adsorption of acridine orange from aqueous solutions, RSC Adv. 6 (2016) 34770-34781. https://doi.org/10.1039/C6RA07625D

[109] F. Liu, Z. Wu, D. Wang, J. Yu, X. Jiang, X. Chen, Magnetic porous silica-graphene oxide hybrid composite as a potential adsorbent for aqueous removal of p-nitrophenol, Colloid Surface A 490 (2016) 207-214. https://doi.org/10.1016/j.colsurfa.2015.11.053

[110] T. Jiao, Y. Liu, Y. Wu, Q. Zhang, X. Yan, F. Gao, A.J.P. Bauer, J. Liu, T. Zeng, B. Li, Facile and scalable preparation of graphene oxide-based magnetic hybrids for fast and highly efficient removal of organic dyes, Sci. Reports 5 (2015) 12451. https://doi.org/10.1038/srep12451

[111] G. Jiang, Q. Chang, F. Yang, X. Hu, H. Tang, Sono-assisted preparation of magnetic ferroferric oxide/graphene oxide nanoparticles and application on dye removal, Chinese J. Chem. Eng. 23 (2015) 510-515. https://doi.org/10.1016/j.cjche.2014.06.037

[112] M. Namvari, H. Namazi, Clicking graphene oxide and Fe_3O_4 nanoparticles together: An efficient adsorbent to remove dyes from aqueous solutions, Int. J. Environ. Sci. Technol. 11 (2014) 1527-1536. https://doi.org/10.1007/s13762-014-0595-y

[113] Z. Guo, J. Huang, Z. Xue, X. Wang, Electrospun graphene oxide/carbon composite nanofibers with well-developed mesoporous structure and their adsorption performance for benzene and butanone, Chem. Eng. J. 306 (2016) 99-106. https://doi.org/10.1016/j.cej.2016.07.048

[114] Q. Wan, M. Liu, Y. Xie, J. Tian, Q. Huang, F. Deng, L. Mao, Q. Zhang, X. Zhang, Y. Wei, Facile and highly efficient fabrication of graphene oxide-based polymer nanocomposites through mussel-inspired chemistry and their environmental pollutant removal application, J Mater. Sci. 52 (2017) 504-518. https://doi.org/10.1007/s10853-016-0349-y

[115] S. Shahabuddin, N.M. Sarih, M.A. Kamboh, H.R. Nodeh, S. Mohamad, Synthesis of polyaniline-coated graphene oxide@SrTiO₃ nanocube nanocomposites for enhanced removal of carcinogenic dyes from aqueous solution, Polymers 8 (2016). https://doi.org/10.3390/polym8090305

[116] Y. Qi, M. Yang, W. Xu, S. He, Y. Men, Natural polysaccharides-modified graphene oxide for adsorption of organic dyes from aqueous solutions, J. Colloid Interf. Sci. 486 (2017) 84-96. https://doi.org/10.1016/j.jcis.2016.09.058

[117] J. Yan, Y. Zhu, F. Qiu, H. Zhao, D. Yang, J. Wang, W. Wen, Kinetic, isotherm and thermodynamic studies for removal of methyl orange using a novel β-cyclodextrin functionalized graphene oxide-isophorone diisocyanate composites, Chem. Eng. Res. Des. 106 (2016) 168-177. https://doi.org/10.1016/j.cherd.2015.12.023

[118] X. Liu, L. Yan, W. Yin, L. Zhou, G. Tian, J. Shi, Z. Yang, D. Xiao, Z. Gu, Y. Zhao, A magnetic graphene hybrid functionalized with β-cyclodextrins for fast and efficient removal of organic dyes, J. Mater. Chem. A 2 (2014) 12296-12303. https://doi.org/10.1039/C4TA00753K

[119] H. Yan, H. Yang, A. Li, R. Cheng, pH-tunable surface charge of chitosan/graphene oxide composite adsorbent for efficient removal of multiple pollutants from water, Chem. Eng. J. 284 (2016) 1397-1405. https://doi.org/10.1016/j.cej.2015.06.030

[120] L. Hu, Z. Yang, L. Cui, Y. Li, H.H. Ngo, Y. Wang, Q. Wei, H. Ma, L. Yan, B. Du, Fabrication of hyperbranched polyamine functionalized graphene for high-efficiency removal of Pb(II) and methylene blue, Chem. Eng. J. 287 (2016) 545-556. https://doi.org/10.1016/j.cej.2015.11.059

[121] L. Wang, L. Jiang, D. Su, C. Sun, M. Chen, K. Goh, Y. Chen, Non-covalent synthesis of thermo-responsive graphene oxide-perylene bisimides-containing poly(N-isopropylacrylamide) hybrid for organic pigment removal, J Colloid Interf. Sci. 430 (2014) 121-128. https://doi.org/10.1016/j.jcis.2014.05.031

[122] L. Van Hoang, J.S. Chung, E.J. Kim, S.H. Hur, The molecular level control of three-dimensional graphene oxide hydrogel structure by using various diamines, Chem. Eng. J. 246 (2014) 64-70. https://doi.org/10.1016/j.cej.2014.01.105

[123] H. Ren, X. Shi, J. Zhu, Y. Zhang, Y. Bi, L. Zhang, Facile synthesis of N-doped graphene aerogel and its application for organic solvent adsorption, J. Mater. Sci. 51 (2016) 6419-6427. https://doi.org/10.1007/s10853-016-9939-y

[124] X. Wang, M. Lu, H. Wang, Y. Pei, H. Rao, X. Du, Three-dimensional graphene aerogels-mesoporous silica frameworks for superior adsorption capability of phenols, Sep. Purif. Technol. 153 (2015) 7-13. https://doi.org/10.1016/j.seppur.2015.08.030

[125] H. Guo, T. Jiao, Q. Zhang, W. Guo, Q. Peng, X. Yan, Preparation of graphene oxide-based hydrogels as efficient dye adsorbents for wastewater treatment, Nanoscale Res. Lett. 10 (2015). https://doi.org/10.1186/s11671-015-0931-2

[126] L. Gan, S. Shang, E. Hu, C. Wah, M. Yuen, S.-x. Jiang, Konjac glucomannan/graphene oxide hydrogel with enhanced dyes adsorption capability for methyl blue and methyl orange, Appl. Surf. Sci. 357 (2015) 866-872. https://doi.org/10.1016/j.apsusc.2015.09.106

[127] W. Cui, J. Ji, Y.F. Cai, H. Li, R. Ran, Robust, anti-fatigue, and self-healing graphene oxide/hydrophobically associated composite hydrogels and their use as recyclable adsorbents for dye wastewater treatment, J. Mater. Chem. A 3 (2015) 17445-17458. https://doi.org/10.1039/C5TA04470G

[128] Z. Cheng, J. Liao, B. He, F. Zhang, F. Zhang, X. Huang, L. Zhou, One-step fabrication of graphene oxide enhanced magnetic composite gel for highly efficient dye adsorption and catalysis, ACS Sustain. Chem. Eng. 3 (2015) 1677-1685. https://doi.org/10.1021/acssuschemeng.5b00383

[129] L. Liu, B. Zhang, Y. Zhang, Y. He, L. Huang, S. Tan, X. Cai, Simultaneous removal of cationic and anionic dyes from environmental water using montmorillonite-pillared graphene oxide, J. Chem. Eng. Data 60 (2015) 1270-1278. https://doi.org/10.1021/je5009312

[130] D. Robati, B. Mirza, M. Rajabi, O. Moradi, I. Tyagi, S. Agarwal, V.K. Gupta, Removal of hazardous dyes-BR 12 and methyl orange using graphene oxide as an adsorbent from aqueous phase, Chem. Eng. J. 284 (2016) 687-697. https://doi.org/10.1016/j.cej.2015.08.131

[131] T. Phatthanakittiphong, G.T. Seo, Characteristic evaluation of graphene oxide for bisphenol a adsorption in aqueous solution, Nanomaterials 6 (2016). https://doi.org/10.3390/nano6070128

[132] X. Tao, X. Wang, Z. Li, S. Zhou, Ultralow temperature synthesis and improved adsorption performance of graphene oxide nanosheets, Appl. Surf. Sci. 324 (2015) 363-368. https://doi.org/10.1016/j.apsusc.2014.10.153

[133] S.-W. Nam, C. Jung, H. Li, M. Yu, J.R.V. Flora, L.K. Boateng, N. Her, K.-D. Zoh, Y. Yoon, Adsorption characteristics of diclofenac and sulfamethoxazole to graphene oxide in aqueous solution, Chemosphere 136 (2015) 20-26. https://doi.org/10.1016/j.chemosphere.2015.03.061

[134] G.Z. Kyzas, A. Koltsakidou, S.G. Nanaki, D.N. Bikiaris, D.A. Lambropoulou, Removal of b-blockers from aqueous media by adsorption onto graphene oxide, Sci. Total Environ. 537 (2015) 411-420. https://doi.org/10.1016/j.scitotenv.2015.07.144

[135] R. Rajesh, S.S. Iyer, J. Ezhilan, S.S. Kumar, R. Venkatesan, Graphene oxide supported copper oxide nanoneedles: An efficient hybrid material for removal of toxic azo dyes, Spectrochim. Acta A 166 (2016) 49-55. https://doi.org/10.1016/j.saa.2016.05.002

[136] S. Swaminathan, A. Muthumanickkam, N.M. Imayathamizhan, An effective removal of methylene blue dye using polyacrylonitrile yarn waste/graphene oxide nanofibrous composite, Int. J. Environ. Sci. Technol. 12 (2015) 3499-3508. https://doi.org/10.1007/s13762-014-0711-z

[137] Z. Ma, D. Liu, Y. Zhu, Z. Li, Z. Li, H. Tian, H. Liu, Graphene oxide/chitin nanofibril composite foams as column adsorbents for aqueous pollutants, Carbohyd. Polym. 144 (2016) 230-237. https://doi.org/10.1016/j.carbpol.2016.02.057

[138] C. Cheng, D. Wang, Hydrogel-assisted transfer of graphene oxides into nonpolar organic media for oil decontamination, Angew Chem. Int. Ed. 55 (2016) 6853-6857. https://doi.org/10.1002/anie.201600221

[139] L.-C. Chen, S. Lei, M.-Z. Wang, J. Yang, X.-W. Ge, Fabrication of macroporous polystyrene/graphene oxide composite monolith and its adsorption property for tetracycline, Chinese Chem. Lett. 27 (2016) 511-517. https://doi.org/10.1016/j.cclet.2016.01.057

[140] D. Chen, H. Zhang, K. Yang, H. Wang, Functionalization of 4-aminothiophenol and 3-aminopropyltriethoxysilane with graphene oxide for potential dye and copper removal, J Hazard Mater. 310 (2016) 179-187. https://doi.org/10.1016/j.jhazmat.2016.02.040

[141] J. Zhang, M.S. Azam, C. Shi, J. Huang, B. Yan, Q. Liu, H. Zeng, Poly(acrylic acid) functionalized magnetic graphene oxide nanocomposite for removal of methylene blue, RSC Adv. 5 (2015) 32272-32282. https://doi.org/10.1039/C5RA01815C

[142] X. Yang, Y. Li, Q. Du, J. Sun, L. Chen, S. Hu, Z. Wang, Y. Xia, L. Xia, Highly effective removal of basic fuchsin from aqueous solutions by anionic polyacrylamide/graphene oxide aerogels, J. Colloid Interf. Sci. 453 (2015) 107-114. https://doi.org/10.1016/j.jcis.2015.04.042

[143] S. Yang, T. Zeng, Y. Li, J. Liu, Q. Chen, J. Zhou, Y. Ye, B. Tang, Preparation of graphene oxide decorated $Fe_3O_4@SiO_2$ nanocomposites with superior adsorption capacity and SERS detection for organic dyes, J. Nanomater. (2015) 817924.

[144] S. Wang, Y. Li, X. Fan, F. Zhang, G. Zhang, β-cyclodextrin functionalized graphene oxide: an efficient and recyclable adsorbent for the removal of dye pollutants, Front. Chem. Sci. Eng. 9 (2015) 77-83. https://doi.org/10.1007/s11705-014-1450-x

[145] D. Wang, L. Liu, X. Jiang, J. Yu, X. Chen, X. Chen, Adsorbent for p-phenylenediamine adsorption and removal based on graphene oxide functionalized with magnetic cyclodextrin, Appl. Surf. Sci. 329 (2015) 197-205. https://doi.org/10.1016/j.apsusc.2014.12.161

[146] Z. Liu, X. Wang, Z. Luo, M. Huo, J. Wu, H. Huo, W. Yang, Removing of disinfection by-product precursors from surface water by using magnetic graphene oxide, Plos One 10 (2015). https://doi.org/10.1371/journal.pone.0143819

[147] K. Liu, H. Li, Y. Wang, X. Gou, Y. Duan, Adsorption and removal of rhodamine B from aqueous solution by tannic acid functionalized graphene, Colloid Surface A 477 (2015) 35-41. https://doi.org/10.1016/j.colsurfa.2015.03.048

[148] J.R. Lee, J.Y. Bae, W. Jang, J.H. Lee, W.S. Choi, H.Y. Koo, Magnesium hydroxide nanoplate/graphene oxide composites as efficient adsorbents for organic dyes, RSC Adv. 5 (2015) 83668-83673. https://doi.org/10.1039/C5RA11184F

[149] R. Hu, S. Dai, D. Shao, A. Alsaedi, B. Ahmad, X. Wang, Efficient removal of phenol and aniline from aqueous solutions using graphene oxide/polypyrrole

composites, J. Mol. Liq. 203 (2015) 80-89.
https://doi.org/10.1016/j.molliq.2014.12.046

[150] H.J. Song, X. Zhang, T. Chen, One step synthesis of beta-FeOOH nanowire bundles/graphene oxide nanocomposites, J. Mater. Sci-Mater. El 25 (2014) 3680-3686. https://doi.org/10.1007/s10854-014-2075-z

[151] M. Namvari, H. Namazi, Synthesis of magnetic citric-acid-functionalized graphene oxide and its application in the removal of methylene blue from contaminated water, Polym. Int. 63 (2014) 1881-1888. https://doi.org/10.1002/pi.4769

[152] Z. Dong, D. Wang, X. Liu, X. Pei, L. Chen, J. Jin, Bio-inspired surface-functionalization of graphene oxide for the adsorption of organic dyes and heavy metal ions with a superhigh capacity, J. Mater. Chem. A 2 (2014) 5034-5040. https://doi.org/10.1039/C3TA14751G

[153] G. Xie, P. Xi, H. Liu, F. Chen, L. Huang, Y. Shi, F. Hou, Z. Zeng, C. Shao, J. Wang, A facile chemical method to produce superparamagnetic graphene oxide-Fe_3O_4 hybrid composite and its application in the removal of dyes from aqueous solution, J. Mater. Chem. 22 (2012) 1033-1039. https://doi.org/10.1039/C1JM13433G

[154] N. Cai, D. Peak, P. Larese-Casanova, Factors influencing natural organic matter sorption onto commercial graphene oxides, Chem. Eng. J. 273 (2015) 568-579. https://doi.org/10.1016/j.cej.2015.03.108

[155] M.T. Raad, H. Behnejad, M. El Jamal, Equilibrium and kinetic studies for the adsorption of benzene and toluene by graphene nanosheets: A comparison with carbon nanotubes, Surf. Interf. Anal. 48 (2016) 117-125. https://doi.org/10.1002/sia.5877

[156] D. Cortes-Arriagada, A. Toro-Labbe, A theoretical investigation of the removal of methylated arsenic pollutants with silicon doped graphene, RSC Adv. 6 (2016) 28500-28511. https://doi.org/10.1039/C6RA03813A

[157] I.S.S. de Oliveira, R.H. Miwa, Organic molecules deposited on graphene: A computational investigation of self-assembly and electronic structure, J. Chem. Phys. 142 (2015). https://doi.org/10.1063/1.4906435

[158] S. Thangavel, G. Venugopal, Understanding the adsorption property of graphene-oxide with different degrees of oxidation levels, Powder Technol. 257 (2014) 141-148. https://doi.org/10.1016/j.powtec.2014.02.046

[159] G. Moussavi, Z. Hossaini, M. Pourakbar, High-rate adsorption of acetaminophen from the contaminated water onto double-oxidized graphene oxide, Chem. Eng. J. 287 (2016) 665-673. https://doi.org/10.1016/j.cej.2015.11.025

[160] Y. Zhou, O.G. Apul, T. Karanfil, Adsorption of halogenated aliphatic contaminants by graphene nanomaterials, Water Res. 79 (2015) 57-67. https://doi.org/10.1016/j.watres.2015.04.017

[161] H. Yan, H. Wu, K. Li, Y. Wang, X. Tao, H. Yang, A. Li, R. Cheng, Influence of the surface structure of graphene oxide on the adsorption of aromatic organic compounds from water, ACS Appl. Mater. Inter. 7 (2015) 6690-6697. https://doi.org/10.1021/acsami.5b00053

[162] I. Ahmed, S.H. Jhung, Remarkable adsorptive removal of nitrogen-containing compounds from a model fuel by a graphene oxide/MIL-101 composite through a combined effect of improved porosity and hydrogen bonding, J. Hazard. Mater. 314 (2016) 318-325. https://doi.org/10.1016/j.jhazmat.2016.04.041

[163] A. Junkaew, C. Rungnim, M. Kunaseth, R. Arroyave, V. Promarak, N. Kungwan, S. Namuangruk, Metal cluster-deposited graphene as an adsorptive material for m-xylene, New J. Chem. 39 (2015) 9650-9658. https://doi.org/10.1039/C5NJ01975C

[164] S. Filice, D. D'Angelo, S. Libertino, I. Nicotera, V. Kosma, V. Privitera, S. Scalese, Graphene oxide and titania hybrid Nafion membranes for efficient removal of methyl orange dye from water, Carbon 82 (2015) 489-499. https://doi.org/10.1016/j.carbon.2014.10.093

[165] Q. Zhou, Y.H. Zhong, X. Chen, J.-H. Liu, X.J. Huang, Y.C. Wu, Adsorption and photocatalysis removal of fulvic acid by TiO2-graphene composites, J. Mater. Sci. 49 (2014) 1066-1075. https://doi.org/10.1007/s10853-013-7784-9

[166] Y. Zhang, G. Li, H. Lu, Q. Lv, Z. Sun, Synthesis, characterization and photocatalytic properties of MIL-53(Fe)-graphene hybrid materials, RSC Adv. 4 (2014) 7594-7600. https://doi.org/10.1039/c3ra46706f

[167] Y. Zhang, Y. Cheng, N. Chen, Y. Zhou, B. Li, W. Gu, X. Shi, Y. Xian, Recyclable removal of bisphenol A from aqueous solution by reduced graphene oxide-magnetic nanoparticles: Adsorption and desorption, J. Colloid Interf. Sci. 421 (2014) 85-92. https://doi.org/10.1016/j.jcis.2014.01.022

[168] Y. Wu, H. Luo, H. Wang, Synthesis of iron(III)-based metal-organic framework/graphene oxide composites with increased photocatalytic performance

for dye degradation, RSC Adv. 4 (2014) 40435-40438.
https://doi.org/10.1039/C4RA07566H

[169] J. Wang, T. Tsuzuki, B. Tang, L. Sun, X.J. Dai, G.D. Rajmohan, J. Li, X. Wang, Recyclable textiles functionalized with reduced graphene oxide@ZnO for removal of oil spills and dye pollutants, Aust. J. Chem. 67 (2014) 71-77.
https://doi.org/10.1071/CH13323

Chapter 2

Applications of Reverse Osmosis for the Removal of Organic Compounds from Wastewater: A state-of-the-art from Process Modelling to Simulation

I.M. Mujtaba [1],* M.A. Al-Obaidi [1,2] and C. Kara-Zaïtri [1]

[1] Chemical Engineering, School of Engineering, Faculty of Engineering and Informatics, University of Bradford. Bradford, West Yorkshire BD7 1DP, UK

[2] Middle Technical University, Iraq – Baghdad

E-mail: I.M.Mujtaba@bradford.ac.uk*

Abstract

This chapter presents the state-of-the-art on distributed models and associated performances of the most recent wastewater treatment methods based on the reverse osmosis (RO) process for the removal of high toxicological organic compounds from wastewater. The chapter presents the key challenges in connection with the removal of such harmful compounds. It provides a comprehensive critique on various models taking into account the impact of variable operating conditions and membrane dimensions and configurations. Whilst the literature review readily shows major RO successes in this aspect of work, it nevertheless highlights the continuing challenge of completely removing N-nitrosodimethylamine-D6 (NDMA) from wastewater.

Keywords

Wastewater Treatment, Reverse Osmosis, Modelling, Simulation, Organic Compound Removal

Contents

1. Wastewater and associated challenges

The fast-growing population and the associated increase in industrialization have resulted in a significant increase in groundwater contamination. The demand for freshwater sources in this changing environment with many draught areas seldom hitherto reported, is posing a real challenge. The trend of disposing large volumes of industrial effluents and sewage into rivers and oceans is set to increase in the short and long terms. It is not surprising to see that there have been various initiatives for implementing sustainable alternative solutions by recycling, reclaiming, and reusing of different types of water. Water reuse is on the increase even in countries with little or no water shortage [1] thus reducing the quantity of wastewater disposed to surface water. A significant volume of research continues to focus on the removal of micropollutants from wastewater as these adversely affect the natural ecosystem and human health. However, this is a great challenge because the organic pollutants found in wastewater can neither be easily nor cheaply removed. In general, wastewater treatment is a much more difficult process than water desalination, not only because of the complex toxicological compounds, which exist in the wastewater but also because such treatment would require advanced and integrated technologies [2].

The key challenges for removing pollutants from wastewater are listed below:

1. The unique regulation of the restricted concentration of new organic compounds such as N-nitrosamine in wastewater is complex.
2. A variety of organic compounds can be found in wastewater, which can lead to harmful chemical reactions and therefore high toxicological products.
3. The existence of some chemicals in the secondary treatment process of effluents with a very low concentration creates more complications during the removal process due to the complex analytical determination.

4. The redesign of existing water and wastewater treatment plants is not easy.

Extensive research work has been performed by solving the above challenges by developing complex treatment systems based on several technologies used alone or in combination.

2. Reuse water applications

Reclaimed and reused waters have been used in a wide variety of applications [1], which include:

1. Industrial reuse: Reuse water of low quality is utilized in cooling towers and power plants due to high-water demand.
2. Agricultural and irrigation reuse: These are considered as the highest consumers of recycled water for irrigating edible and non-edible agricultural crops. This is due to low cost of wastewater with nutrient content that eliminates the use of fertilizers.
3. Indirect potable reuse: Examples of these include facilities in Orange County, California, Windhoek, Namibia and NEWater in Singapore.

Table 1 reflects the tendency and the purpose of using recycling/reclaiming water in some countries [3].

Table 1: Purposes and rates of using wastewater in some countries [3].

Country	Purpose	Rate	Notes
Pakistan	Agricultural	96%	Non-treated wastewater
Tunisia	Agricultural	86%	Treated wastewater
Namibia	Municipal	29%	Namibia and Singapore have the most
Singapore	Municipal	45%	important water reuse for human
	Industrial	51%	consumption reclamation projects
USA	Industrial	45%	USA and Germany have a larger number of
Germany	Industrial	69%	recycling and reuse projects across various industries

3. Pollutants complexity challenges

Phenol and phenolic compounds (aromatic compounds) are considered as micropollutants that can be found in a variety of concentrations in wastewater effluents of many industrial processes such as refineries (6-500 ppm), coal (9-6800 ppm), petrochemical (28-1220 ppm) and pharmaceutical production (0.1-1600 ppm) [4,5]. These organic compounds are particularly harmful and cause an adverse impact on human health due to high toxicity even at low concentrations [6]. The oral tolerable daily intake of phenols has been limited down 0.5 mg/kg/day as confirmed by the European Food Standards Agency [7]. Also, the existence of trace amounts of phenol in industrial effluents can prevent the reuse of water in many applications [8]. There are currently a number of methods for removing these pollutants including adsorption on powdered activated carbon, UV/H_2O_2, catalyst wet air oxidation and membrane technology. The chemical oxidation and reverse osmosis processes have exhibited better performances for removing phenol from wastewater, which in turn aids to produce good quality water for several industries [5,9,10].

N-nitrosamines are considered as carcinogenic organic chemical compounds found in wastewater and formed due to the disinfection of secondary effluents, and the existence of inhibitors in the water [11]. The US Environmental Protection Agency (US EPA, 2009) [12] has restricted their concentration in recycled water to low levels of 0.7 ng/L. There are several treatment methods used to eliminate the risk of disposing NDMA from the reuse water, such as adsorption on activated carbon, ozonation and exposure to ultraviolet radiation [13]. However, the successful method is the combination of several treatment technologies including coagulation with ferric chloride, disinfection by chloramination, ultrafiltration UF, reverse osmosis (RO), and UV/H_2O_2 [14,15]. However, the main drawback of this advanced technology is the high cost of treatment. This resulted in diverting much attention to membrane technology for reducing treatment costs.

4. Process modelling complexity challenges

The modelling of an industrial process has always played an important role in the process design. Many challenges exist including a better understanding of the processing mechanism, rigorous evaluation of various designs via simulation and the optimisation of the process performance by calculating optimal operating conditions and process control parameters. These challenges are more complicated in industrial treatment processes involving reverse osmosis, but they have nevertheless been attempted with some success in a wide range of applications including textile, paper, electrochemical and biochemical industries, and seawater desalination. RO has been used in such applications due to its ability to separate impurities effectively commensurate with environmental demands. Further recent attempts have yielded improved optimisation models for removing organic compounds from wastewater [16,17].

5.   Simulation models challenges

The performance of many industrial processes can be analyzed using two simulation methods. Dynamic models use variable flow and loads, whereas steady-state models use constant flows and loads with respect to time. Dynamic simulation models are therefore much more complex than steady-state models because of their ability to predict the process response for a range of operating conditions in time.

6.   Wastewater treatment methods

A number of treatment technologies are currently available for removing organic pollutants from wastewater. However, the performance of applied technologies dependent on the nature and complexity of contaminants in the wastewater. These treatment technologies fall into one of the following categories [18]:

1.   Physicochemical treatment: coagulation-flocculation process, adsorption by activated carbon and advanced oxidation (i.e. ozone, UV) methods are used for the treatment of wastewater. Specifically, the oxidation method uses chlorine or ozone and classified as a highly-graded treatment technology.

2.   Biological treatment: activated sludge and biological trickling filters are used as sewage treatment system, which can renovate the organic pollutants into easily separated biomass using clarifiers.

3.   Advanced treatments: ultra-violet (UV) photolysis, ion exchange, membrane bioreactor process and membrane filtration technology such as reverse osmosis (RO) and nanofiltration (NF) treatment methods. These methods are always used to embed the microbiological safety of the reclaimed water.

The performance of one example of the physicochemical treatment and the advanced treatment of the RO process used for removing organic compounds from wastewater is discussed below supporting with comprehensive literature review.

6.1   The physicochemical oxidation treatment method

The process of the catalyst wet air oxidation (CWAO) is regarded as a well-known wastewater treatment method for detoxifying the organic compounds in the aqueous phase. Fig. 1 represents the schematic diagram of the underlying process, which is characterized by oxidizing the organic compounds using pure oxygen at a solid fixed catalyst packed inside a trickle bed reactor (TBR). The specifications of the catalyst ($Pt/\gamma-Al_2O_3$) and the characterization of the reactor used to remove phenol from wastewater in a pilot-scale size experiment are shown in Tables 2 and 3. The feed wastewater is collected in a feed tank and pumped to the process, while a compressor is used to compress O_2 which is concurrently fed to the reactor with the wastewater. The preferable operating conditions of the reactor are 4.93 to 197.38 atm and 100–350°C of pressure and temperature respectively. However, the other operating conditions such as initial phenol concentration, oxygen partial pressure, wastewater hourly space velocity and gas flow rate have to be managed to ensure the process stability. The hot outlet reactor stream is sent to a heat exchanger to regulate its temperature, while a separator is used to separate the liquid and gas for further analysis [5].

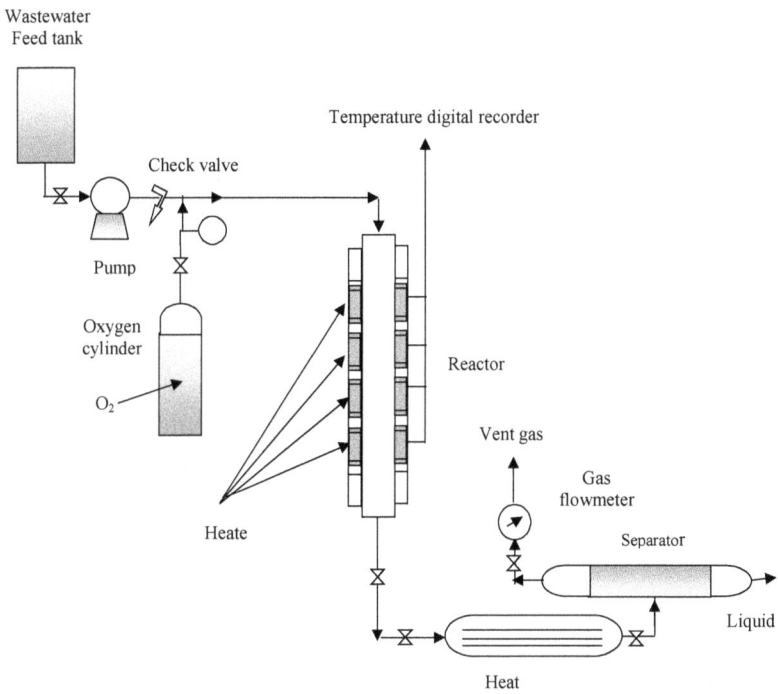

Figure 1. Schematic diagram of the CWAO process (Adapted from [5]).

Table 2: Characterisation of catalyst (Pt/ɣ–Al₂O₃) [19] (Adapted from [5]).

Parameter	Value
Particle shape	Sphere
Support	ɣ–Al2O3
Active phase	(0.48 wt%) Pt
Diameter particle(d_{pe})	1.6 (mm)
Specific surface area of particle (S_a)	259.9 (m²/g)
Pore volume (V_a)	0.308 (cm³/g)
Bulk density (ρ_{cat})	0.647 (g/cm³)
Calcination temperature	400°C

Table 3: Characterisation of the trickle bed reactor of CWAO process [19]) (Adapted from [5]).

Parameter	Value
Construction material	Stainless steel
Volume of catalyst in bed (V_{cat})	85 cm³
Length of bed catalyst (L_r)	30 cm
Inner reactor diameter (D_r)	1.9 cm
Length of reactor	77 cm

6.1.1 The model of Mohammed *et al.* [5]

Mohammed *et al.* developed a distributed model [5] consisting of a set of algebraic and differential equations and physical properties equations for the CWAO process considering the removal of phenol from wastewater using pure oxygen and catalyst (Pt/ɣ–Al₂O₃) type. The model developed is given in details in Table A.1 of Appendix A. The model is totally based on the work of Safaa [19], where it is carried out in a laboratory scale of TBR reactor.

6.1.1.1   Assumptions
1. The plug flow is assumed inside the reactor.
2. The validity of the film theory of Al-Dahhan *et al.* [20].
3. The phenol disappearance was described using the kinetic equation of Langmuir-Hinshelwood [21].

6.1.1.2   Parameter estimation technique

The unknown parameters of the membrane elements and the operating conditions are normally determined before solving model equations. In this simulation study, experimental data are used to predict the best values of unknown parameters, which are then used with the known parameters to check the behavior of the unit against variable operating conditions. Mohammed *et al.* [5] used linear and non-linear regression methods and implemented these using the parameter estimation tool of the gPROMS (general Process Modelling System) software suite to estimate the unknown model parameters. These include the reaction orders of phenol concentration (n), activation energy (EA) and oxygen partial pressure (m) as well as the pre-exponential factor (A^0) based on the experimental data of Safaa [19]. These methods of parameter estimation minimize the sum of square errors between the model predicted results and experimental data. The results of the parameter estimation method are listed in Table 4.

Table 4: Parameter estimation results of the CAWO process (Adapted from [5]).

Parameter	Symbol	Unit	Value
Order of phenol concentration	n	(-)	2.1066
Activation energy	EA	$(\dfrac{J}{mole})$	16315.735
Order of oxygen partial pressure	m	(-)	0.6112
Pre-exponential factor	A^0	$(sec^{-1}(\dfrac{cm^3}{mole})^{-1.11})$	668879.2

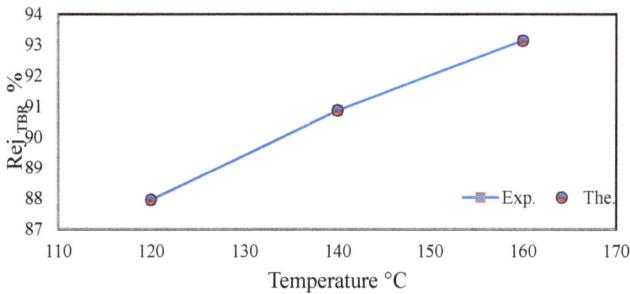

Figure 2. Model predictions and experimental results of phenol rejection verse operating temperature (Adapted from [5]).

6.1.1.3   Model validation

Fig. 2 readily shows full consistency between model predictions and experimental results of the variation of rejection parameter with the operating temperature.

6.1.1.4   The performance of the CWAO process for the removal of phenol

Mohammed *et al.'s* CWAO process model [5] is used to predict the behavior of the process according to different operating conditions of initial phenol concentration, reaction temperature, liquid hourly space velocity, oxygen partial pressure and gas flow rate based on the estimated values of the model' unknown parameters.

The process simulation is carried out by considering the following operating conditions.

- Initial phenol concentration ($C_{ph,L(in)}$): 1000, 3000 and 5000 mg/L.

- Gas flow rates (-): 20, 40, 80 and 100%.

- Oxygen partial pressure (P): 7.895, 9.869 and 11.843 atm.

- Reaction temperature (T): 120, 140 and 160°C.

- Liquid hourly space velocity ($LHSV$): 1, 2 and 3 1/h.

The simulation results showed that phenol rejection varied between 60 and 94%, indicating many dependencies on the operating conditions.

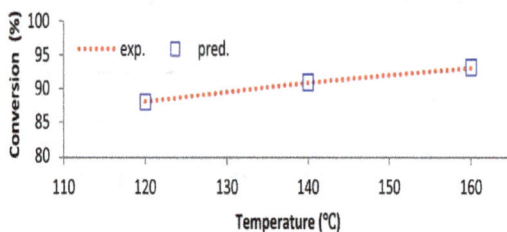

Figure 3. Effect of operating temperature on phenol conversion at constant liquid hourly space 1 1/h, reaction conditions (initial phenol concentration = 5 g/L, oxygen partial pressure = 0.8 MPa and gas flowrate = 80%) (Adapted from [5]).

The impact of feed temperature on phenol rejection is shown in Fig. 3. It is not difficult to see that higher conversion can be obtained using higher reaction temperatures. In other words, the reaction rate constant is directly proportional to the temperature and the activation energy is inversely proportional to the temperature. The impact of liquid hourly space velocity on phenol rejection is illustrated in Fig. 4, where an increase in liquid velocity causes a decrease in rejection parameter. This phenomenon can be ascribed to reducing the residence time of wastewater inside the reactor due to an increase in the liquid velocity, which in turn reduces the phenol conversion.

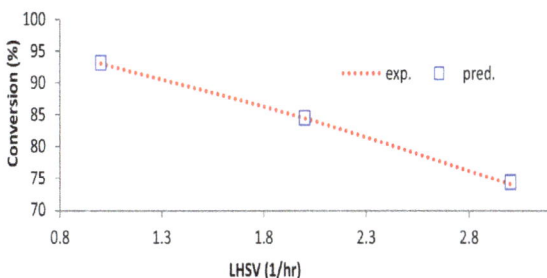

Figure 4. Effect of liquid hourly space on phenol conversion at constant operating temperature of 160°C, reaction conditions (initial phenol concentration = 5 g/L, oxygen partial pressure = 0.8 MPa and gas flowrate = 80%) (Adapted from [5]).

Fig. 5 shows the impact of oxygen partial pressure on phenol rejection at constant operating conditions, which indicates that an increase in the operating oxygen partial pressure from 0.8 to 1.2 MPa results in an increase in phenol rejection from 93 to 94.6%. This is because of the increasing gas density and solubility due to an increase in the gas partial pressure, which lifts the rejection parameter. Finally, the impact of initial phenol concentration on the performance of the process is shown in Fig. 6 at constant other operating conditions. The maximum phenol rejection of 94.75% is conducted at phenol concentration of 5 g/L, which indicates that increasing the phenol concentration can reinforce the rejection parameter due to increasing phenol molecules that cover a bigger catalyst area compared with the lower concentrations.

45

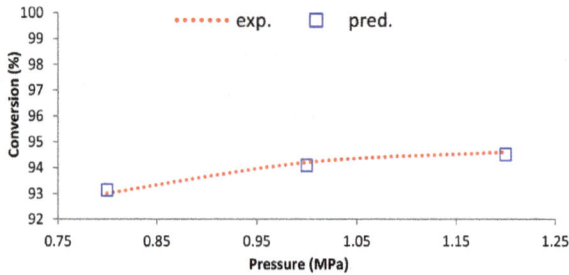

Figure 5. Effect of oxygen partial pressure on phenol conversion at reaction conditions of 160°C and initial phenol concentration = 5 g/L and gas flowrate = 80%) (Adapted from [5]).

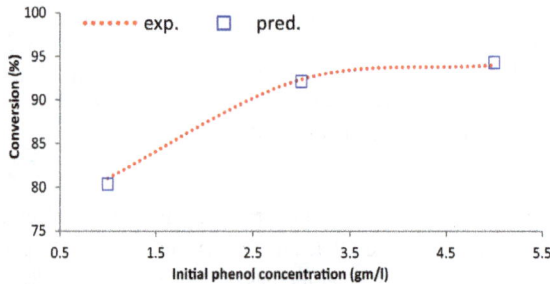

Figure 6. Effect of phenol concentration on phenol conversion at reaction conditions of 160°C and oxygen partial pressure 1.2 MPa and gas flowrate = 80%) (Adapted from [5]).

6.2   Reverse osmosis process

Reverse osmosis (RO) is a pressure driven process used to remove salts/pollutants from water so that it can be used for various purposes such as human consumption, as well as for agricultural and industrial uses [22]. RO has many immediate advantages including high packing density, minimum thermal damage and lower energy consumption [23]. More specifically, the spiral wound membrane modules are often preferred in both desalination and industrial processes since these offer particular characteristics of accepted permeation rates, ease of operation, replace, high membrane area to volume ratio, and low water production costs [24]. Furthermore, spiral wound RO modules are less liable to membrane fouling and are easier to clean [25]. Therefore, the modelling of the spiral wound module will be discussed in detail below to highlight the performance of the RO process for the removal of organic compounds from wastewater.

Spiral wound modules contain one or more membrane leaves based on the module type and diameter. Two back-to-back glued membrane sheets with a feed spacer are used to form one leaf, which are rolled around a perforated central tube of the produced water. The feed spacer provides a channel for the brine, which is forced to flow along the membrane length and exit from the opposite end as a concentrated brine. The fresh water penetrates the membrane and is collected in the permeate side, where it flows towards the centred tube.

Fig. 7 presents a schematic diagram of the spiral wound module and the two opposite directions of the brine and produced water flow inside the module. The x-axis represents the axial coordinate along the membrane length L. While, the y-axis represents the tangential coordinate in the spiral direction starting from the sealed end of the leaf to the end of the membrane width W. The effective membrane area can be written as $A = LW$. Also, t_f and t_p are denoted the height of the feed and permeate spacers channels respectively. The membrane area is equally divided into several sub-sections of equal Δx and Δy, where Δx and Δy are the lengths of sub-section in the x and y coordinates respectively. Therefore, the area of the sub-section is given by Eq. (1).

$$A_{sub-sction} = \Delta x \, \Delta y \tag{1}$$

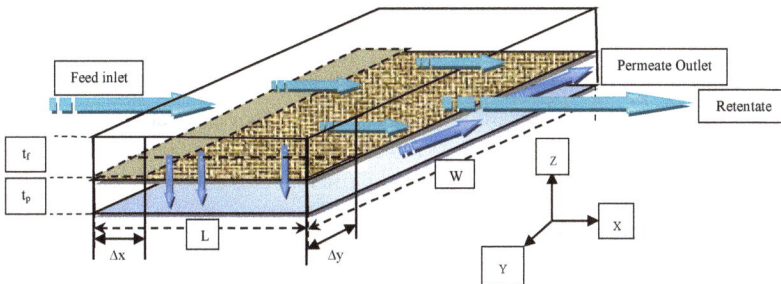

Figure 7. Schematic diagram of the spiral-wound module (Adapted from [26]).

The RO technology has seen a significant growth in water recycling and wastewater treatment in several industries [27]. For example, RO is widely considered for wastewater/effluent treatments in (a) textile industry [28], (b) dairy industry [29,30], (c) tannery industry [31] and (d) pharmaceutical industry [32]. These have stimulated continued research with an ultimate objective of maximizing the performance of the unit and, at the same time, reducing the cost of filtration.

6.2.1 The feasibility of the RO process for wastewater

Reverse osmosis has seen a significant increase in water desalination and treatment applications. This section discusses the state-of-the-art in respect of the feasibility and reliability of the RO process for removing organic compounds from wastewater.

Chai *et al.* [33] used a filtration system of two spiral wound non-cellulose composite membranes (total area of 10 m²) type ATFRO (APV, Denmark) linked in a series configuration in addition to a UF membrane for the pre-treatment of an original electroplating wastewater containing copper. Experimental results showed that the maximum copper rejection was around 96.4%.

Several membrane types (HR95PP, SEPA-MS05, DESAL-3Band DESAL-3LP) were tested by Bódalo-Santoyo *et al.* [34] in a pilot-scale filtration batch system of an individual flat cell RO module supplied by INDEVEN (Spain) to treat a multi-component synthetic effluent stream of inorganic species of sulfate, ammonium and cyanide compounds. The sulfate rejection varied between 96 and 99.5% where HR95PP and DESAL-3LP showed the highest rejections. However, the ammonium was retained in lower values than sulfate, where HR95PP showed a maximum of 84% and a range of 74 to 78% for the rest of the membranes. Surprisingly, the cyanide rejection was around 19% for HR95PP with negative rejections (higher permeate concentration compared to feed concentration), as found for other tested membranes. Bódalo-Santoyo *et al.* [35] used the same apparatus and the effluent stream to test the performance of membrane DESAL-3 (polyethersulfone) of 3E-3 m². The results showed that the rejection of ammonium, cyanide, and sulfate in the range of 92.7-93.5%, 88-90% and 93.8-93.9% respectively, which were dependent on the use of different feed concentrations.

Abo-Qdais and Moussa [36] tested the performance of a bench scale membrane module containing tubular spiral wound RO membrane of 2.5 m² to treat synthetic wastewater samples of different concentrations of copper and cadmium. The process efficiencies of removing Cu and Cd pollutants from wastewater were 98 and 99% respectively.

Mohsen-Nia *et al.* [37] studied the performance of a laboratory scale RO thin film (1.95 m²) composite spiral wound membrane (RE2012-100) for removing copper and nickel from the mixed salt system. The rejection values were around 98% with the slightly higher removal of copper due to its larger size. These results confirmed that the underlying process can remove these pollutants effectively.

Gómez et al. [38] investigated the impact of the nature of membrane and operational variables on the performance of a bench scale of flat sheet membrane (3E-3 m²) for removing aniline from aqueous solutions. This study concluded that membrane (HR98PP) has the highest rejection of 91.8% while the lowest rejection of 79% was obtained with MS05. Also, the aniline rejection was slightly increased due to an increase in operating pressure.

The pilot-plant scale of the RO process of a spiral wound module type 2521 TF (Korean CSM company) was used by Mohammadi et al. [39] to study the performance of the process for removing chromium (Cr^{+6}) from electroplating industry's effluents. Under optimum operating conditions, the highest removal of chromium was about 99%.

Sagne et al. [40] investigated the impact of operating variables and membrane type (CPA2 and ESPA2 from Hydanautics and BW30 from DOW) (2.6 m²) for eliminating small organic solutes of carboxylic acids and alcohols from distillery condensates using a recycle mode RO pilot-plant of an individual spiral wound module. The experiments showed that caproic acid and 2.3-butanediol were completely removed. Moreover, the membrane type BW30 achieved maximum removal of more than 80% of butyric acid, valeric acid, and 2-phenylethanol solutes compared to other membranes tested.

Madaeni and Koocheki [41] studied the efficiency of the cross-flow filtration system of two thin film composite spiral wound membranes type Filmtec TW30HP-4641 (11.89 m²) in a series configuration to remove phosphate, sulfite, nitrite, and nitrate from wastewater. The experiment results showed maximum rejection of 91, 93, 95 and 98-99% for nitrite, nitrate, sulfite and phosphate ions, respectively. However, the rejection of nitrite and nitrate ions can be improved to 99% by the addition of KH_2PO_4 into the solution.

Tabassi et al. [42] investigated the performance of a pilot plant for an individual commercial polyamide thin film composite RO membrane spiral wound type SG 2514TF (Osmonics company) of 0.6 m² in removing phenol from synthetic aqueous solutions of different concentrations. The effect of the operating parameters on the process performance was studied where the maximum achieved rejection exceeded to 80%.

Khazaali et al. [43] confirmed that the maximum rejection of bisphenol A (BPA) from aqueous solutions using cross-flow filtration system of a low-pressure polyamide thin film composite spiral wound RO membrane type TW30-1812-100(0.446 m²) was around

87%. The research was directed to pinpoint the optimum operating pressure that commensurates with the highest rejection.

Finally, Thirugnanasambandham *et al.* [44] evaluated the performance of a bench scale filtration system of a spiral wound membrane (2 m²) to treat wine industry wastewater under several operating conditions. The results revealed that all the operating conditions have a significant impact on the process performance and the optimum conditions, were explored for the maximum rejections of 91, 93 and 97% of color, COD, and TDS respectively.

6.2.2 The RO process modelling

There are a number of models available to research the membrane performance of a spiral wound module with different features and applications based on some assumptions. These models have been applied to seawater and brackish water and validated against wastewater experimental data. However, the majority of these models did not take into account the impact of operating parameters variation along the membrane dimensions. To the best of the authors' knowledge, the available literature includes a few attempts which explored such operating parameters and their impact on the membrane length and width, especially for the water treatment industry.

The next section focuses on an extensive literature review on distributed modelling of the RO processes considering both desalination and water treatment RO processes. This section will highlight the latest distributed models developed for the removal of organic compounds from wastewater. It is also aimed to evaluate the importance of using distributed modelling as well as the efficiency of the RO process for the removal of organic compounds from wastewater.

6.2.3 Distributed modelling of the spiral wound RO process

The development of distributed models is based on a thorough understanding of the flow pattern inside the module for enhancing the process design. It can be argued that this is different in lumped models.

Al-Obaidi and Mujtaba [45] confirmed that most of the available models are based on the solution-diffusion model and the three-parameter irreversible thermodynamics model. The available current literature is discussed in the following section.

Analytical models were developed by Rautenbach and Dahm [46] for a spiral wound module and worked out by Evangelista [24], for high rejecting membranes and by Avlonits *et al.* [47] and Boudinar *et al.* [48] for both Roga and FilmTech membrane types respectively. These models considered the validity of the solution-diffusion model with a fully axial flow of brine solution but they neglected the components of the tangential feed flow and the axial permeate flow. Also, these models assumed constant density and viscosity and ignored the concentration polarization impact. In addition to this, some of these models did not consider the pressure drop in the brine and permeate compartments.

Based on the irreversible thermodynamic model, Spiegler and Kedem [49], Senthilmurugan *et al.* [50] and Mane *et al.* [51] developed models for turbulent flow by considering the pressure drop in both channels. Whereas, Mane *et al.* [51] considered two dimensions (x and y) for the feed flow rate and stimulated the rejection of boron by the RO process. Geraldes *et al.* [52] developed a one-dimensional model for spiral wound RO membranes by ignoring the diffusion flow in the feed channel and also the pressure drop in the permeate channel. Sagne *et al.* [53] investigated an improved dynamic one-dimensional model based on the solution-diffusion methodology to simulate the removal of five volatile organic compounds from brackish water albeit by neglecting the concentration polarization impact and degrading the solute flux. Avlonits *et al.* [54] showed a two-dimensional model by assuming only convective flux and neglecting diffusive mass transport and ignoring the variance of permeate concentration along the axial and spiral directions. Oh *et al.* [55] developed a one-dimensional model based on the solution-diffusion model for spiral wound RO system. They assumed constant mass transfer coefficient and constant water flux in the case of changing the inlet feed flow rate. They also ignored the pressure drop in the permeate channel. Kaghazchi *et al.* [56] proposed a one-dimensional model based on the solution-diffusion methodology where the bulk flow rate was calculated as an average value of inlet and outlet feed flow rates.

All the above models were validated against seawater and brackish water experimental data. Observations made on these proposed models, which have been validated against wastewater treatment data are summarised in the next section.

Sundaramoorthy *et al.* [10,57] suggested a one-dimensional model based on the solution-diffusion model by assuming constant values for both the pressure and permeate concentration at the permeate channel. The experimental data of chlorophenol and dimethylphenol removal from wastewater were used to validate the model developed. Fujioka *et al.* [58] developed a one-dimensional model based on the Spiegler and Kedem model and considered the variation of the operating parameters and assuming zero permeate pressure. The model is validated against experimental data of N-nitrosamine rejection. Al-Obaidi and Mujtaba [45] and Al-Obaidi *et al.* [26,59] developed one and two-dimensional steady-state and dynamic models for simulating the removal of organic compounds from wastewater using an individual element of RO membrane. The models were developed based on the combined principles of the solution-diffusion model, the Spiegler and Kedem model and the concentration polarisation mechanism. The models were validated against experimental data of chlorophenol, N-nitrosamine and dimethylphenol removal from wastewater.

From the above, it can be said that the solution-diffusion and the irreversible thermodynamic models have been used more widely to understand the mechanism of transport phenomena through the membrane. Both Murthy and Gupta [64] and Mujtaba [65] affirmed the accuracy of the Spiegler and Kedem and the solution-diffusion models to express the mechanism of water and solute transport in the RO process.

Several distributed models, which were developed for the spiral wound RO process for the removal of phenolic and N-nitrosamine compounds from wastewater, will be discussed in the following paragraphs.

6.2.4 The model of Sundaramoorthy *et al.* [10]

A one-dimensional model has been developed for an individual spiral wound RO process by Sundaramoorthy *et al.* [10]. The model was able to predict the spatial variation of operating parameters along the membrane length. The proposed model equations are listed in Table A.2. of Appendix A.

6.2.4.1   Assumptions

1. The validity of the solution-diffusion model to elucidate the transport phenomena of water and solute.
2. Constant permeate pressure of 1 atm at the permeate channel.
3. The validity of the Vant Hoff's relation to express the osmotic pressure.
4. The validity of the film theory model to characterize the concentration polarisation.
5. Constant values of water and solute transport parameters and friction factor.
6. The constant solute concentration at the permeate channel.

6.2.4.2   Experimental method

Srinivasan *et al.* [9] and Sundaramoorthy *et al.* [57] used a pilot-scale RO filtration system of a commercial thin film composite membrane packed into a spiral wound module in their experimental work to remove chlorophenol and dimethylphenol from aqueous solutions of different concentrations. The solutes concentrations varied from 0.778E-3 to 6.548E-3 kmol/m³. The feed was pumped at three different flow rates of 2.166E-4, 2.33E-4 and 2.583E-4 m³/s with a set of inlet feed pressures varying from 5.83 to 13.58 atm for each flow rate. Fig. 8 shows a schematic diagram of the corresponding RO filtration system.

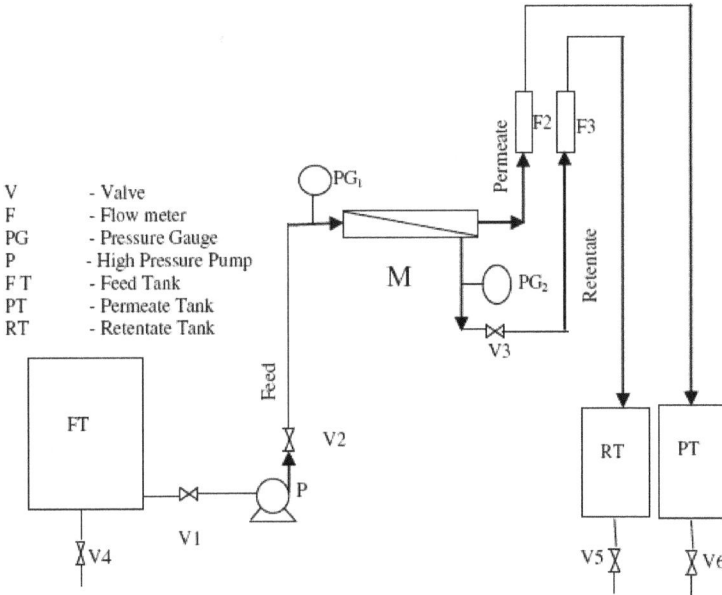

Figure 8. Schematic diagram of an individual RO process (Adapted from [9]).

6.2.4.3   Parameters estimation

Sundaramoorthy *et al.* [10] also developed a graphical method for estimating the model unknown parameters including water A_w and solute B_s transport parameters and the friction factor b. Table 5 shows the parameters considering the membrane type Ion Exchange, India for the experiments of chlorophenol and dimethylphenol removal from wastewater. While, Table 6 shows the characteristics of the RO membrane used.

Table 5: Results of parameter estimation (Adapted from [57] and [9]).

Solute	Parameter	Value
Chlorophenol	b	$8529.45\left(\frac{atm\ s}{m^4}\right)$
	A_w	$9.5188E\text{-}7\left(\frac{m}{atm\ s}\right)$
	B_s	$8.468E\text{-}8\left(\frac{m}{s}\right)$
Dimethylphenol	b	$9400.9\left(\frac{atm\ s}{m^4}\right)$
	A_w	$9.7388E\text{-}7\left(\frac{m}{atm\ s}\right)$
	B_s	$1.5876E\text{-}8\left(\frac{m}{s}\right)$

Table 6: Input data: Membrane characteristics and geometry (Adapted from [57]).

Company	Ion Exchange, India
Membrane material and configuration	TFC Polyamide, spiral-wound
Feed spacer thickness (t_f)	0.8 *mm*
Permeate channel thickness (t_p)	0.5 *mm*
Number of turns	30
Module width (W)	8.4 *m*
Module length (L)	0.934 *m*
Module diameter	3.25 *inches*

6.2.4.4 Model validation

The model developed has been validated against experimental data of chlorophenol [57] and dimethylphenol [9] removal from wastewater. A maximum error of 15% was found in estimating the permeate concentration with a large range of very low errors associated with the other operating parameters of outlet feed flow rate and rejection parameter tested. Therefore, it can be said that the model is able to estimate the performance of an individual spiral wound module with sufficient accuracy.

6.2.4.5 The performance of the RO process for the removal of phenolic compounds

The experimental results of Sundaramoorthy *et al.* [57] and Srinivasan *et al.* [9] confirmed that an individual spiral wound RO module is able to remove chlorophenol and dimethylphenol from wastewater at a maximum level of 83 and 97.3% respectively depending on the range of operating parameters used.

6.2.5 The model of Fujioka *et al.* [58]

A comprehensive work to investigate the performance of a full-scale RO plant was carried out by Fujioka *et al.* [58] who developed a specific one-dimensional model based on the irreversible thermodynamics principles and hydrodynamic calculations to investigate the total N-nitrosamine rejection from wastewater. The model equations are given in Table A.3 in Appendix A.

6.2.5.1   Assumptions

1. Spiegler and Kedem's model was used to describe the mass transport through the module.
2. The pressure drop in the feed section was identified using the Schock and Miquel model.
3. Local permeate pressure at the permeate channel was taken as zero.
4. The underlying process was assumed to be isothermal.

6.2.5.2   Materials and methods

Fujioka *et al.* [58] used a full-scale RO filtration system of three 4-inch glass-fibre pressure vessels shown in Fig. 9 in the experiments of eight N-nitrosamine solutes rejection with a molecular weight in the range of 74 to158 g/mol as summarised in Table 8 (note NDMA, NPYR and NEMA only were selected here). The filtration experiments were carried out by introducing the stock solution of N nitrosamine compounds in Milli-Q water conditioning with 20 Mm NaCl, 1 Mm CaCl$_2$ and 1 Mm NaHCO$_3$ to obtain approximately 250 ng/L of each target Nitrosamine compound. Specifically, one spiral wound element type ESPA2-4040 is stuffed in each pressure vessel, where a series connections is used. Therefore, the concentrated feed solution of the first vessel was transferred to the second vessel followed by the third one. The permeate was then collected from the stages and recycled back with the retentate solution into the feed tank to maintain constant feed concentration. The feed was pumped at constant volumetric flow rate of 2.43E-3 m³/s, while the average permeate flux was fixed at 2.78E-6,5.56E-6 and 8.33E-6 m/s, which were equivalent to 10, 20 and 30 Lm²/h respectively, during the experiments by increasing the feed pressure from 4, 6.5 and 10.1 atm, which are equivalent to 0.4, 0.659 and 1.02 MPa respectively. The feed temperature was adjusted at 20 ± 0.1 °C during the experiments. The specifications of the spiral wound membrane element are giving in Table 7.

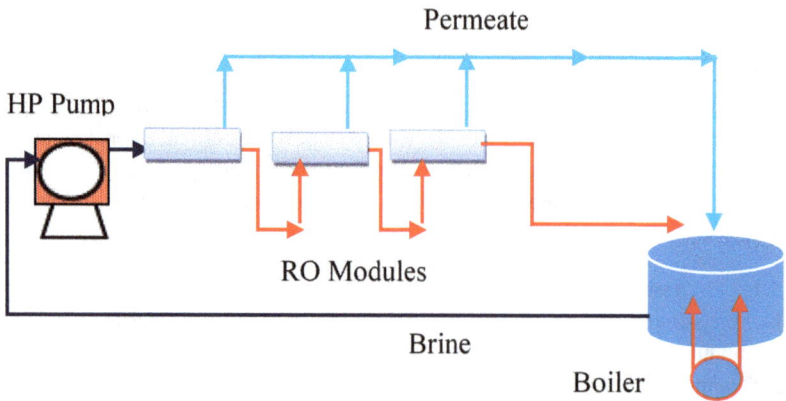

Figure 9. Schematic diagram of full-scale three elements RO plant (Adapted from [58]).

6.2.5.3 Parameter estimation

The unknown friction parameter of the model was estimated using an iteration procedure to minimise the difference between the model prediction and the observed feed pressure. This method yielded the values of 3.9, 4.3 and 5.5 at permeate flow rate of 10, 20 and 30 Lm²/h respectively. Water and solute permeability constants and the reflection coefficient of each solute are provided in Table 8.

Table 7: Specifications of the spiral-wound membrane element (Adapted from [59]).

Make	Hydranautics, Oceanside, CA., USA
Membrane type and configuration	ESPA2-4040, Spiral-wound, Composite Polyamide
Feed and permeate spacer thickness t_f and t_p (m)	6.6E-4
Membrane sheet area (m²)	7.9
Maximum feed flow rate (m³/s)	4.44E-3
Membrane sheet length L (m)	0.9
Membrane sheet width W (m)	8.7778

Table 8: Physical and transport parameters of the selected N-nitrosamines (Adapted from [61,58]).

Name	Abbreviation	Molecular weight (g/mol)	Permeability coefficient, B_s (m/s)	Reflection coefficient, σ (dimensionless)
N-nitrosodimethylamine-D6	NDMA	74.05	5.35E-6	0.953
N-nitrosopyrrolidine-D8	NPYR	100.06	5.12E-7	0.973
N-nitrosomethylethylamine-D3	NMEA	88.06	1.14E-6	0.958
Pure water permeability at 20°C L_p = 5.2 ($\mp$ 0.2) L/m²h bar				

6.2.5.4   Model validation

Figs. 10 and 11 demonstrate the comparison between the model predictions and the experimental data of the feed pressure variation along the membranes length (series of three membranes) and the rejection parameters of the three tested N-nitrosamine compounds. The model prediction was almost consistent with the experimental results, especially at low and medium feed pressures. However, the model overestimated the rejections especially of NDMA, which was probably caused due to inaccurately estimating its solute transport parameter value.

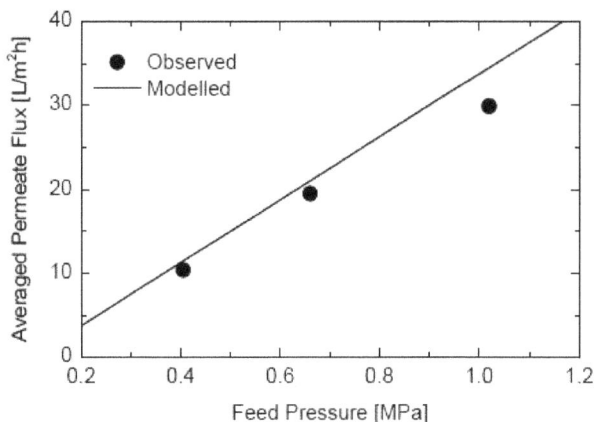

Figure 10. Model prediction and observed feed pressure (Adapted from [58]).

6.2.5.5   Effects of processing parameters on the RO process performances

Fig. 11 shows the relation between the rejection of three compounds of N-nitrosamine and the permeate flux. It is not difficult to see that an increase in the rejection of N-nitrosamine occurs due to an increase in the permeate flux, which is already associated with an increase in the feed pressure. Finally, the simulation results of NDMA rejection showed a range of 40 to 61% for a system of three membranes in a series configuration compared to other N-nitrosamine compounds, which confirm a poor performance of the RO process for the removal of NDMA from wastewater. The same findings are confirmed by Steinle-Darling et al. [14] who noticed that the NDMA removal using a pilot-scale of three flat-sheet of commercial RO membranes was between 54 – 70%. While Plumlee et al. [62] and Krauss et al. [63] affirmed the NDMA removal ranges between 24 – 56% and 40 – 70% respectively of the RO process.

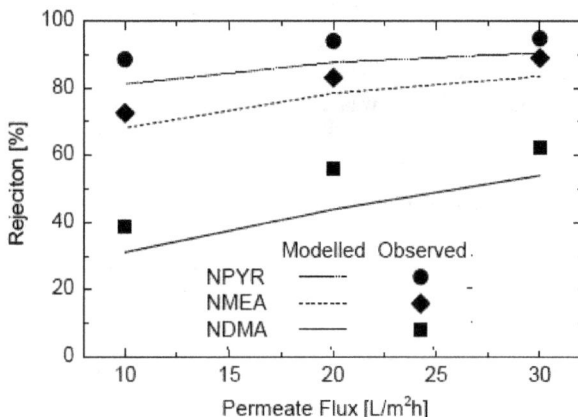

Figure 11. Model prediction and observed permeate flux of three membranes (Adapted from [58]).

6.2.6   The model of Al-Obaidi et al. [59]

Al-Obaidi et al. [59] developed a one-dimensional mathematical model of integrated differential and algebraic equations to investigate the separation mechanism of N-nitrosamine from wastewater in a spiral wound RO process. The model was based on Spiegler and Kedem's principle on mass transport, Darcy's law and the concentration polarization to investigate the pressure drop and mass transfer coefficient in the module feed channel respectively. The physical properties equations were considered identical to the water equations proposed by Koroneos *et al.* [60] due to the use of dilute aqueous solutions of N-nitrosamine in the experiments of Fujioka *et al.* [58]. The model equations are provided in Table A.4 in Appendix A.

6.2.6.1   Assumptions

The following assumptions were made to develop the proposed process model:

1. Spiegler and Kedem's principles were used to elucidate the transport phenomena.
2. The Darcy's law is valid where the friction parameter was used to characterize the pressure drop in the feed and permeate channels.
3. Constant solute concentration in the permeate channel and the average value were calculated from the inlet and outlet calculated concentrations.
4. Isothermal filtration process.

6.2.6.2   Determination of transport parameters

The friction and the model transport parameters (b, L_p, B_s) for each run were calculated using the gEST parameter estimation of the gPROMS software based on the experiments conducted by Fujioka *et al.* [58]. The values used for the friction parameters were 62, 194 and 395 (atm s/m^4) for the average permeate fluxes 2.78E-6, 5.56E-6 and 8.33E-6 m/s respectively. The selected membrane transport parameters of the three N-nitrosamine solutes (NDMA, NEMA and NPYR) and the water transport parameter L_p are provided in Table 9. The estimated values presented in Table 9 are close to those reported by Fujioka *et al.* [58].

Table 9: Physical and transport parameters of the selected N-nitrosamines (Adapted from [59]).

Name	Molecular weight (g/mol)	Water permeability coefficient, (L_p) (m/s atm) at 20 °C	Permeability coefficient, B_s (m/s)	Reflection coefficient, σ (dimensionless)
NDMA	74.05	1.06E-6 – 1.27E-6	4.75E-6 – 5.36E-6	0.951-0.985
NEMA	88.06	1.05E-6 – 1.21E-6	1.15E-6 – 1.23E-6	0.936-0.989
NPYR	100.06	1.05E-6 – 1.23E-6	4.18E-7 – 4.39E-7	0.983-0.998

6.2.6.3   Model validation

The model developed by Al-Obaidi *et al.* [59] has been validated against the actual experimental results of Fujioka *et al.* [58]. Fig. 12 compares the experimental results and the model prediction of the three selected N-nitrosamines rejections at three different overall permeate fluxes. Generally, the clear corroboration with experimental data readily shows the correctness of the model to measure the observed rejection data with a marginal error within high operating pressures. While, at low operating pressure (low average permeate flux), Fig. 12 shows that the model is able to simulate the observed data with a minor deviation. This might be ascribed to the inaccurate estimation of the transport membrane parameters at such pressure. In addition, Figs. 13 and 14 show the linear fittings with a regression coefficient R^2 (close to 1) for both the experimental and model prediction of the outlet feed flow rate and pressure respectively.

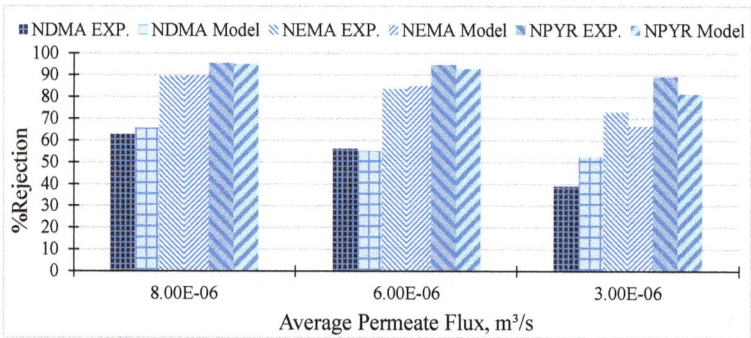

Figure 12. Experimental and model rejections of NDMA, NEMA and NDEA with average permeate flux (Adapted from [59]).

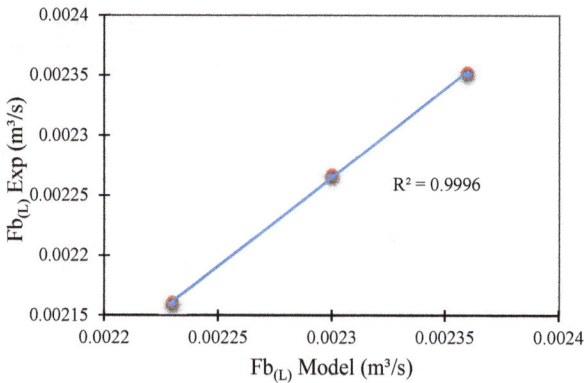

Figure 13. Linear fitting of experimental and model outlet feed flow rate (Adapted from [59]).

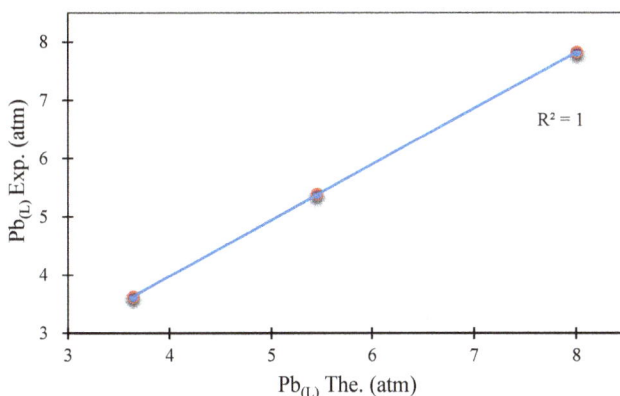

Figure 14. Linear fitting of experimental and model outlet feed pressure (Adapted from [59]).

6.2.7 The model of Al-Obaidi *et al.* [26]

The novel concept Adapted by Al-Obaidi *et al.* [26] lies in the development of an explicit simple two-dimensional spiral wound RO model for the removal of organic compounds from wastewater. This model was based on the solution-diffusion model and relaxed the assumptions of constant physical properties, constant pressure and concentration of the fresh water on the permeate side was considered in the past by many researchers. In addition, the impact of the plug-flow and diffusion flow was taken into account to express variation of the brine concentration along the membrane length and width. Moreover, the model characterizes the consideration of the tangential feed flow and axial permeate flow rates which were not considered in most existing spiral wound published models. The model developed was useful to study the effect of various operating parameters on the performance of the RO process.

6.2.7.1   Assumptions

A number of reasonable assumptions and simplifications were used in order to develop this model:

1. The module was made up of a non-porous flat sheet with spacers and negligible leaf curvature.

2. The validity of the solution-diffusion model for the transport of the solvent and solute through the membrane.
3. The validity of Darcy's law for the feed and permeate channels, which assumes that the pressure drop is proportional to the feed and permeate flow rate in the case of laminar flow conditions and the friction parameter is applied to characterize the pressure drop.
4. The validity of the film model theory to estimate the concentration polarization impact.
5. The main value of the permeate concentration for all the increments has been taken as the total freshwater concentration since the flowing of outlet water is in the spiral direction towards the perforated collected pipe.

6.2.7.2   Governing equations

Table A.5 in Appendix A presents the model equations of Al-Obaidi et al. [26] used to estimate the performance of an individual spiral wound RO process to remove organic compounds from wastewater. The physical properties equations were considered identical to water equations proposed by Koroneos et al. [60] due to the use of dilute aqueous solutions of dimethylphenol in the experiments of Srinivasan et al. [9].

6.2.7.3   Parameter estimation

The parameters of the model were estimated using the proposed graphical method of the linear fit of Sundaramoorthy et al. [10]. This method was used to determine the values of solvent transport coefficient A_w, solute transport coefficient B_s and the feed channel friction parameter b. The details of these parameters are provided in Table 10. The estimated values of transport parameters showed some difference in comparison with the values suggested by Srinivasan et al. [9] shown in Table 5.

Table 10: Parameter estimation results (Adapted from [26]).

Parameter	Value
A_w	9.42009E-7$\left(\frac{m}{atm\ s}\right)$
B_s (dimethylphenol)	2.22577E-8$\left(\frac{m}{s}\right)$
b	9400.9 $\left(\frac{atm\ s}{m^4}\right)$

6.2.7.4   Model validation

For aqueous solutions of dimethylphenol, Tables A.6 to A.8 in Appendix A depict the experimental results of Srinivasan *et al.* [9] and the model predictions for three groups of feed flow rates; (each group holding five different feed concentrations under five different feed pressures) with estimating percentage of error between the experimental and the model predictions. Tables A.6 to A.8 in Appendix A compare experimental and model prediction results for a range of operating parameters including the retentate concentration, retentate pressure, retentate flow rate, average permeate concentration and average solute rejection with a set of different inlet feed flow rates, pressures, and concentrations. Generally, a good agreement between the model prediction and experimental results can be seen over the ranges of pressures, feed flow rates, and concentrations. Fig. 15 shows this for solute rejection for the whole data within 2.1% error. The model is able to predict the permeate concentration within a maximum of 15% error (Tables A.6 to A.8 in Appendix A) and less than 4% error for about 76% of retentate flow rate readings. Finally, 79% of retentate pressure readings are within 4% error. The model has been used for simulation as reported in the following section.

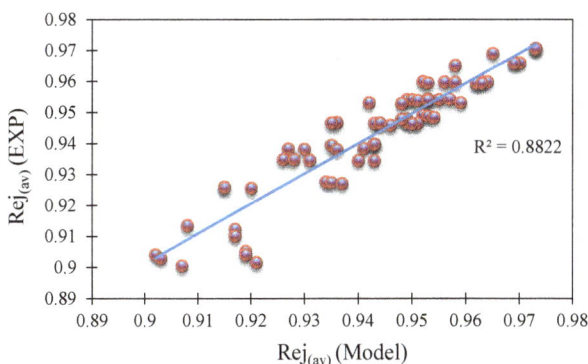

Figure 15. Comparison of experimental and model prediction of solute rejection (Adapted from [26]).

6.2.7.5   The effect of processing parameters on the RO process performances

The model was used to simulate the process, explore the sensitivity of the model to different parameters of the process, and to check the membrane performance under the impact of varying the process parameters, such as inlet feed flow rate, inlet feed concentration, inlet feed pressure, and inlet feed temperature.

The two-dimensional results presented in Figs. 16 and 17 show the output data and provide a good insight into the steady state feed flow rate and pressure in the two dimensions throughout the membrane sheet. The feed flow rate decreases quite quickly as water passes through the membrane with a rapid increase in the osmotic pressure due to an increase in solute concentration. It was also noted that the feed pressure drops at the end of the membrane length due to increasing pressure loss, in turn, due to the friction of the wall of the membrane. Fig. 18 clearly shows the spatial progress of feed concentration in two dimensions in the feed channel due to retained solute along the membrane wall. The net result was an increase of the concentration polarization and osmotic pressure caused by the build-up of the solute on the membrane thus resulting in a reduction in water flux.

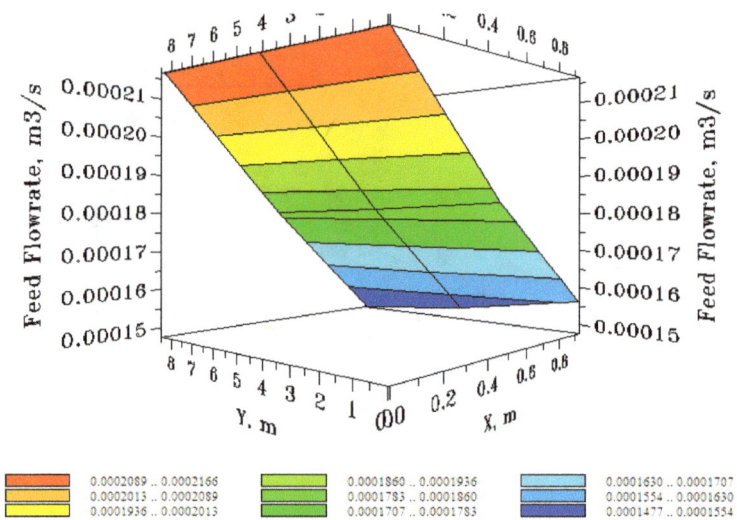

Figure 16. Steady-state feed flow rate along the two-dimensions at inlet feed conditions (2.166E-4 m³/s, 6.548E-3 kmol/m³, 13.58 atm and 31.5°C) (Adapted from [26]).

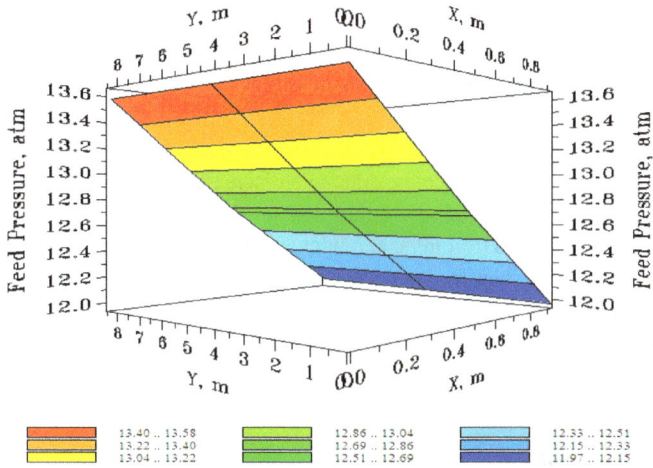

Figure 17. Steady-state feed pressure along the two-dimensions at inlet feed conditions (2.166E-4 m³/s, 6.548E-3 kmol/m³, 13.58 atm and 31.5°C (Adapted from [26]).

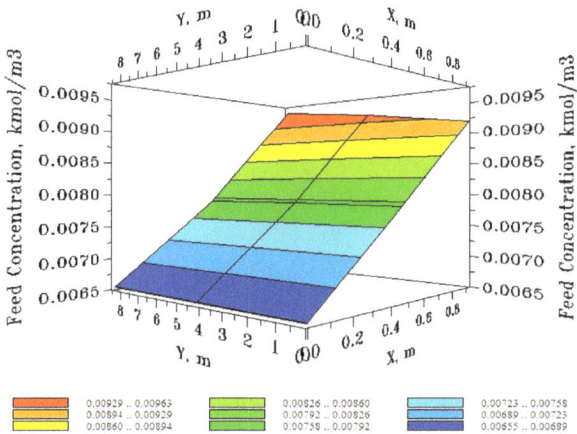

Figure 18. Steady-state solute concentration along the two-dimensions at inlet feed conditions (2.166E-4 m³/s, 6.548E-3 kmol/m³, 13.58 atm and 31.5°C) (Adapted from [26]).

Fig. 19 shows the response of solute rejection for the variation in both inlet feed pressure from (5.83 to 13.58 atm) and inlet feed concentration from (0.819E-3 to 6.548E-3 kmol/m³) with constant values of inlet feed flow rate and temperature (2.583E-4 m³/s and 31°C) respectively. It is worthy to note that the dimethylphenol rejection varies between 88 and 96.8%. Also, the solute rejection increases because of an increase in the inlet feed pressure and concentration. This is due to an increase in the water flux by increasing the inlet feed pressure, which reduces solute concentration in the permeate channel. Also, the membrane rejection intensity increases due to an increase in the inlet feed concentration. Fig. 20 depicts the response of solute rejection for the variation in both inlet feed flow rate from 2.166E-4 to 2.583E-4 m³/s, and inlet feed pressure from 5.83 to 13.58 atm with constant values of high inlet feed concentration and temperature (6.548E-3 kmol/m³ and 31°C) respectively.

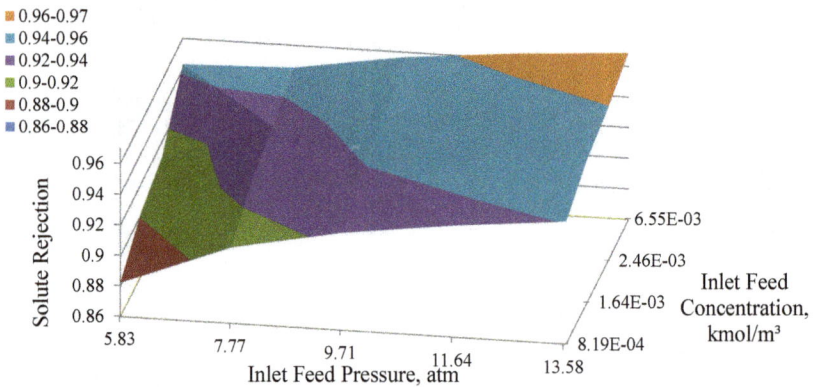

Figure 19. Impact of variation in inlet feed pressure and concentration on solute rejection at fixed inlet feed flow rate and temperature (2.583E-4 m³/s and 31°C) (Adapted from [26]).

It is noted that under high concentrations and pressures conditions, increasing inlet feed flow rate has a comparable impact on the solute rejection compared with using high concentrations and low pressures conditions. The reason for this can be explained as follows. At low-pressure conditions, increasing inlet feed flow rate results in decreasing the concentration polarization, which reduces the solute concentration along the wall membrane and the solute flux through the membrane. Then, this decreases the permeate solute concentration and increases the solute rejection. However, at high-pressure conditions, it seems that there is a conflict between the operating variables. Firstly, high pressure increases water flux due to a decrease in the osmotic pressure. However, the water flux somewhat decreases by increasing the friction with the membrane wall. This might be explained by a lower impact of the increasing inlet feed flow rate on the solute rejection response at high inlet feed concentrations. On the other hand, using lower inlet feed concentration, the experimental data (Tables A.6 to A.8 in Appendix A) show somehow an increase in the average solute rejection by increasing the inlet feed flow rate. This is due to the absence of concentration polarization impact at lower feed concentrations, which increases the mass transfer coefficient and total water flux. Furthermore, increasing the applied pressure results in reducing the concentration of the permeated water, which in turn increases the solute rejection due to an increase in water flux.

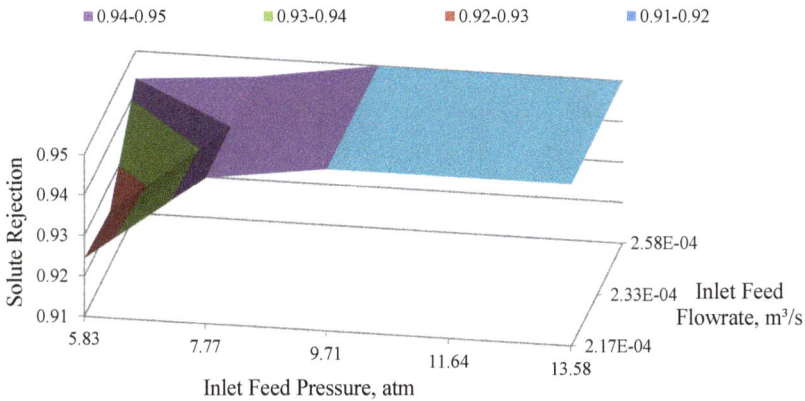

Figure 20. Impact of variation in inlet feed pressure and flow rate on solute rejection at fixed inlet feed concentration and temperature (6.548E-3 kmol/ m³ and 31°C) (Adapted from [26]).

Fig. 21 clearly shows the response of solute rejection for the variation in inlet feed concentration from 0.819E-3 to 6.548E-3 kmol/m³, and inlet feed flow rate from 2.166E-4 to 2.583E-4 m³/s with constant values of inlet feed pressure and temperature of 13.58 atm and 31°C respectively. It is also clear that the impact of variation in inlet feed concentration on solute rejection is comparable to the inlet feed flow rate in the case of using high inlet feed pressure conditions (Fig. 21). The solute rejection increases because of an increase in the inlet feed concentration and this may be owing to an increase in the membrane solute isolation intensity. The increasing inlet feed flow rate at high operating pressure results in a little increase in solute rejection for all the operating concentrations (Tables A.6 to A.8 in Appendix A). Also, it is easy to see that the temperature has a considerable impact on the solute rejection (Fig. 22).

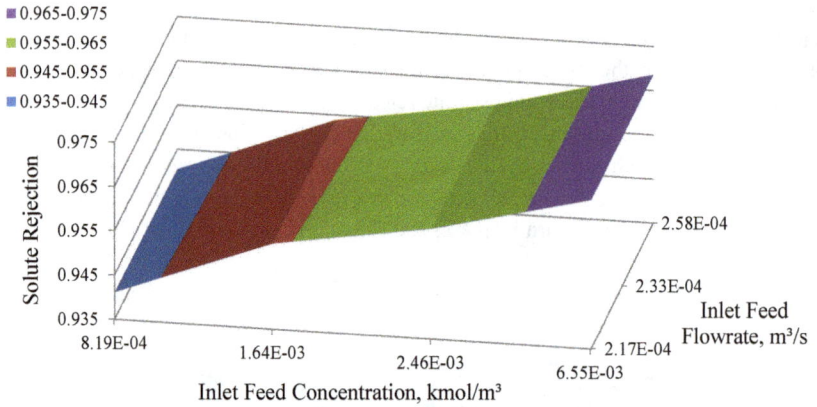

Figure 21. Impact of variation in inlet feed concentration and flow rate on solute rejection at fixed inlet feed pressure and temperature (13.58 atm and 31°C) (Adapted from [26]).

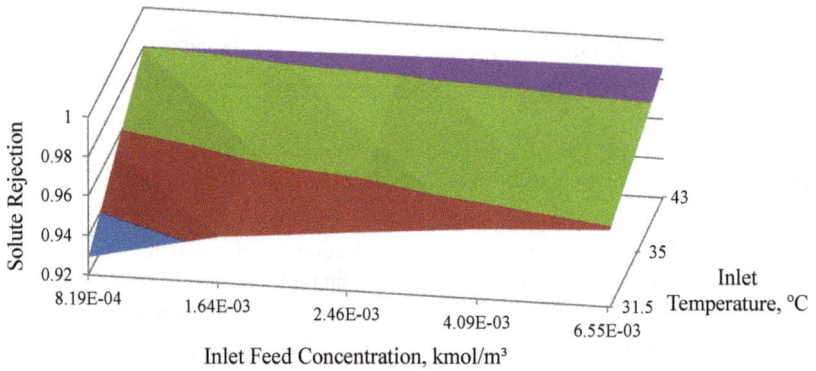

Figure 22. Impact of variation in inlet feed concentration and temperature on solute rejection at fixed inlet feed pressure and flow rate (9.71 atm and 2.166E-4 m³/s) (Adapted from [26]).

Increasing the operating temperature results in decreasing the viscosity of brine, increasing water flux and reducing the concentration at the permeate channel. Figs. 23 to 26 exhibit the variation of the total water recovery in respect of the inlet feed parameters such as flow rate, concentration, pressure, and temperature respectively. Fig. 23 shows an increasing total water recovery as a result of an increase in the operating pressure, which lifts the quantity of water flux. The total water recovery decreases as a result of an increase in operating concentration due to the reduction of water flux caused by increasing osmotic pressure, which already reduces the driving force of water flux.

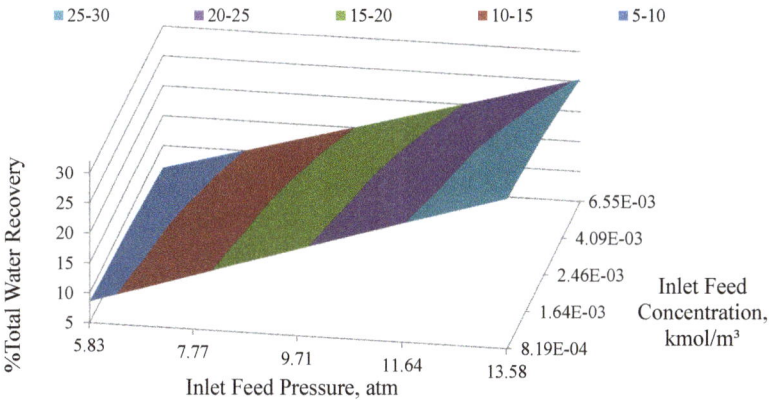

Figure 23. Impact of variation in inlet feed pressure and concentration on %total water recovery at fixed inlet feed flow rate and temperature (2.583E-4 m³/s and 31°C) (Adapted from [26]).

Figs. 24 and 25 reveal a reduction in total water recovery as a result of an increase in the inlet feed flowrate in spite of the gain of osmotic pressure reduction. It seems that increasing inlet feed flow rate leads to an increase in the frictional pressure drop along the membrane that creates a low driving force for the flow of fresh water in addition to a decrease in the residence time of feed inside the unit. Also, total water recovery decreases as a result of an increase in operating concentration. This is due to an increase in the osmotic pressure, which in turn reduces the water flux. Lastly, increasing operating temperature has a significant impact on total water recovery by increasing the quantity of water flux, as depicted in Fig. 26.

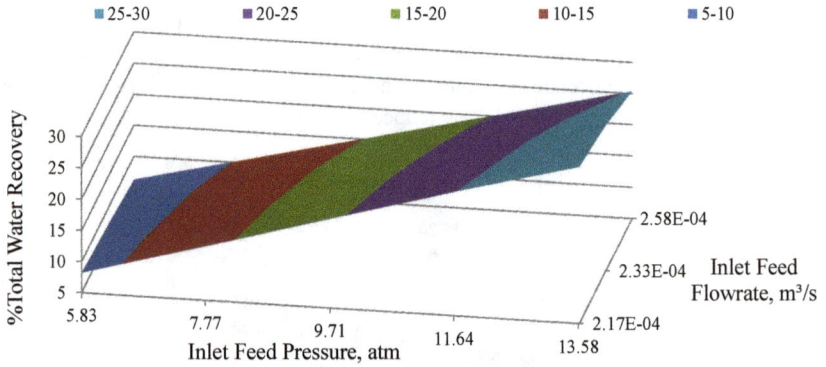

Figure 24. Model simulation of %total water recovery at varying inlet feed flow rate and inlet feed pressure at fixed inlet feed concentration temperature (6.548E-3 kmol/m³ and 31ºC) (Adapted from [26]).

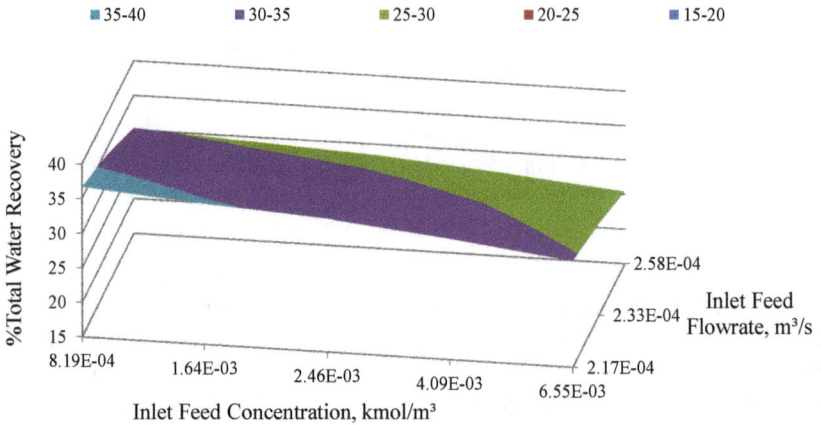

Figure 25. Impact of variation in inlet feed concentration and flow rate on %total water recovery at fixed inlet feed pressure and temperature (13.58 atm and 31ºC) (Adapted from [26]).

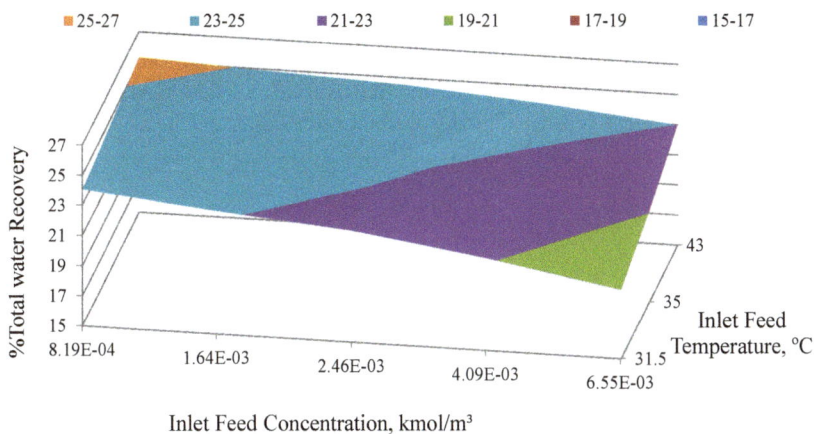

Figure 26. Impact of variation in inlet feed concentration and temperature on %total water recovery at fixed inlet feed pressure and flow rate (9.71 atm and 2.166E-4 m³/s) (Adapted from [26]).

Conclusions

This chapter provides a state-of-the-art of distributed models and associated performance of wastewater treatment methods using the reverse osmosis process for the removal of highly toxic organic compounds of phenol and N-nitrosamine. The RO process has confirmed its applicability and suitability for treating secondary effluents at low cost. However, the performance of RO process to remove N-nitrosodimethylamine-D6 (NDMA) has been a challenge. Whilst the concentration of these micro-pollutants is relatively low in wastewater, it will continue to challenge future research for developing an improved separation process. The direction of travel for providing a sustainable solution for treating these highly toxic compounds will continue to attract RO researchers for many reasons, not least because of the tightening regulation for lower recommended concentrations in both drinking water and wastewater. Whilst the treatment of phenol from wastewater in other applications is very specific, this chapter elucidates one example only of catalyst wet air oxidation process and focuses on highlighting successful modelling methodologies.

To the best of the authors' knowledge, the abilities and indeed possibilities of the RO process for removing NDMA are yet to be fully explored or realized with many opportunities and challenges for optimising the underlying operating conditions, superstructure and membrane synthesis. It is fair to say that the RO process remains the most promising treatment method for the removal of a wide range of hazardous chemicals albeit largely dependent on the operating conditions and the associated process configuration.

Appendix A

Table A.1. Modelling of CWAO process of Mohammed et al. [5]

Model Equations	Specifications	Eq. no.
$\dfrac{dC_{ph,L}}{dZ} = -\left(\dfrac{\eta_{LS}K_{LS}a_{LS}}{u_l}\right)\left(C_{ph,L} - C_{ph,L-s}\right)$	The mass balance equations for the concentrations of phenol in the liquid phase.	1
$\dfrac{dC_{O2,G}}{dZ} = -\left(\dfrac{K_{GL}a_{GL}}{u_g}\right)\left(\dfrac{C_{O2,G}}{H_{O2}} - C_{O2,L}\right)$	The concentration of oxygen and the mass transfer across the gas–liquid interface.	2
$\dfrac{dC_{O2,G}}{dZ} = -\left(\dfrac{K_{GL}a_{GL}}{u_l}\right)\left(\dfrac{C_{O2,G}}{H_{O2}} - C_{O2,L}\right)$ $-\left(\dfrac{\eta_{LS}K_{LS}a_{LS}}{u_l}\right)\left(C_{O2,L} - C_{O2,L-s}\right)$	The mass balance equations for the concentrations of oxygen in the liquid phase.	3
$K_{LS}a_{LS}\left(C_{O2,L} - C_{O2,L-S}\right)$ $= 7\eta_0(1 - \varepsilon_B)R_{ph}$	The oxygen chemical reaction.	4
$K_{LS}a_{LS}\left(C_{ph,L} - C_{ph,L-S}\right)$ $= \eta_0(1 - \varepsilon_B)R_{ph}$	The phenol chemical reaction.	5
$R_{ph} = \rho_{cat}K_{het}\dfrac{C_{ph}^n C_{O2}^m}{\left(1 + K_{ph}C_{ph,L}\right)^2}$	The kinetic equation of Langmuir–Hinshelwood type that accounts for phenol disappearance.	6
$K_{ph} = exp\left(-\dfrac{364.47}{T} - 2.3854\right)$	Calculates the adsorption equilibrium constant of phenol (K_{ph}).	7
$K_{het} = A^0\, exp\left(-\dfrac{EA}{RT}\right)$	Calculates the Reaction rate constant (K_{het}).	8
$\dfrac{K_{O2}^L a_L}{D_{O2}^L} = 7\left(\dfrac{\rho_{ph}u_l}{\mu_{ph}}\right)^{0.4}\left(\dfrac{\mu_{ph}}{\rho_{ph}D_{O2}^L}\right)^{0.5}$	Calculates the gas–liquid mass transfer coefficient of phenol.	9
$\dfrac{K_{ph}^s}{D_{ph}^L a_{LS}} = 1.8\left(\dfrac{\rho_{ph}u_l}{a_{LS}\mu_{ph}}\right)^{0.5}\left(\dfrac{\mu_{ph}}{\rho_{ph}D_{ph}^L}\right)^{1/3}$	Calculates the liquid–solid mass transfer coefficient of phenol.	10

Table A.1. Modelling of CWAO process of Mohammed et al. [5] (continued)

Model Equations	Specifications	Eq. no.
$\dfrac{K_{O2}^s}{D_{O2}^L a_{LS}} = 1.8\left(\dfrac{\rho_{ph}u_L}{a_{LS}\mu_{ph}}\right)^{0.5}\left(\dfrac{\mu_{ph}}{\rho_{ph}D_{O2}^L}\right)^{1/3}$	Calculates the liquid–solid mass transfer coefficient of oxygen.	11
$D_{O2}^L = 8.93x10^{-8}\dfrac{v_L^{0.267}\,T}{v_{O2}^{0.267}\,\mu_{ph}}$	Calculates the molecular diffusivity oxygen.	12
$D_{ph}^L = 8.93x10^{-8}\dfrac{v_L^{0.267}\,T}{v_{ph}^{0.267}\,\mu_{ph}}$	Calculates the molecular diffusivity of phenol.	13
$v_L = 0.285\,(v_c^L)^{1.048}$	Calculates The molar volume of liquid (v_L).	14
$v_{O2} = 0.285(v_c^{O2})^{1.048}$	Calculates The molar volume of oxygen (v_{O2}).	15
$v_{ph} = 0.285(v_c^{ph})^{1.048}$	Calculates The molar volume of phenol (v_{ph}).	16
$H_{O2} = \left(6088.8 - 871.2\ln T - \dfrac{326284}{T}\right)$	Calculates the Henry's constant for oxygen (H_{O2}).	17
$\rho_{ph} = \dfrac{MW_{ph}P_c}{R\,T_c Z_c(1 - T_r)^{2/7}}$	Calculates the density of phenol.	18
$T_r = \dfrac{T}{T_c}$	Calculates the reduced temperature.	19
$\emptyset = \dfrac{1 - T_r}{1 - T_{br}}$	Calculates the Volume fraction of molecule ($\emptyset$).	20
$\rho_{O2} = \dfrac{P\,MW_{O2}}{Z_{O2}\,R\,T}$	Calculates the density of oxygen.	21
$\mu_{ph} = exp\left(\ln(\alpha\,x\,\mu_{ph,b})x\left(\dfrac{\ln(\mu_{ph,b})}{\ln(\alpha\,x\,\mu_{ph,b})}\right)^{\emptyset}\right)$	Calculates the viscosity of phenol.	22
$T_{br} = \dfrac{T_b}{T_c}$	Calculates the reduced boiling point temperature.	23

Table A.1. Modelling of CWAO process of Mohammed et al. [5] (continued)

Model Equations	Specifications	Eq. no.
$\eta_0 = \dfrac{3(\varphi \coth\varphi - 1)}{\varphi^2}$	Calculates the effectiveness factor of for the sphere particle.	24
$\varphi = \dfrac{V_p}{S_p} \sqrt{\left(\dfrac{n+1}{2}\right)\left(\dfrac{K_{het} C_{ph}^{n-1} \rho_p}{D_{ei}}\right)}$	Calculates the Thiel modulus (φ).	25
$\rho_p = \dfrac{\rho_{cat}}{1 - \varepsilon_B}$	Calculates the particle density (ρ_p).	26
$V_p = \dfrac{4}{3}\pi(rp)^2$	Calculates the external volume (V_p) of the spherical shape of particle.	27
$S_p = 4\pi(rp)^2$	Calculates the surface area (S_p) of the spherical shape of particle.	28
$a_{L,S} = \dfrac{S_p(1 - \varepsilon_B)}{V_p}$	Calculates the surface area of particle per unit volume of the bed.	29
$D_{ei} = \dfrac{\varepsilon_S}{\tau} \dfrac{1}{\dfrac{1}{D_{mo,i}} + \dfrac{1}{D_{kn,i}}}$	Calculates The effective diffusivity (D_{ei}).	30
$\varepsilon_S = \rho_P V_g$	Calculates the catalyst particle porosity (ε_S).	31
$\varepsilon_B = 0.38 + 0.073 \left(1 + \dfrac{\left(\dfrac{d_t}{d_{pe}} - 2\right)^2}{\left(\dfrac{d_t}{d_{pe}}\right)^2}\right)$	Calculates the bed porosity (ε_B).	32
$D_{kn,i} = 9700\, r_g \sqrt{\dfrac{T}{MW_{ph}}}$	Calculates the Knudsen diffusivity ($D_{kn,i}$).	33
$r_g = 2\dfrac{V_g}{S_g}$	Calculate the mean pore radius (r_g).	34
$Rej_{TBR} = \dfrac{C_{ph,L(in)} - C_{ph,L(out)}}{C_{ph,L(in)}} x100$	Calculate the rejection of phenol	35

Table A.2. Equations describing the spiral-wound RO modelling of Sundaramoorthy et al. [10]

Title	The Mathematical Expression	Eq. no.
Calculate water flux at each point along the x-axis	$J_{w(x)} = \dfrac{\emptyset}{A \sinh \emptyset}\left[\left(F_{b(0)}\cosh \emptyset\left(1-\dfrac{x}{L}\right)\right.\right.$ $\left.\left. - F_{b(L)}\cosh\dfrac{\emptyset\, x}{L}\right)\right]$	1
Calculate the parameter $\emptyset$ in Eq. (1)	$\emptyset = L\left(\dfrac{W b A_w}{\left(1 + A_w\dfrac{R}{B_s}T\,C_p\right)}\right)^{0.5}$	2
Calculate the feed pressure at any point along the x-axis	$P_{b(x)} = P_{b(0)} - \dfrac{bL}{\emptyset \sinh \emptyset}\left[F_{b(L)}\left(\cosh\dfrac{\emptyset x}{L}-1\right)\right.$ $\left. - F_{b(0)}\left(\cosh \emptyset\left(1-\dfrac{x}{L}\right)\right.\right.$ $\left.\left. - \cosh \emptyset\right)\right]$	3
Calculate the outlet feed pressure	$P_{b(L)} = P_{b(0)} - \dfrac{bL}{\emptyset \sinh \emptyset}\left[\left(F_{b(0)}\right.\right.$ $\left.\left. + F_{b(L)}\right)(\cosh \emptyset - 1)\right]$	4
		5
Calculate the osmotic pressure at each point along the x-axis	$\Delta\pi_{s(x)} = R(T + 273.15)\left(C_{b(x)} - C_p\right)$	6
Calculate the progress of feed concentration at each point along the x-axis	$C_{b(x)} = C_p + \dfrac{F_{b(0)}\left(C_{b(0)} - C_p\right)}{F_{b(x)}}$	7
Calculate the rejection parameter at each point along the membrane length	$Rej = 1 - \dfrac{C_p}{C_{b(L)}}$	8
Calculate mass transfer coefficient at any point along the x-axis	$k_{(x,y)}d e_b$ $= 246.9\, D_{b(x,y)}\,Re_{b(x,y)}^{0.101}\,Re_{p(x,y)}^{0.803}\,C_{m(x,y)}^{0.129}$	9
Calculate the permeate concentration at each point along the permeate channel	$C_p = \dfrac{C_{b(x)}}{\left(1 + \dfrac{J_{w(x)}}{B_s}\over e^{\frac{J_{w(x)}}{k_{(x)}}}\right)}$	10

Table A.3. Equations describing the spiral-wound RO modelling of Fujioka et al. [58]

Title	The Mathematical Expression	Eq. no.
Calculate water flux at each point along the x-axis	$J_{w(x)} = L_p [(P_{b(x)} - P_p - \sigma \Delta \pi_{(x)})]$	1
Calculate the total permeate flux per each slide	$Q_{p(x)} = J_{w(x)} \Delta S$	2
Calculate the total permeate flux of membrane	$Q_{p,t} = \sum Q_{p(x)}$	3
Calculate the osmotic pressure at each point along the x-axis	$\pi_{(x)} = 1.19(T + 273.15) \sum C_{b(x)}$	4
Calculate the progress of feed concentration at each point along the x-axis	$C_{b(x+1)} = \dfrac{F_{b(x)} C_{b(x)} - C_{p(x)} Q_{p(x)}}{F_{b(x+1)}}$	5
Calculate the sub-section feed flow rate at each point along the x-axis	$F_{b(x+1)} = F_{b(x)} - Q_{p(x)}$	6
Calculate the pressure drop at each point along the membrane length	$\Delta P_{b(x)} = 0.5\, b\, \rho_{b(x)} U_{b(x)}^2 \dfrac{\Delta x}{d_h}$	7
Calculate the total pressure drop per each element	$\Delta P_{b,t} = \sum \Delta P_{b(x)}$	8
Calculate the progress of feed pressure at each sub-section	$P_{b(x+1)} = P_{b(x)} - \Delta P_{b(x)}$	9
Calculate the density parameter	$\rho_{b(x)} = 498.4\, m_f \sqrt{\left[248400\, m_f^2 + 752.4\, m_f \cdot C_{b(x)} \times 18.0153\right]}$	10
Calculate the parameter M_f in Eq. (8)	$m_f = 1.0069 - 2.757.10^{-4} T_b$	11
Calculate the rejection parameter at each point along the membrane length	$Rej_{(x)} = \dfrac{\sigma(1 - F_{(x)})}{(1 - \sigma F_{(x)})}$	12
Calculate the observed rejection parameter	$Rej_{obs(x)} = \dfrac{Rej_{(x)}}{(1 - Rej_{(x)}) x\, exp\left(\dfrac{J_{w(x)}}{k_{(x)}}\right) + Rej_{(x)}}$	13
Calculate the parameter $(F_{(x)})$ in Eq. (10)	$F_{(x)} = exp\left[-\dfrac{(1 - \sigma)}{B_s} J_{w(x)}\right]$	14

Title	The Mathematical Expression	Eq. no.
Calculate mass transfer coefficient. K, $\mu_{b(x)}$ and $\rho_{b(x)}$ are the efficiency of mixing ($K = 0.5$)	$k_{(x)} = 0.753 \left(\dfrac{K}{2\text{-}K}\right)^{0.5} \left(\dfrac{D_{b(x)}}{t_f}\right) \left(\dfrac{\mu_{b(x)}\rho_{b(x)}}{D_{b(x)}}\right)^{0.1666} \left(\dfrac{2\,t_f^{2}U_{b(x)}}{D_{b(x)}\,\Delta L}\right)^{0.5}$	15
Calculate the viscosity parameter	$\mu = 2.141E - 5x\ 10^{\frac{247.8}{T-140}}$	16
Calculate the permeate concentration at each point along the permeate channel	$C_{p(x)} = C_{b(x)}(1 - Re_{j_{obs}})$	17
Calculate the overall permeate concentration	$C_{p(av)} = \dfrac{\sum C_{p(x)}Q_{p(x)}}{\sum Q_{p(x)}}$	18

Title	The Mathematical Expression	Eq. no.
Calculate water flux	$J_{w(x)} = \dfrac{L_p}{\varnothing_{(x)}}\left\{\Delta P_{b(0)} - (b\times F_{b(0)}) - \left(b\times \left(\dfrac{W\,L_p}{b\,\varnothing_{(x)}}\right)^{0.5}\left(\Delta P_{b(x)}-\Delta P_{b(0)}\right)\right) - \left[b\times\left(2\left(F_{b(0)}\text{-}F_{b(x)}\right)\left(F_{b(x)}\text{-}F_{b(0)}\right)\right)^{0.5}\right] + \left[b\times\left(F_{b(0)}\text{-}F_{b(x)}\right)\right]\right\}$	1
Calculate the parameter $\varnothing_{(x)}$ in Eq. (1)	$\varnothing_{(x)} = 1 + \dfrac{\sigma\,C_{p(av)}\,L_p}{\omega} - \dfrac{\tilde{C}_{b(x)}(1-\sigma)\,\sigma\,L_p}{\omega}$	

77

Calculate solute flux	$$J_{s(x)} = \left(J_{w(x)}(1-\sigma)\tilde{C}_{b(x)}\right) + \left(\omega\, R\, T_b \left(C_{b(av)} - C_{p(av)}\right) e^{\frac{J_{w(x)}}{k_{(x)}}}\right)$$	2
Calculate the mean solute concentration in the feed side	$$\tilde{C}_{b(x)} = \frac{C_{b(x)} - C_{p(av)}}{\ln\left(\dfrac{C_{b(x)}}{C_{p(av)}}\right)}$$	3
Calculate feed flow rate	$$F_{b(x)}^2 = F_{b(0)}^2 + \frac{W\,L_p}{b\,\varnothing_{(x)}}\left(\Delta P_{b(x)}^2 - \Delta P_{b(0)}^2\right) + 2\left(F_{b(0)} - F_{b(x)}\right)\left(F_{b(x)} - F_{b(0)}\right)$$	4
Calculate feed pressure	$$P_{b(x)}^2 = \left[\left[P_{b(0)} - (b \times F_{b(0)}) - \left(b \times \left(\frac{W\,L_p}{b\,\varnothing_{(x)}}\right)^{0.5}\left(\Delta P_{b(x)} - \Delta P_{b(0)}\right)\right)\right]^2 - \left[b^2 x^2 \left(2\left(F_{b(0)} - F_{b(x)}\right)\left(F_{b(x)} - F_{b(0)}\right)\right)\right]\right]$$	5
Calculate trans-membrane pressure	$$\Delta P_{b(x)}^2 = \left[\left[\Delta P_{b(0)} - (b \times F_{b(0)}) - \left(b \times \left(\frac{W\,L_p}{b\,\varnothing_{(x)}}\right)^{0.5}\left(\Delta P_{b(x)} - \Delta P_{b(0)}\right)\right)\right]^2 - \left[b^2 x^2 \left(2\left(F_{b(0)} - F_{b(x)}\right)\left(F_{b(x)} - F_{b(0)}\right)\right)\right]\right] + \left[b^2 x^2 \left(F_{b(0)} - F_{b(x)}\right)^2\right]$$	
Calculate feed velocity	$$U_{b(x)} = F_{b(x)} / (t_f\,W)$$	6
Calculate feed solute concentration	$$\frac{d\left(\dfrac{C_{b(x)} F_{b(x)}}{t_f\,W}\right)}{dx} = -\frac{J_{w(x)} C_{p(av)}}{t_f} + \frac{J_{w(x)} C_{b(x)}}{t_f} + \frac{d}{dx}\left(D_{b(x)} \frac{dC_{b(x)}}{dx}\right)$$	7
Calculate mass transfer coefficient. The efficiency of mixing ($K = 0.5$)	$$k_{(x)} = 0.753 \left(\frac{K}{2-K}\right)^{0.5} \left(\frac{D_{b(x)}}{t_f}\right) \left(\frac{\mu_{b(x)} \rho_{b(x)}}{D_{b(x)}}\right)^{0.1666} \left(\frac{2\,t_f^2\,U_{b(x)}}{D_{b(x)}\,\Delta L}\right)^{0.5}$$	8

Table A.4. Equations describing the spiral-wound RO modelling of Al-Obaidi et al. [59] (continued)

Title	The Mathematical Expression	Eq. no.
Calculate permeate flow rate	$$F_{p(x)}^2 = F_{p(0)}^2 + \left(\frac{W\,L_{p\,x}}{\emptyset_{(x)}}\Delta P_{b(0)}\right)^2 - \left(\frac{W\,L_{p\,b}}{\emptyset_{(x)}}F_{b(0)}\left(\frac{x^2}{2}\right)\right)^2$$ $$- \left\{\left(\frac{W\,L_p}{\emptyset_{(x)}}\right)^{1.5} b^{0.5}\left(\frac{x^2}{2}\right)(\Delta P_{b(x)}-\Delta P_{b(0)})\right\}^2$$ $$- \left\{\left(\frac{W\,L_{p\,b}}{\emptyset_{(x)}}\right)\left(\frac{x^2}{2}\right)\left(2\left(F_{b(0)}-F_{b(x)}\right)\left(F_{b(x)}-F_{b(0)}\right)\right)\right\}^2$$ $$+ \left(\left(\frac{W\,L_{p\,b}}{\emptyset_{(x)}}\right)\left(\frac{x^2}{2}\right)\left(F_{b(0)}-F_{b(x)}\right)\right)^2$$	9
Calculate permeate pressure	$$P_{p(x)}^2 = P_{p(0)}^2 - \left(b\,x\,F_{b(0)}\right)^2 - \left(\left(\frac{W\,L_{p\,b}}{\emptyset_{(x)}}\right)\left(\frac{x^2}{2}\right)\Delta P_{b(0)}\right)^2$$ $$+ \left(\left(\frac{W\,L_p}{\emptyset_{(x)}}\right)\left(\frac{x^3}{6}\right)b^2 F_{b(0)}\right)^2$$ $$+ \left[b^{1.5}\left(\frac{x^3}{6}\right)\left(\frac{W\,L_p}{\emptyset_{(x)}}\right)^{1.5}(\Delta P_{b(x)}-\Delta P_{b(0)})\right]^2$$ $$+ \left(b^2\left(\frac{W\,L_p}{\emptyset_{(x)}}\right)\left(\frac{x^3}{6}\right)\left(2\left(F_{b(0)}-F_{b(x)}\right)\left(F_{b(x)}-F_{b(0)}\right)\right)\right)^2$$ $$- \left(\left(\frac{x^3}{6}\right)\left(\frac{W\,L_{p\,b}}{\emptyset_{(x)}}\right)\left(F_{b(0)}-F_{b(x)}\right)\right)^2$$	10
Calculate solute diffusion parameter	$$D_{b(x)} = 6.725E-6\,\exp\left\{0.1546\times10^{-3}C_{b(x)}\,(18.01253)-\frac{2513}{T_b+273.15}\right\}$$	12
Calculate viscosity parameter	$$\mu_{b(x)} = 1.234E-6\,\exp\left\{0.0212\,C_{b(x)}\,(18.0153)+\frac{1965}{T_b+273.15}\right\}$$	13
Calculate density parameter	$$\rho_{b(x)} = 498.4\,m_{f(x)} + \sqrt{[248400\,m_{f(x)}^2 + 752.4\,m_{f(x)}C_{b(x)}\,(18.0153)]}$$	14
Calculate wall solute concentration	$$\frac{C_{w(x)}-C_{p(av)}}{C_{b(x)}-C_{p(av)}} = \exp\left(\frac{J_{w(x)}}{k_{(x)}}\right)$$	15
Calculate permeate solute concentrations $(C_{p(x)})$ at x=0 and x=L. B_s is solute transport coefficient (m/s) used for the case of assuming $(\sigma = 1)$	$$C_{p(0)} = \frac{B_s\,R\,T_b\,C_{b(0)}\,e^{\frac{J_{w(0)}}{k_{(0)}}}}{J_{w(0)}+B_s\,R\,T_b\,e^{\frac{J_{w(0)}}{k_{(0)}}}}\quad\text{and}\quad C_{p(L)} = \frac{B_s\,R\,T_b\,C_{b(L)}\,e^{\frac{J_{w(L)}}{k_{(L)}}}}{J_{w(L)}+B_s\,R\,T_b\,e^{\frac{J_{w(L)}}{k_{(L)}}}}$$	16, 17

| Calculate the average permeate solute concentration | $C_{p(av)} = \dfrac{C_{p(0)} + C_{p(L)}}{2}$ | 18 |
| Calculate the total rejection and water recovery rate | $Rej = \dfrac{C_{b(L)} - C_{p(av)}}{C_{b(L)}} x100$ and $Rec_{(Total)} = \dfrac{F_{p(Total)}}{F_{b(0)}} x\,100$ | 19, 20 |

Table A.5. Equations describing the spiral-wound RO modelling of Al-Obaidi et al. [26]

Title	The Mathematical Expression	Eq. no.
Axial and vertical water flux	$J_{w(x,y)} = A_w \left((P_{b(x,y)} - P_{p(x,y)}) - RT_{b(x,y)} (C_{w(x,y)} - C_{p(x,y)}) \right)$	1
Axial and vertical solute flux	$J_{s(x,y)} = B_s \exp\left(\dfrac{J_{w(x,y)}}{k_{(x,y)}}\right)(C_{b(x,y)} - C_{p(x,y)})$	2
Axial and vertical membrane wall concentration	$C_{w(x,y)} = \left(C_{p(x,y)} + \exp\left(\dfrac{J_{w(x,y)}}{k_{(x,y)}}\right)(C_{b(x,y)} - C_{p(x,y)}) \right)$	3
Pressure difference along the two dimensions of the membrane	$\Delta P_{b(x,y)} = (P_{b(x,y)} - P_{p(x,y)})$	4
Axial and vertical feed flow rate	$\left(\dfrac{dF_{b(x,y)}}{dx}\right) + \left(\dfrac{dF_{b(x,y)}}{dy}\right) = [-(\Delta x + \Delta y)(J_{w(x,y)})]$	5
Axial and vertical feed pressure	$\left(\dfrac{dP_{b(x,y)}}{dx}\right) + \left(\dfrac{dP_{b(x,y)}}{dy}\right) = (-b\,F_{b(x,y)})$	6
Axial and vertical permeated pressure	$\left(\dfrac{dP_{p(x,y)}}{dx}\right) + \left(\dfrac{dP_{p(x,y)}}{dy}\right) = (-b\,F_{p(x,y)})$	7
Axial and vertical permeated flow rate	$F_{p(x,y)} = J_{w(x,y)}\Delta x\,\Delta y$	8
Axial and vertical molar flux of feed	$\dfrac{C_{b(x,y)}}{t_f \Delta y}\dfrac{dF_{b(x,y)}}{dx} + \dfrac{F_{b(x,y)}}{t_f.\Delta y}\dfrac{dC_{b(x,y)}}{dx} + \dfrac{C_{b(x,y)}}{t_f \Delta x}\dfrac{dF_{b(x,y)}}{dy}$ $+ \dfrac{F_{b(x,y)}}{t_f \Delta x}\dfrac{dC_{b(x,y)}}{dy}$ $= \dfrac{d}{dx}\left[D_{b(x,y)}\dfrac{dC_{b(x,y)}}{dx}\right]$ $+ \dfrac{d}{dy}\left[D_{b(x,y)}\dfrac{dC_{b(x,y)}}{dy}\right] - \dfrac{J_{s(x,y)}}{t_f}$	9

Axial and vertical molar flux of permeate

$$\frac{C_{p(x,y)}}{t_f \Delta y}\frac{dF_{p(x,y)}}{dx} + \frac{F_{p(x,y)}}{t_f \Delta y}\frac{dC_{p(x,y)}}{dx} - \frac{d}{dx}\left[D_{p(x,y)}\frac{dC_{p(x,y)}}{dx}\right]$$
$$+ \frac{C_{p(x,y)}}{t_f \Delta x}\frac{dF_{p(x,y)}}{dy} + \frac{F_{p(x,y)}}{t_f \Delta x}\frac{dC_{p(x,y)}}{dy}$$
$$- \frac{d}{dy}\left[D_{p(x,y)}\frac{dC_{p(x,y)}}{dy}\right] = \frac{J_{s(x,y)}}{t_f} \qquad 10$$

Axial and vertical feed temperature

$$0 = \left[\frac{J_{w(x,y)}\left(T_{b(x,y)} - T_{p(x,y)}\right)}{t_f}\right] \qquad 11$$

Total permeated flow rate

$$F_{p(Total)} = \Sigma F_{p(x,y)} \qquad \text{From } (0,0) \text{ to } (L, W) \qquad 12$$

%Total water recovery

$$Rec_{(Total)} = \frac{F_{p(Total)}}{F_{b(0,y)}}x100 \qquad 13$$

Average %solute rejection

$$Rej_{(av)} = \frac{C_{b(x=L,y)} - C_{p(av)}}{C_{b(x=L,y)}}x100 \qquad 14$$

Average permeated concentration

$$C_{p(av)} = \frac{\Sigma C_{p(x,y)}}{(\text{no of subdivisions})} \qquad 15$$

Axial and vertical mass transfer coefficient

$$k_{(x,y)}de_b = 246.9\ D_{b(x,y)}\ Re_{b(x,y)}^{0.101}\ Re_{p(x,y)}^{0.803}\ C_{m(x,y)}^{0.129} \qquad 16$$

Axial and vertical Dimensionless solute concentration

$$C_{m(x,y)} = \frac{C_{b(x,y)}}{\rho_w} \qquad 17$$

Axial and vertical feed diffusivity

$$D_{b(x,y)}$$
$$= 6.725.\,10^{-6}\ exp\left\{0.1546.\,10^{-3}C_{b(x,y)}x18.01253\right.$$
$$\left. - \frac{2513}{T_{b(x,y)} + 273.15}\right\} \qquad 18$$

Axial and vertical permeated diffusivity

$$D_{p(x,y)}$$
$$= 6.725.\,10^{-6}\ exp\left\{0.1546.\,10^{-3}C_{p(x,y)}x18.01253\right.$$
$$\left. - \frac{2513}{T_{p(x,y)} + 273.15}\right\} \qquad 19$$

Axial and vertical feed viscosity

$$\mu_{b(x,y)} = 1.234.\,10^{-6}\ exp\left\{0.0212\ C_{b(x,y)}x18.0153\right.$$
$$\left. + \frac{1965}{T_{b(x,y)} + 273.15}\right\} \qquad 20$$

Axial and vertical permeated viscosity

$$\mu_{p(x,y)} = 1.234.\,10^{-6}\ exp\left\{0.0212\ C_{p(x,y)}x18.0153\right.$$
$$\left. + \frac{1965}{T_{p(x,y)} + 273.15}\right\} \qquad 21$$

| Axial and vertical feed density | $\rho_{b(x,y)}$ $= 498.4\, m_{f(x,y)} \sqrt{\left[248400\, m_{f(x,y)}^2 + 752.4\, m_{f(x,y)} \cdot C_{b(x,y)} x18.0153\right]}$ | 22 |
| Axial and vertical permeated density | $\rho_{p(x,y)}$ $= 498.4\, m_{p(x,y)} \sqrt{\left[248400\, m_{p(x,y)}^2 + 752.4\, m_{p(x,y)} C_{p(x,y)} x18.0153\right]}$ | 23 |

Table A.5. Equations describing the spiral-wound RO modelling of Al-Obaidi et al. [26] (continued)

Title	The Mathematical Expression	Eq. no.
Axial and vertical variable in Eq.22	$m_{f(x,y)} = 1.0069 - 2.757.10^{-4} T_{b(x,y)}$	24
Axial and vertical variable in Eq. 23	$m_{p(x,y)} = 1.0069 - 2.757.10^{-4} T_{p(x,y)}$	25
Axial and vertical feed channel Reynolds number, dimensionless	$Re_{b(x,y)} = \dfrac{\rho_{b(x,y)}\, de_b F_{b(x,y)}}{t_f\, W\, \mu_{b(x,y)}}$	26
Axial and vertical permeate channel Reynolds number, dimensionless	$Re_{p(x,y)} = \dfrac{\rho_{p(x,y)} de_p J_{w(x,y)}}{\mu_{p(x,y)}}$	27
The equivalent diameter of feed channel	$de_b = 2t_f$	28
The equivalent diameter of permeated channel	$de_p = 2t_p$	29

Table A.6. Model validation with experimental results for inlet feed flow rate of $(F_{b(0,y)} = 2.166E\text{-}4\ m^3/s)$ (Adapted from [26])

No	Pb (inlet) atm	Tb (inlet) °C	Cb(inlet) x10³ (kmol/m³)	Cb(outlet)x10³ (kmol/m³)			Pb(outlet), atm			Cp(av)x10³ (kmol/m³)			Rej(av)			Fb(outlet)x10⁴ (m³/s)		
				Exp.	The.	%Error	Exp.	The.	%Error	Exp.	The.	%Error	Exp.	The.	%Error	Exp.	The.	%Error
1	5.83	32.5	0.819	0.9500	0.9230	2.84	4.46	4.06	8.96	0.0931	0.0885	4.84	0.902	0.904	-0.22	1.8	1.916	-6.44
2	7.77	32.5	0.819	1.0164	0.9909	2.51	6.31	6.06	3.96	0.0864	0.0734	15.0	0.915	0.9259	-1.19	1.67	1.786	-6.94
3	9.71	32.5	0.819	1.0821	1.0710	1.03	8.14	8.06	0.98	0.0790	0.0662	16.2	0.927	0.9382	-1.20	1.59	1.656	-4.15
4	11.64	32.5	0.819	1.1562	1.1650	-0.75	9.98	10.05	-0.70	0.0740	0.0623	15.8	0.936	0.9466	-1.13	1.5	1.528	-1.86
5	13.58	32.5	0.819	1.2568	1.2800	-1.83	11.8	12.05	-2.11	0.0729	0.0600	17.6	0.942	0.9531	-1.17	1.37	1.399	-2.11
6	5.83	31	1.637	1.8839	1.8300	2.86	4.41	4.05	8.16	0.1526	0.1730	-13.3	0.919	0.9051	1.51	1.851	1.932	-4.37
7	7.77	31	1.637	2.0227	1.9580	3.20	6.27	6.05	3.50	0.1335	0.1405	-5.24	0.934	0.9276	0.68	1.736	1.807	-4.08
8	9.71	31	1.637	2.1210	2.1100	0.52	8.09	8.05	0.49	0.1209	0.1272	-5.21	0.943	0.9397	0.34	1.63	1.681	-3.12
9	11.64	31	1.637	2.2882	2.2830	0.22	9.93	10.03	-1.00	0.1167	0.1189	-1.88	0.949	0.9479	0.11	1.523	1.557	-2.23
10	13.58	31	1.637	2.4255	2.4940	-2.82	11.76	12.03	-2.29	0.1140	0.1140	0.00	0.953	0.9543	-0.13	1.416	1.433	-1.20
11	5.83	31	2.455	2.7989	2.7310	2.42	4.37	4.042	7.50	0.2575	0.2367	8.07	0.908	0.9133	-0.58	1.868	1.942	-3.96
12	7.77	31	2.455	2.9783	2.9170	2.06	6.22	6.038	2.92	0.2204	0.1900	13.7	0.926	0.9348	-0.95	1.761	1.819	-3.29
13	9.71	31	2.455	3.1192	3.1350	-0.50	8.05	8.034	0.19	0.1778	0.1680	5.51	0.943	0.9464	-0.36	1.666	1.696	-1.80
14	11.64	31	2.455	3.3529	3.3880	-1.04	9.89	10.02	-1.31	0.1710	0.1557	8.94	0.949	0.954	-0.52	1.566	1.576	-0.63
15	13.58	31	2.455	3.5062	3.6900	-5.24	11.72	12.016	-2.52	0.1683	0.1482	11.9	0.952	0.9598	-0.81	1.478	1.453	1.69
16	5.83	30	4.092	4.6600	4.5070	3.28	4.32	4.03	6.71	0.3029	0.2730	9.87	0.935	0.9393	-0.45	1.898	1.962	-3.37
17	7.77	30	4.092	4.8066	4.7870	0.40	6.17	6.024	2.36	0.2884	0.3130	-8.52	0.94	0.9344	0.59	1.808	1.848	-2.21
18	9.71	30	4.092	5.1470	5.1160	0.60	8	8.017	-0.21	0.2625	0.2740	-4.38	0.949	0.9467	0.24	1.681	1.731	-2.97
19	11.64	30	4.092	5.2933	5.4950	-3.80	9.84	10	-1.62	0.2382	0.2525	-6.00	0.955	0.954	0.10	1.65	1.617	2.00

No	Pb(inlet), atm	Tb(inlet), °C	Cb(inlet) x10³ (kmol/m³)	Cb(outlet)x10³ (kmol/m³) Exp	The	%Error	Pb(outlet), atm Exp	The	%Error	Cp(av)x10³ (kmol/m³) Exp	The	%Error	Re|(av) Exp	The	%Error	Fb(outlet)x10⁴ (m³/sec) Exp	The	%Error
20	13.58	30	4.092	5.6648	5.9410	-4.87	11.67	11.99	-2.74	0.2096	0.2380	-13.5	0.963	0.9597	0.34	1.536	1.502	2.21
21	5.83	31.5	6.548	*	7.1620	*	*	4.025	*	*	0.5141	*	*	0.9282	*	*	1.978	*
22	7.77	31.5	6.548	7.7583	7.6060	1.96	6.13	6.017	1.84	0.3724	0.3878	-4.13	0.952	0.949	0.31	1.828	1.863	-1.91
23	9.71	31.5	6.548	8.1052	8.1220	-0.20	7.96	8.01	-0.62	0.3080	0.3299	-7.11	0.962	0.9593	0.28	1.75	1.747	0.17
24	11.64	31.5	6.548	8.6566	8.7160	-0.68	9.79	9.993	-2.07	0.2597	0.2970	-14.3	0.97	0.9659	0.42	1.641	1.633	0.48
25	13.58	31.5	6.548	8.9111	9.4110	-5.60	11.62	11.98	-3.09	0.2406	0.2760	-14.7	0.973	0.9706	0.24	1.575	1.517	3.68

Note: (*) the experimental data not reported

Table A.7. Model validation with experimental results for inlet feed flow rate of ($F_{b(0,y)} = 2.33E\text{-}4\ m^3/s$) (Adapted from [26])

No	Pb(inlet), atm	Tb(inlet), °C	Cb(inlet) x10³ (kmol/m³)	Cb(outlet)x10³ (kmol/m³) Exp	The	%Error	Pb(outlet), atm Exp	The	%Error	Cp(av)x10³ (kmol/m³) Exp	The	%Error	Re|(av) Exp	The	%Error	Fb(outlet)x10⁴ (m³/sec) Exp	The	%Error
1	5.83	32.5	0.819	0.9432	0.9127	3.24	4.39	3.92	10.7	0.0915	0.0887	3.06	0.903	0.9028	0.02	1.957	2.085	-6.54
2	7.77	32.5	0.819	1.0058	0.9740	3.16	6.23	5.916	5.04	0.0855	0.0723	15.4	0.915	0.9257	-1.16	1.86	1.955	-5.10
3	9.71	32.5	0.819	1.0600	1.0450	1.41	8.06	7.916	1.78	0.0742	0.0647	12.8	0.93	0.9381	-0.87	1.742	1.825	-4.76
4	11.64	32.5	0.819	1.1246	1.1300	-0.47	9.9	9.907	-0.07	0.0731	0.0605	17.2	0.935	0.9465	-1.22	1.639	1.694	-3.35
5	13.58	32.5	0.819	1.1961	1.2280	-2.66	11.73	11.908	-1.51	0.0622	0.0580	6.75	0.948	0.9527	-0.49	1.542	1.566	-1.55
6	5.83	31	1.637	1.8753	1.8113	3.41	4.34	3.91	9.90	0.1519	0.1730	-13.8	0.919	0.904	1.63	2.01	2.1	-4.47
7	7.77	31	1.637	1.9830	1.9280	2.77	6.19	5.899	4.70	0.1289	0.1390	-7.83	0.935	0.9274	0.81	1.894	1.9758	-4.31
8	9.71	31	1.637	2.0929	2.0620	1.48	8.02	7.905	1.43	0.1193	0.1240	-3.93	0.943	0.9397	0.34	1.794	1.848	-3.01
9	11.64	31	1.637	2.2019	2.2170	-0.68	9.86	9.89	-0.30	0.1123	0.1150	-2.40	0.949	0.9478	0.12	1.684	1.723	-2.31
10	13.58	31	1.637	2.3617	2.4000	-1.62	11.68	11.88	-1.71	0.1110	0.1103	0.63	0.953	0.954	-0.10	1.594	1.5988	-0.30
11	5.83	31	2.455	2.7734	2.7060	2.43	4.29	3.898	9.13	0.2302	0.2370	-2.95	0.917	0.9123	0.51	2.022	2.109	-4.30

No	Pb(inlet)	Tb(inlet)	Cb(inlet)	Cb(outlet) Exp	Cb(outlet) The	%Error	Pb(outlet) Exp	Pb(outlet) The	%Error	Cp(av) Exp	Cp(av) The	%Error	Rej(av) Exp	Rej(av) The	%Error	Fb(outlet) Exp	Fb(outlet) The	%Error
12	7.77	31	2.455	2.9513	2.8750	2.58	6.14	5.894	4.00	0.2125	0.1874	11.8	0.928	0.9348	-0.73	1.907	1.986	-4.14
13	9.71	31	2.455	3.1000	3.0700	0.96	7.97	7.89	1.00	0.1736	0.1645	5.24	0.944	0.9464	-0.25	1.815	1.863	-2.64
14	11.64	31	2.455	3.2800	3.2940	-0.42	9.81	9.876	-0.67	0.1640	0.1515	7.62	0.95	0.954	-0.42	1.707	1.74	-1.93
15	13.58	31	2.455	3.5022	3.5580	-1.59	11.64	11.88	-2.06	0.1541	0.1435	6.87	0.956	0.9596	-0.37	1.591	1.618	-1.69
16	5.83	30	4.092	4.5546	4.4680	1.90	4.25	3.887	8.54	0.2915	0.2770	4.97	0.936	0.938	-0.21	2.072	2.129	-2.75
17	7.77	30	4.092	4.7964	4.7240	1.51	6.1	5.88	3.60	0.2734	0.3100	-13.3	0.943	0.9343	0.92	1.974	2.015	-2.07
18	9.71	30	4.092	4.9938	5.0220	-0.56	7.92	7.87	0.63	0.2447	0.2690	-9.93	0.951	0.9464	0.48	1.887	1.897	-0.52
19	11.64	30	4.092	5.1790	5.3610	-3.51	9.76	9.85	-0.92	0.2227	0.2460	-10.4	0.957	0.9541	0.30	1.805	1.783	1.21
20	13.58	30	4.092	5.4361	5.7550	-5.86	11.59	11.85	-2.24	0.1957	0.2310	-18.0	0.964	0.9597	0.44	1.722	1.664	3.36
21	5.83	31.5	6.548	*	7.1040	*	*	3.88	*	*	0.5190	*	*	0.9269	*	*	2.144	*
22	7.77	31.5	6.548	7.5553	7.5100	0.59	6.05	5.873	2.92	0.3551	0.3844	-8.25	0.953	0.9488	0.44	1.987	2.029	-2.11
23	9.71	31.5	6.548	7.8131	7.9977	-2.36	7.88	7.87	0.12	0.2969	0.3240	-9.12	0.962	0.9593	0.28	1.902	1.913	-0.57
24	11.64	31.5	6.548	8.1806	8.5100	-4.02	9.72	9.85	-1.33	0.2536	0.2901	-14.3	0.969	0.9659	0.31	1.815	1.798	0.93
25	13.58	31.5	6.548	8.6740	9.1260	-5.21	11.54	11.84	-2.59	0.2342	0.2680	-14.4	0.973	0.9705	0.25	1.734	1.681	3.05

Note: (*) the experimental data not reported

Table A.8. Model validation with experimental results for inlet feed flow rate of $(F_{b(0,y)} = 2.583E\text{-}4\ m^3/s)$ (Adapted from [26])

No	Pb (inlet) atm	Tb(inlet) °C	Cb(inlet) x10³ (kmol/m³)	Cb(outlet)x10³ (kmol/m³) Exp.	The.	%Error	Pb(outlet), atm Exp.	The.	%Error	Cp(av)x10³ (kmol/m³) Exp.	The.	%Error	Rej(av) Exp.	The.	%Error	Fb(outlet)x10⁴(m³/s) Exp.	The.	%Error
1	5.83	32.5	0.819	0.9290	0.8997	3.15	4.27	3.69	13.5	0.0864	0.08958	-3.68	0.907	0.9004	0.72	2.199	2.345	-6.63
2	7.77	32.5	0.819	0.9975	0.9533	4.43	6.11	5.69	6.87	0.0798	0.07102	11.0	0.92	0.9255	-0.59	2.075	2.21	-6.50
3	9.71	32.5	0.819	1.0610	1.0160	4.24	7.94	7.69	3.14	0.0626	0.06267	-0.11	0.941	0.9383	0.28	1.953	2.08	-6.50

4	11.64	32.5	0.819	1.1160	1.0840	2.86	9.78	9.67	1.12	0.0558	0.0582	-4.30	0.95	0.9462	0.40	1.838	1.955	-6.36
5	13.58	32.5	0.819	1.2073	1.1783	2.40	11.61	11.68	-0.60	0.0495	0.0551	-11.3	0.959	0.9531	0.61	1.72	1.807	-5.05
6	5.83	31	1.637	1.8481	1.7880	3.25	4.22	3.68	12.7	0.1460	0.1750	-19.8	0.921	0.9016	2.10	2.261	2.359	-4.33
7	7.77	31	1.637	1.9523	1.8890	3.24	6.07	5.68	6.42	0.1230	0.1370	-11.3	0.937	0.9271	1.05	2.148	2.23	-3.81
8	9.71	31	1.637	2.0456	2.0050	1.98	7.89	7.67	2.78	0.1166	0.1210	-3.77	0.943	0.9396	0.36	2.042	2.107	-3.18
9	11.64	31	1.637	2.1461	2.1360	0.47	9.73	9.67	0.61	0.1116	0.1149	-2.95	0.948	0.9478	0.02	1.947	1.982	-1.79
10	13.58	31	1.637	2.2204	2.2880	-3.04	11.56	11.66	-0.86	0.1088	0.1056	2.94	0.951	0.9538	-0.29	1.85	1.855	-0.27
11	5.83	31	2.455	2.7457	2.6720	2.68	4.17	3.675	11.8	0.2279	0.2400	-5.30	0.917	0.91	0.76	2.29	2.368	-3.40
12	7.77	31	2.455	2.8985	2.8200	2.71	6.02	5.671	5.79	0.2000	0.1847	7.65	0.931	0.9345	-0.37	2.173	2.245	-3.31
13	9.71	31	2.455	2.9821	2.9880	-0.19	7.85	7.668	2.31	0.1670	0.1602	4.07	0.944	0.9464	-0.25	2.08	2.121	-1.97
14	11.64	31	2.455	3.1659	3.1790	-0.41	9.66	9.654	0.06	0.1488	0.1463	1.68	0.953	0.9539	-0.09	1.97	1.997	-1.37
15	13.58	31	2.455	3.3142	3.3990	-2.55	11.51	11.65	-1.21	0.1392	0.1376	1.14	0.958	0.9595	-0.15	1.868	1.874	-0.32
16	5.83	29	4.092	*	4.4080	*	*	3.66	*	*	0.3163	*	*	0.9282	*	*	2.393	*
17	7.77	29	4.092	*	4.6300	*	*	5.65	*	*	0.2320	*	*	0.9498	*	*	2.278	*
18	9.71	29	4.092	4.9000	4.8820	0.36	7.8	7.65	1.92	0.2303	0.1981	13.9	0.953	0.9594	-0.67	2.113	2.162	-2.31
19	11.64	29	4.092	5.0476	5.1640	-2.30	9.61	9.63	-0.20	0.2120	0.1803	14.9	0.958	0.965	-0.73	2.07	2.047	1.11
20	13.58	29	4.092	5.3657	5.4860	-2.24	11.47	11.62	-1.30	0.1878	0.1698	9.58	0.965	0.969	-0.41	1.972	1.93	2.12
21	5.83	31.5	6.548	7.1666	7.0360	1.82	4.08	3.66	10.2	0.3870	0.3810	1.55	0.946	0.9458	0.02	2.337	2.401	-2.73
22	7.77	31.5	6.548	7.5021	7.3880	1.52	5.93	5.65	4.72	0.3451	0.3812	-10.4	0.954	0.9483	0.59	2.253	2.287	-1.50
23	9.71	31.5	6.548	7.8270	7.7960	0.39	7.75	7.64	1.41	0.2896	0.3172	-9.53	0.963	0.9593	0.38	2.17	2.1703	-0.01
24	11.64	31.5	6.548	8.0064	8.2550	-3.10	9.57	9.637	-0.70	0.2482	0.2810	-13.2	0.969	0.9658	0.33	2.09	2.053	1.77
25	13.58	31.5	6.548	8.5037	8.7780	-3.22	11.42	11.62	-1.75	0.2296	0.2589	-12.7	0.973	0.9705	0.25	2.011	1.936	3.72

Note: (*) the experimental data not reported

References

[1]   G. Wade Miller, Integrated concepts in water reuse: managing global water needs, Desalination 187 (2006) 65-75. https://doi.org/10.1016/j.desal.2005.04.068

[2]   Henze, M.; Loosdrecht, M.C.M. van; Ekama, G.A.; Brdjanovic, D., 2008. Biological Wastewater Treatment - Principles, Modelling and Design. IWA Publishing. Retrieved from: http://app.knovel.com/hotlink/toc/id:kpBWTPMD04/biological-wastewater/biological-wastewater

[3]   Jiménez B.; Asano T.; 2008. Water reuse, an international survey of current practice, issues and needs. London: IWA Publishing.

[4]   G. Busca, S. Berardinelli, C. Resini, L. Arrighi, Technologies for the removal of phenol from fluid streams: A short review of recent developments, J. Hazard Mater. 160 (2008) 265-288. https://doi.org/10.1016/j.jhazmat.2008.03.045

[5]   A.E. Mohammed, A. Jarullah, S. Gheni, I.M. Mujtaba, Optimal design and operation of an industrial three phase reactor for the oxidation of phenol, Comput. Chem. Eng. 94 (2016) 257-271. https://doi.org/10.1016/j.compchemeng.2016.07.018

[6]   A.A. Gami, M.Y. Shukor, K. Abdul Khalil, F.A. Dahalan, A. Khalid, S.A. Ahmad, Phenol and its toxicity, J. Envir. Microb. Tox. 2 (2014) 11-24.

[7]   EFSA, 2013. Scientific opinion on the toxicological evaluation of phenol. EFSA Journal, 11(4), 3189. Available at: http://www.efsa.europa.eu/en/efsajournal/doc/3189.pdf

[8]   P.A. Mangrulkar, A.K. Bansiwal, S.S. Rayalu, Adsorption of phenol and chlorophenol on surface altered fly ash based molecular sieves, Chem Eng J. 138 (2008) 73-77. https://doi.org/10.1016/j.cej.2007.05.030

[9]   G. Srinivasan, S. Sundaramoorthy, D.V.R. Murthy, Validation of an analytical model for spiral wound reverse osmosis membrane module using experimental data on the removal of dimethylphenol, Desalination 281 (2011) 199-208. https://doi.org/10.1016/j.desal.2011.07.053

[10]   S. Sundaramoorthy, G. Srinivasan, D.V.R. Murthy, An analytical model for spiral wound reverse osmosis membrane modules: part I—model development and parameter estimation, Desalination, 280 (1–3) (2011) 403-411. https://doi.org/10.1016/j.desal.2011.03.047

[11]   J.W.A. Charrois, J.M. Boyd, K. L. Froese, S.E. Hrudey, Occurrence of N-nitrosamines in Alberta public drinking-water distribution systems, J. Environ. Eng. Sci. 6 (2007) 103-114. https://doi.org/10.1139/s06-031

[12]   US EPA, 2009. United States Environmental Protection Agency IRIS database. Available at: https://www.epa.gov/iris.

[13]   C. Lee, W. Choi, Y.G. Kim, J. Yoon, UV photolytic mechanism of N-nitrosodimethylamine in water: Dual pathways to methylamine versus dimethylamine, Environ. Sci. Tech. 39 (2005) 2101-2106. https://doi.org/10.1021/es0488941

[14]   E. Steinle-Darling, M. Zedda, M.H. Plumlee, H.F. Ridgway, M. Reinhard, Evaluating the impacts of membrane type, coating, fouling, chemical properties and water chemistry on reverse osmosis rejection of seven nitrosoalklyamines, including NDMA, Water Res. 41 (2007) 3959-3967. https://doi.org/10.1016/j.watres.2007.05.034

[15]   T. Fujioka, L.D. Nghiem, S.J. Khan, J.A. McDonald, Y. Poussade, J.E. Drewes, Effects of feed solution characteristics on the rejection of N-nitrosamines by reverse osmosis membranes, J. Membr. Sci. 409 (2012) 66-74. https://doi.org/10.1016/j.memsci.2012.03.035

[16]   Dochain D.; Vanrolleghem P.A., 2001. Dynamical Modelling and Estimation in Wastewater Treatment Processes. London: IWA Publishing, ProQuest Ebook Central.

[17]   M.A. Al-Obaidi, C. Kara-Zaitri, I.M. Mujtaba, Scope and limitations of the irreversible thermodynamics and the solution diffusion models for the separation of binary and multi-component systems in reverse osmosis process, Comput. Chem. Eng. 100 (2017) 48-79. https://doi.org/10.1016/j.compchemeng.2017.02.001

[18]   N. Bolong, A.F. Ismail, M.R. Salim, T. Matsuura, A review of the effects of emerging contaminants in wastewater and options for their removal, Desalination 239 (2009) 229-246. https://doi.org/10.1016/j.desal.2008.03.020

[19]   M. Safaa, 2009. Catalytic Wet Air Oxidation of Phenolic Compounds in Wastewater in a Trickle Bed Reactor at High Pressure. MSc. Thesis. University of Tikrit, Iraq.

[20]   M.H. Al-Dahhan, L. Faical, P. Milorad, A.L. Dudukovic, High pressure trickle bed reactors: a review, Ind. Eng. Chem. Res. 36 (1997) 3292-3314. https://doi.org/10.1021/ie9700829

[21]   W. Qiang, H. Xijun, L. Po, J. Yue, X. Feng, H. Chen, S. Zhang, Modeling of a pilot-scale trickle bed reactor for the catalytic oxidation of phenol, Sep. Purif. Technol. 67 (2009) 158-165. https://doi.org/10.1016/j.seppur.2009.03.021

[22]   K.M. Sassi, I.M. Mujtaba, Optimal design and operation of reverse osmosis desalination process with membrane fouling, Chem. Eng. J. 171 (2011) 582-593. https://doi.org/10.1016/j.cej.2011.04.034

[23]   A. Reverberi, B. Fabiano, C. Cerrato, V. Dovì, Concentration polarization in reverse osmosis membranes: Effect of membrane splitting, Chem. Eng. Trans. 39 (2014) 763-768.

[24]   F. Evangelista, An improved analytical method for the design of spiral-wound modules, Chem. Eng. J. 38 (1988) 33-40. https://doi.org/10.1016/0300-9467(88)80051-0

[25]   L. Song, S. Hong, J.Y. Hu, S.L. Ong, W.J. Ng, Simulations of Full-Scale Reverse Osmosis Membrane Process, J. Environ. Eng. 128 (10) (2002) 960-966. https://doi.org/10.1061/(ASCE)0733-9372(2002)128:10(960)

[26]   M.A. Al-Obaidi, C. Kara-Zaïtri, I.M. Mujtaba, Wastewater treatment by spiral wound reverse osmosis: Development and validation of a two dimensional process model, J. Cleaner Prod. 140 (2017) 1429-1443. https://doi.org/10.1016/j.jclepro.2016.10.008

[27]   S. Lee, R.M. Lueptow, Rotating reverse osmosis: a dynamic model for flux and rejection, J. Membr. Sci. 192 (2001) 129-143. https://doi.org/10.1016/S0376-7388(01)00493-8

[28]   N.B. Amar, N. Kechaou, J. Palmeri, A. Deratani, A. Sghaier, Comparison of tertiary treatment by nanofiltration and reverse osmosis for water reuse in denim textile industry, J. Hazard Mater. 170 (2009) 111-117. https://doi.org/10.1016/j.jhazmat.2009.04.130

[29]   I. Koyuncu, M.T. Topacik, A. Ates, Application of low pressure nanofiltration membranes for the recovery and reuse of dairy industry effluents, Water Sci. Tech. 41 (2000) 213-221.

[30]   S. Álrez, F.A. Riera, R. Álvarez, J. Coca, Concentration of apple juice by reverse osmosis at laboratory and pilot-plant scales, Ind. Eng. Chem. Res. 41 (2002) 6156-6164. https://doi.org/10.1021/ie020013g

[31]   P. Bhattacharya, A. Roy, S. Sarkar, S. Ghosh, S. Majumdar, S. Chakraborty, S. Mandal, A. Mukhopadhyay, S. Bandyopadhyay, Combination technology of ceramic microfiltration and reverse osmosis for tannery wastewater recovery, Water Resour. Ind. 3 (2013) 48-62. https://doi.org/10.1016/j.wri.2013.09.002

[32]   R.M. Mitra-Gholami, R.R. Kalantary, A. Sabzali, F. Gatei, Performance evaluation of reverse osmosis technology for selected antibiotics removal from synthetic pharmaceutical wastewater, Iran. J. Environ. Health Sci. Eng. 9 (2012) 19. https://doi.org/10.1186/1735-2746-9-19

[33]   X. Chai, G. Chen, P.-L. Yue, Y. Mi, Pilot scale membrane separation of electroplating waste water by reverse osmosis, J. Membr. Sci. 123 (1997) 235-242. https://doi.org/10.1016/S0376-7388(96)00217-7

[34]   A. Bódalo-Santoyo, J.L. Gómez-Carrasco, E. Gómez-Gómez, F. Máximo-Martin,
        A.M. Hidalgo-Montesinos, Application of reverse osmosis to reduce pollutants
        presents in industrial wastewater, Desalination 155 (2003) 101-108.
        https://doi.org/10.1016/S0011-9164(03)00287-X

[35]   A. Bódalo-Santoyo, J.L. Gómez-Carrasco, E. Gómez- Gómez, G. León, M. Tejera,
        Sulfonated polyethersulfone membrane in the desalination of aqueous,
        Desalination 168 (2004) 277-282. https://doi.org/10.1016/j.desal.2004.07.009

[36]   H. Abo-Qdais, H. Moussa, Removal of heavy metals from wastewater by
        membrane processes: a comparative study, Desalination 164 (2004) 105-110.
        https://doi.org/10.1016/S0011-9164(04)00169-9

[37]   M. Mohsen-Nia, P. Montazeri, H. Modarress, Removal of Cu^{+2} and Ni^{+2} from
        wastewater with chelating agent and reverse osmosis processes, Desalination 217
        (2007) 276-281. https://doi.org/10.1016/j.desal.2006.01.043

[38]   J.L. Gómez, G. León, A.M. Hidalgo, M. Gómez, M.D. Murcia, G. Griñán,
        Application of reverse osmosis to remove aniline from wastewater, Desalination
        245 (2009) 687-693. https://doi.org/10.1016/j.desal.2009.02.038

[39]   H. Mohammadi, M. Cholami, M. Rahimi, Application and optimization in
        chromium-contaminated wastewater treatment of the reverse osmosis technology,
        Desal. Water Treat. 9 (2009) 229-233. https://doi.org/10.5004/dwt.2009.808

[40]   C. Sagne, C. Fargues, R. Lewandowski, M.-L. Lameloise, M. Gavach, M.
        Decloux, A pilot scale study of reverse osmosis for the purification of condensate
        arising from distillery stillage concentration plant, Chemical Engineering and
        Processing: Process Intensification 49 (2010) 331-339.
        https://doi.org/10.1016/j.cep.2010.03.002

[41]   S.S. Madaeni, S. Koocheki, Influence of di-hydrogen phosphate ion performance
        of polyamide reverse osmosis membrane for nitrate and nitrite removal. J. Porous
        Mater. 17 (2010) 163-168. https://doi.org/10.1007/s10934-009-9276-5

[42]   D. Tabassi, A. Mnif, B. Hamrouni, Influence of operating conditions on the
        retention of phenol in water by reverse osmosis SG membrane characterized using
        Speigler-Kedem model, Desal. Water Treat. 52 (2014) 1792-1803.
        https://doi.org/10.1080/19443994.2013.807049

[43]   F. Khazaali, A. Kargari, M. Rokhsaran, Application of low-pressure reverse
        osmosis for effective recovery of bisphenol A from aqueous wastes, Desal. Water
        Treat. 52 (2014) 7543-7551. https://doi.org/10.1080/19443994.2013.831795

[44]   K. Thirugnanasambandham, V. Sivakumar, K. Loganathan, R. Jayakumar, K.
        Shine, Pilot scale evaluation of feasibility of reuse of wine industry wastewater
        using reverse osmosis system: modeling and optimization, Desal. Water Treat. 57
        (2016) 1-11. https://doi.org/10.1080/19443994.2016.1154894

[45]   M.A. Al-Obaidi, I.M. Mujtaba, Steady state and dynamic modeling of spiral wound wastewater reverse osmosis process, Comput. Chem. Eng. 90 (2016) 278-299. https://doi.org/10.1016/j.compchemeng.2016.04.001

[46]   R. Rautenbach, W. Dahm, Design and optimization of spiral-wound and hollow fiber RO-modules, Desalination 65 (1987) 259-275. https://doi.org/10.1016/0011-9164(87)90138-X

[47]   S. Avlonits, W.T. Hanbury, M.B. Boudinar, Spiral wound modules performance. An analytical solution - part I, Desalination 81 (1991) 191-208. https://doi.org/10.1016/0011-9164(91)85053-W

[48]   M.B. Boudinar, W.T. Hanbury, S. Avlonits, Numerical simulation and optimisation of spiral-wound modules, Desalination 86 (1992) 273-290. https://doi.org/10.1016/0011-9164(92)80038-B

[49]   K.S. Spiegler, O. Kedem, Thermodynamics of hyperfiltration (reverse osmosis): criteria for efficient membranes, Desalination 1 (1966) 311-326. https://doi.org/10.1016/S0011-9164(00)80018-1

[50]   S. Senthilmurugan, A. Ahluwalia, S.K. Gupta, Modeling of a spiral-wound module and estimation of model parameters using numerical techniques, Desalination 173 (2005) 269-286. https://doi.org/10.1016/j.desal.2004.08.034

[51]   P.P. Mane, P.K. Park, H. Hyung, J.C. Brown, J.H. Kim, Modeling boron rejection in pilot- and full-scale reverse osmosis desalination processes, J. Membr. Sci. 338 (2009) 119-127. https://doi.org/10.1016/j.memsci.2009.04.014

[52]   V. Geraldes, N. Escórcio Pereira, M. Norberta de Pinho, Simulation and optimization of medium-sized seawater reverse osmosis processes with spiral wound modules, Ind. Eng. Chem. Res. 44 (2005) 1897-1905. https://doi.org/10.1021/ie049357s

[53]   C. Sagne, C. Fargues, B. Broyart, M. Lameloise, M. Decloux, Modeling permeation of volatile organic molecules through reverse osmosis spiral wound membranes, J. Membr. Sci. 330 (2009) 40-50. https://doi.org/10.1016/j.memsci.2008.12.038

[54]   S.A. Avlonits, M. Pappas, K. Moutesidis, A unified model for the detailed investigation of membrane modules and RO plants performance, Desalination 203 (2007) 218-228. https://doi.org/10.1016/j.desal.2006.04.009

[55]   H. Oh, T. Hwang, S. Lrr, A simplified model of RO systems for seawater desalination, Desalination 238 (2009) 128-139. https://doi.org/10.1016/j.desal.2008.01.043

[56]   T. Kaghazchi, M. Mehri, M. Takht Ravanchi, A. Kargari, A mathematical modeling of two industrial seawater desalination plants in the Persian Gulf region, Desalination 252 (2010) 135-142. https://doi.org/10.1016/j.desal.2009.10.012

[57]   S. Sundaramoorthy, G. Srinivasan, D.V.R. Murthy, An analytical model for spiral wound reverse osmosis membrane modules: part II—experimental validation, Desalination 277 (1–3) (2011) 257-264.
https://doi.org/10.1016/j.desal.2011.04.037

[58]   T. Fujioka, S.J. Khan, J.A. Mcdonald, A. Roux, Y. Poussade, J.E. Drewes, L.D. Nghiem, Modelling the rejection of N-nitrosamines by a spiral-wound reverse osmosis system: Mathematical model development and validation, J. Membr. Sci. 454 (2014) 212-219. https://doi.org/10.1016/j.memsci.2013.12.008

[59]   M.A. Al-Obaidi, C. Kara-Zaïtri, I.M. Mujtaba, Development and validation of N-nitrosamine rejection mathematical model using a spiral-wound reverse osmosis process, Chem. Eng. Trans. 52 (2016) 1129-1134.

[60]   C. Koroneos, A. Dompros, G. Roumbas, Renewable energy driven desalination systems modelling, J. Cleaner Prod. 15 (2007) 449-464.
https://doi.org/10.1016/j.jclepro.2005.07.017

[61]   T. Fujioka, S.J. Khan, J.A. Mcdonald, A. Roux, Y. Poussade, J.E. Drewes, L.D. Nghiem, N-nitrosamine rejection by nanofiltration and reverse osmosis membranes: The importance of membrane characteristics, Desalination 316 (2013) 67–75. https://doi.org/10.1016/j.desal.2013.01.028

[62]   M.H. Plumlee, M. Lo´pez-Mesas, A. Heidlberger, K.P. Ishida, M. Reinhard, N-nitrosodimethylamine (NDMA) removal by reverse osmosis and UV treatment and analysis via LC–MS/MS, Water Res. 42 (2008) 347-355.
https://doi.org/10.1016/j.watres.2007.07.022

[63]   M. Krauss, P. Longrée, E. Van Houtte, J. Cauwenberghs, J. Hollender, Assessing the fate of nitrosamine precursors in wastewater treatment by physicochemical fractionation, Environ. Sci. Technol. 44 (2010) 7871-7877.
https://doi.org/10.1021/es101289z

[64]   Z.V.P. Murthy, S.K. Gupta, Sodium cyanide separation and parameter estimation for reverse osmosis thin film composite polyamide membrane, J. Membr. Sci. 154 (1999) 89-103. https://doi.org/10.1016/S0376-7388(98)00280-4

[65]   I.M. Mujtaba, The Role of PSE Community in Meeting Sustainable Freshwater Demand of Tomorrow's World via Desalination, In Computer Aided Chemical Engineering- 31, I. A. Karimi and Rajagopalan Srinivasan (Editors) Vol 31 (2012) 91-98, Elsevier.

Chapter 3

Treatment of Refractory Organic Pollutants using Ionic Liquids

I.M. AlNashef *[1], R. Sulaiman [1], S.S. AlSaleem [2]

[1] Khalifa University of Science and Technology, Masdar Institute, Department of Chemical Engineering, P. O. Box 54224, Abu Dhabi, United Arab Emirates

[2] College of Engineering, Civil Engineering Department, Qassim University, Buraidah, Saudi Arabia

* enashf@masdar.ac.ae

Abstract

In recent years, the pollutants that are refractory to treat by conventional biological, physical, and chemical methods together with the stricter restrictions imposed by new legislation have caused many researchers to look for alternative treatment processes. In addition, there is uncertainty regarding the formation of toxic by products following conventional chemical oxidation. Halogenated hydrocarbons (HHCs), which are priority chemicals, have been used extensively in a number of industrial processes. However, it was discovered that many of these halogenated hydrocarbons are carcinogens and have serious negative impact on the eco system. Ionic liquids (ILs), a new class of solvents have many favorable characteristics, e.g. low vapor pressure, non-flammability, ability to dissolve polar and non-polar compounds, and thermal stability. In this chapter, the use of ILs in the treatment of water contaminated with halogenated hydrocarbons pollutants is presented.

Keywords

Ionic Liquids, Water Treatment, Refractory Pollutants, Halogenated Hydrocarbons, Superoxide Ion, Green Engineering

Contents

1.   Introduction

A lot of chemicals, organic and inorganic, are currently used in different industrial applications. Many of these compounds are considered as potent pollutants and thus must be treated carefully so that they do not contaminate water resources. Water is one of the most used solvents in industry. In addition, it is used as transport or reaction medium [1]. Also, stricter restrictions have been imposed by new legislations in most countries, for example, Table 1 shows the maximum prescribed value of miscellaneous organic compounds in drinking water in the UAE. This led to a tremendous increase in the interest of researchers for finding proper processes that minimize the effect of these chemicals on the eco-system. Contaminated fresh water resources and wastewater of many industrial activities may contain toxic organic compounds that cannot be treated using conventional methods. However, these contaminated water stream resources are not allowed to be discharged, if they do not meet the specifications for discharge effluents that change from country to country [2].

Conventional technologies that include but are not limited to thermal, physicochemical, and biological treatments have been used for removing pollutants from water. However, many of these processes have serious drawbacks. For example, the biological methods require longer residence time for microorganisms to degrade the pollutants and are very sensitive to toxic contaminants. Incineration is a well-known and widely used thermal treatment process. However, it consumes a large amount of energy. In addition, because of the high temperature of the process, other hazardous compounds such as dioxins and furan could be generated [3]. Some other techniques such as flocculation, precipitation, adsorption, air stripping, and reverse osmosis require a post-treatment to dispose of the pollutants from the newly contaminated environment [4]. The aforementioned limitations of conventional methods have encouraged the researchers to develop more efficient and environmentally friendly systems for wastewater treatment. One available option is wet air oxidation (WAO), which was proposed and developed by F. Zimmerman [5]. WAO is one of the most economically and technologically viable advanced oxidation processes (AOP) for wastewater treatment. WAO is suitable to a high organic loading at high flow rate and can partially cover the application range of incineration and biological methods. By using WAO, the organic pollutants are either partially oxidized into biodegradable intermediates or mineralized to carbon dioxide, water, and benign end products under elevated temperatures (125–320°C) and pressures (0.5–20 MPa) using a gaseous source of oxygen (either pure oxygen or air) as the oxidant [6]. However, the necessity of relatively high temperature and high-pressure conditions for WAO is the clear disadvantage of this process. Aside from WAO, other advanced oxidation processes involve the production of hydroxyl radicals (OH$^{\bullet}$), which are strong, nonspecific reactants capable of the complete oxidation of most organic compounds into carbon dioxide, water, and mineral acids [7]. Several technologies are available to produce hydroxyl radicals in the aqueous phase [8] e.g. ozone+UV, ozone+H_2O_2, H_2O_2+UV+ferrous salts, and UV+TiO_2. AOPs differ from other treatment processes (such as adsorption, ion exchange, or stripping) because organic compounds in water are degraded rather than concentrated or transferred into different phases. Furthermore, compounds that are not adsorbable or only partially adsorbable may be destroyed by a reaction with hydroxyl radical [7].

One limitation of using hydroxyl radicals is the presence of free radical scavengers. Free radical scavengers are species in water that consume the hydroxyl radicals leaving fewer of them available to treat organics of interest. Examples of free radical scavengers are CO_3^{2-}, HCO_3^-, Cl^-, NO_3^-, bromide, and cyanide [8,9,10]. Also, in some cases, AOPs have high degree of complexity and require special reactors, control of water pH, and off-gas control. These disadvantages affect AOP performance and thus increase the treatment cost dramatically. Although, hydroxyl radicals are capable of oxidizing almost all reduced compounds present in water without restriction, but these radicals can not degrade oxidized compounds, like halogenated hydrocarbons (HHCs), which are considered essentially nonreactive [8]. Contrary to the hydroxyl radical, superoxide ion radical ($O_2^{\bullet-}$) can destruct highly oxidized compounds, such as halogenated hydrocarbons [11]. Table 2 shows the physical properties of $O_2^{\bullet-}$. The superoxide ion was electrochemically or chemically produced in different convential aprotic solvents, Table 3. AlNashef et al. [12] were the first to report the electrochemical production of a stable $O_2^{\bullet-}$ in ILs. Later on, many other groups investigated the generation of $O_2^{\bullet-}$ in different ILs.

Halogenated hydrocarbons, which are priority chemicals, have been used extensively in a number of industrial processes. carbon tetrachloride, for instance, finds primary applications in dry cleaning, metal degreasing, fire extinguishers, grain fumigation, and as a reaction medium in the industrial synthesis of chemicals. chloroform, which is commonly found in chlorinated drinking water, has been used as a dry cleaning spot remover and for cleaning machines. Other applications include industrial intermediates, insecticide, and the purification of alkaloids and vitamins. Bromoform, which is also commonly found in chlorinated drinking water especially in desalinated seawater, is used as a fluid for mineral ore separation in geological processes, as a laboratory reagent, and in the electronics industry [10]. Carbon tetrachloride, chloroform, and bromoform have been detected in the effluent after treatment of seawater by using reverse osmosis and multistage flash processes [11]. In addition, HHCs include polychlorinated biphenyls (PCBs), hexachlorohexane (HCB), and most chlorinated organics. Many of HHCs are still in use, e.g. trichlorethylene, perchloroethylene, trichloroacetylene, and other chlorinated solvents are being used in dry cleaning, critical cleaning, paint stripping, and similar operations [13]. Some of HHCs have been identified as animal carcinogens and can cause birth defects, and their continual release into the ecosystem has a deleterious effect on animal life [14]. For example, HCB, which is a byproduct of the poly(chloroethylene) solvent industry, is as environmentally persistent as PCBs and is a human carcinogen [15,16].

Chlorophenols have become important pollutants in the environment and their treatment, disposal, and general management have become serious challenges to environmental agencies in most parts of the world. Chlorophenols are usually introduced into the environment as a result of chemical and pharmaceutical industrial activities [15,16]. The prevalence of these pollutants in the environment is associated with the production, use and degradation of numerous pesticides, which can easily find their ways into the environment [17]. The main potential environmental sources of chlorophenols include: (a) direct soil application as biocides; (b) synthesis during engineering processes where chlorine is used; (c) leaching and vaporizing from treated wood items; (d) release or discharge from factories into air and water; and (e) combustion of organic material in the presence of chloride.

Ionic Liquids are organic salts resulting from the combination of bulky, unsymmetrical and flexible organic cations and various inorganic or organic anions that usually melt below 100°C [18,19]. Different ILs can be formed by appropriately combining cations and anions or by chemically modifying either the cation or the anion. Theoretically, at least 10^6 ILs can be synthesized, while there are only hundreds of traditional organic solvents commercially available, which means there is ample space for development [20,21]. Interest in ILs comes from their potential application as "green solvents" due to their negligible vapor pressure and other properties such as non-flammability, high ionic conductivity, high thermostability [18,22] and tunable physicochemical properties such as miscibility and polarity [18,23]. These properties are significant when addressing the health and safety concerns associated with many conventional solvent applications [23].

In addition, ILs can dissolve a wide range of organic and inorganic pollutents. Due to the modular nature of ILs, their properties can readily be tuned for a wide range of task specific applications in different important chemical processes including liquid–pounds [24-27], (Fig. 1).

Table 1: Maximum prescribed value of miscellaneous organic compounds in drinking water. *

Halogenaed Compound	Maximum Prescribed Value (µg/L)
DDT	1
Chlordane	0.2
Trichloroethene	20
Tetrachloromethane	3
Tetrachloroethene	40
1,2-Dichloroethane	30
Dichloromethane	20
Chlorobenzene	300
1,2-Dichloroethene	50
1,2-Dichlorobenzene	1000
1,4-Dichlorobenzene	300
Vinyl Chloride	0.3

* Abu Dhabi water Quality Standards & Regulations Summary, August 2010.

Ionic liquids have been investigated for many applications including, but not limited to: Liquid extraction, oxidation-reduction processes, fuel and solar cells, organometallic synthesis, electrochemical devices, capacitors, lubricants, stationary phases for chromatography, matrices for mass spectrometry, supports for the immobilization of enzymes, separation technologies, azeotropic mixtures separations, liquid crystals, templates for synthesis nanomaterials and materials tissue preservation, preparation of polymer–gel catalytic membranes, biphasic catalysis, and generation of high conductivity materials [17,22,24,25,27].

Interestingly, ILs not only behave as solvents, but also as catalysts in many chemical reactions [21,22]. In recent years, interest has grown in application of ILs in liquid–liquid extraction instead of traditional organic solvent. For example, extraction of heavy metal ions, aromatic hydrocarbons, organic acids, amino acids, sulfur compounds in diesel oil, organic compounds in plants and organics in water using ILs as extractants has received increasing attention [28,29]. Table 2 shows the use of ILs in the extraction of organic compounds from different types of water.

In addition to these advantages, a superoxide ion can be produced easily in ILs [23-25] and some studies have shown that the superoxide ion remaining stable in many ILs for a relatively long time (i.e., 120 min) [21]. The ability of ILs to dissolve many organic compounds and the generation of a stable superoxide ion in some ILs can facilitate the destruction of emerging contaminants after extracting from waste or wastewater.

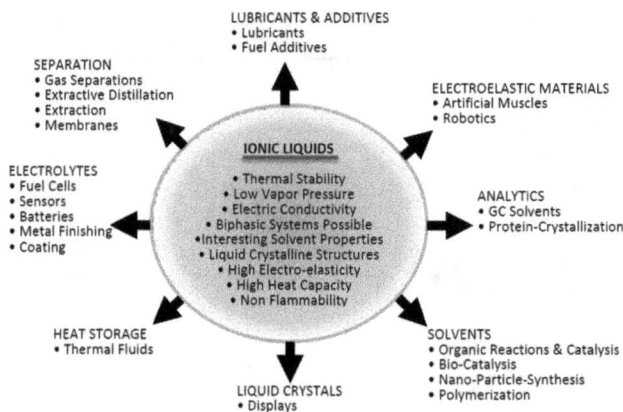

Figure 1. ILs applications , adapted from Ref [27].

These favorable features open a window to a good insight to use ILs as powerful alternative solvent-extractors in the field industrial wastewater treatment and, in general, water treatment.

Table 2: Physical properties of $O_2^{•-}$, adapted from [23].

Property	Value	Ref.
Paramagnetic	one-unpaired electron	[31,32]
O-O bond distance (KO_2)	1.28 °A	[33]
$O_2^{•-}$ (aqueous)	245 nm	[34]
$O_2^{•-}$ (AcN)	255 nm	[35]
Gas-phase basicity	$F^- > O_2^{•-} > Cl^- > Br^-$	[36]
IR absorption spectra, KO_2, (O-O stretch; bond order 1.5)	1145 1/cm 1140 1/cm	[37,38]
$O_2^{•-}$ vibrational frequency	1090 1/cm 1108 1/cm (±20 1/cm)	[39-40]

ILs can overcome the negative environmental and health impacts of commonly used volatile organic which are discharged into the atmosphere each year as a results of industrial processing operations [30].

In the following sections, the solubility of HHCs in ILs, extraction of HHCs from water using ILs, and different methods used for the degradation of the extracted HHCs will be discussed.

Table 3: Conventional aprotic solvents used as media for $O_2^{\bullet-}$, adapted from [23].

Aprotic solvent	Abbreviation	Ref.
Dimethyl formamide	DMF	[41,42,43,44,45]
Dimethyl sulfoxide	DMSO	[46,47,48]
Acetonitrile	AcN	[23,49,50,51,52]
Propylene carbonate	-	[52]
Pyridine	-	[13,54]
Methylene chloride	-	[53,43]
Dimethoxyethane	DME	[44]
Acetone	-	[43]
Diethyl ether	DEE	[44]

2. Solubility of HHCs in ILs

Measurement and prediction of the solubility of water pollutants in hydrophobic ILs is an important issue for assessment of the potential use of these ILs in water treatment processes. The available data in the literature are scarce. AlSaleem et al. [55] reported the solubility of three HHCs, namely carbon tetrachloride (CCl_4), chloroform ($CHCl_3$), and bromoform ($CHBr_3$) in different types of hydrophobic ILs at 25, 35 and 45°C. The three halogenated hydrocarbons were chosen because these are members of priority water contaminants. The authors investigated the effect of the structure of IL cation paired with bis(trifluoromethylsulfonyl)imide [Tf_2N] and tris(pentafluoroethyl)trifluorophosphate [FAP] anions, temperature, polarity, and length of the alkyl chain on the cation on the solubility of the selected HHCs in ILs. It was found that carbon tetrachloride and bromoform were partially miscible in all tested ILs while chloroform exhibited full miscibility. For ammonium based ionic liquids, the solubility increases with the increase of the cation molecular weight and alkyl chain length. The results indicated substantial the solubility of the studied halogenated hydrocarbons in methyltrioctylammonium bis(trifluoromethylsulfonyl)imide, octyltriethylammonium bis(trifluoromethylsulfonyl)imide, and 1-octyl-1-methylpyrrolidinium bis(trifluoromethylsulfonyl)-imide . These results confirmed the potential use of ILs as powerful alternative solvents for wastewater treatment. In addition, the authors used COnductor-like Screening MOdel for Realistic Solvents (COSMO-RS) for predicting the solubility of the three HHCs in the tested ILs. It was found that there was an excellent qualitative agreement with experimental data both for temperature dependence and for cations structure effect.

AlNashef et. al. [56] reported the solubility of HCB, 1,2-dichlorobenzene (1,2DCB), and 1,3-dichlorobenzene (1,3DCB) in ECOENG 500 (Cocosalkyl pentaethoxi methyl ammonium methosulfate), 1-ethyl-3-methylimidazolium ethylsulfate, AMMOENG 110 (quaternary ammonium chloride) and AMMOENG111 (Poly[oxy(methyl-1,2-ethanediyel)]-alpha-[2-diethyl-hydroethylammonio)ethyl]-acetate as a function of temperature. It was found that the structure of the IL has a tremendous effect on the solubility of the studied HHCs. For example, it was found that while 1,2DCB and 1,3DCB were totally miscible with ECOENG 500 at 65°C, but the solubilities in 1-ethyl-3-methyl-imidazolium ethylsulfate were 0.0041 and 0.0071 wt.%, respectively. In addition, the solubility of HCB at 65°C was 29.6 and 0.003 wt.% in ECOENG 500 and 1-ethyl-3-methyl-imidazolium ethylsulfate, respectively. Makowska et al. [57] reported the phase behavior of liquid mixtures of the ILs 1-ethyl-3-methylimidazolium bis(trifluoromethylsulfonyl) imide, 1-butyl-3-methylimidazolium bis(trifluoromethylsulfonyl) imide, 1-pentyl-3-methylimidazolium bis(trifluoromethylsulfonyl) imide with chloroform and chloroform/carbon tetrachloride mixed solvent. The authors used the cloud point method to obtain the phase diagrams that showed very complicated miscibility behavior. Phase diagrams with the upper critical solution temperature (UCST) were obtained in most cases however under some conditions the phase diagrams with the lower critical solution temperature (LCST) or even with the both (UCST and LCST) were also observed. The "hourglass" type of the phase diagrams constitutes another type of the phase behavior were observed in their study.

Siporska and Szydłowski [58] experimentally determined the (liquid + liquid) miscibility temperatures as a function of composition for the binary systems comprising of imidazolium based ionic liquids with bis(trifluoromethylsulfonyl)imide ($[C_nmIm][Tf_2N]$: n = 3 to 10) and fluorobenzene, chlorobenzene, bromobenzene, iodobenzene or 1,2-dichlorobenzene. All the measured systems show limited miscibility with the upper critical solution temperature behavior. Similar to the other systems with the imidazolium cation the increase of the alkyl chain length in this cation improves the miscibility. The effect of the halogenobenzene was also clear. The miscibility was increasing in the order:

iodobenzene < bromobenzene < chlorobenzene < fluorobenzene. This arrangement corresponds to the decreasing molar volume of the substituted benzenes. Generally, the trend observed in the miscibility behavior was the same as that observed by Shiflett et al. for [emIm][Tf$_2$N].

3. ILs as extractants

Many research groups investigate the use of ILs in liquid–liquid extraction instead of traditional organic solvent for the extraction of organics in water, [55]. Table 4 shows the use of ILs in the extraction of organic compounds from different types of water.

Table 4. Extraction of organic compounds from different types of water, adapted from [19,28].

Source	Target compounds	Ionic liquid
Tap, bottled, fountain, well, river, rainwater, treated and raw wastewater	18 Polycyclic Aromatic Hydrocarbons	[omIm][PF$_6$]
Snow, river and brook water	Four Aromatic Amines	[hmIm][PF$_6$]
Tap, lake and fountain water	Fipronil, Chlorfenapyr, Buprofezin, And Hexythiazox	[hmIm][PF$_6$]
Tap, well, rain and Yellow River water	Organophosphorus Pesticides	[omIm][PF$_6$]
Everglade, river, reservoir and snow	Dichlorodiphenyltrichloroethane And Its Metabolites	[hmIm][PF$_6$]
Tap, well, river, and creek water	13 Aromatic Compounds	[bmIm][Tf$_2$N]
Tap water, well water and rain	1-Naphthylamine, N,N Dimethylaniline, Diphenylamine	[bmIm][PF$_6$]
Storm water	Aliphatic And Aromatic Hydrocarbons	[bmIm][PF$_6$]
East Lake and waste water	Phenols	[omIm][PF$_6$]
River water and effluent water	Chlorobenzenes	[bmIm][PF$_6$]
Water	Aromatic And Aliphatic Hydrocarbons	[bmIm][PF$_6$]
Water	4-Nonylphenol, 4-Tert-Octylphenol	[bmIm][PF$_6$]
Water	Phorate, Parathion And Poxim	[bmIm][PF$_6$]
Water	(Benzene, Toluene ,Ethylbenzene, and Xylene), Polycyclic Aromatic Hydrocarbons , Phthalates ,Phenols, Aromatic Amines, Herbicides , Organotin, and Organomecury	[bmIm][PF$_6$] [omIm][PF$_6$]
Water	Triazines	[hmIm][PF$_6$]
Water	Chloroform, Bromodichloromethane, Dibromochloromethane And Bromoform	[omIm][PF$_6$]
Water	Trihalomethanes	[bmIm][PF$_6$]

Very recentlty, AlSaleem et al. [59] reported the evaluation of six different types of hydrophobic ILs for the extraction of three HHCs from water, namely CCl_4, $CHCl_3$, and $CHBr_3$. The ILs were selected based on their: i) affinity to solubilize contaminants of concern, ii) moderate viscosity, iii) hydrophobicity, and iv) stability in the presence of the superoxide ion that the authors suggested to be used for the destruction of HHCs *in situ*. The authors reported that the used ILs successfully removed HHCs from water with extraction efficiencies ranging from 83 to 99%. It was found that octyltriethyl-ammonium and -pyrrolidinium cation liquids paired with bis(trifluoromethylsulfonyl)imide [Tf_2N] anion exhibited high performance for extracting CCl_4, $CHCl_3$, and $CHBr_3$. The authors also investigated the effect of process parameters, including contact time, temperature, pH, hydrogen bond donor (α), hydrogen bond acceptor (β), dipolarity/polarizability (π^*), and octanol/water partition coefficient (Kow) on the extraction efficiency. Although liquid-liquid extraction is considered to be time-consuming but the extraction method described in this work was found to be effective and high extraction efficiencies were achieved within 10 min for the selected ILs. These features provided the opportunity to use ILs as a powerful alternative method in the field of wastewater treatment. The extraction of HHCs also provided a potential analytical approach to identify them in water by chromatography technique.

In addition, ANOVA results revealed that pH and contact time can significantly affect the extraction efficiency of selected ILs with very low P-values less than 0.005; whereas, there was no significant impact with temperature variations between 25 and 45°C on the extraction efficiency.

One of the most interesting characteristics of ILs is the opportunity to fine-tune its physicochemical properties by means of modifying the alkyl chains or the identity of the cation/anion pair. Therefore, using α, β and π^* for the selection or tailoring of ILs can significantly improve and expand the current applications of ILs in extraction.

The authors indicated that further investigation should be conducted to emphasize the relationship of polarity parameters of ILs with extraction. This can help the specialists to predict the kinetic rates and improve the knowledge of the driving force in such a process.

Different methods for the regeneration of ILs from HHCs can be used. For example, the superoxide ion can be used for the *in situ* decomposition of HHCs. In addition, different methods (distillation, oxidation, etc.) for the separation of refractory organics from ILs after extraction, can be viable processes.

Deng et al. [60] reported the use of a hydrophobic magnetic room temperature ionic liquid (MIL), trihexyltetradecylphosphonium tetrachloroferrate (III) ($[3C_6PC_{14}][FeCl_4]$), as a possible separation agent for solvent extraction of phenolic compounds from aqueous solution. Due to its strong paramagnetism, [3C$_6$PC$_{14}$][FeCl$_4$] responded to an external neodymium magnet and hence was employed in the design of a novel magnetic extraction technique. The conditions for extraction, including extraction time, volume ratio between MIL and aqueous phase, pH of aqueous solution, and structures of phenolic compounds were investigated and optimized. The magnetic extraction of phenols achieved equilibrium within 20 min and the phenolic compounds were found to have higher distribution ratios under acidic conditions. In addition, it was observed that phenols containing higher number of chlorine or nitro substituents exhibited higher distribution ratios. For example, the distribution ratio of phenol was 107. In contrast, 3,5-dichlorophenol distribution ratio had a much higher value of 6372 under identical extraction conditions. When compared with four selected traditional non-magnetic room temperature ILs, [3C$_6$PC$_{14}$][FeCl$_4$] exhibited significantly higher extraction efficiency under the same experimental conditions. Pentachlorophenol, a major component in the contaminated soil sample obtained from a superfund site, was successfully extracted and removed by use of [3C$_6$PC$_{14}$][FeCl$_4$] with high extraction efficiency. Pentachlorophenol concentration was dramatically reduced from 7.8 to 0.2 μg/mL after the magnetic extraction by use of [3C$_6$PC$_{14}$][FeCl$_4$]. The authors reported that they were attempting the development of an efficient approach to recovery and recycling of the MIL and they have explored centrifugation for separation of the magnetic IL from bulk aqueous phenolic solution. Another possible approach was to use a strong magnet for forcing MIL to the bottom of the glassware. The bulk solution can then be decanted out. The recovered MIL then can be recycled and tested for the next extraction of phenolic compounds.

Fan et al. [61] synthesized a series of hydroxyl-, benzyl-, and dialkyl-functionalized ILs, and investigated their extraction abilities for phenol, resorcinol, p-nitrophenol, guaiacol, and o-cresol. Results showed that the extraction efficiencies of the five phenols were significantly influenced by the pH value, added salt, phase ratio, and chemical structure of the IL. The authors found that phenols present in non-ionized forms were preferable to transfer into IL phases. The anion/cation hydrogen-bonding characters of ILs were the main structural factors affecting the extraction efficiency. It was reported that the extraction efficiencies of all tested compounds depended on the pH values of the water phase. In addition, the extraction efficiencies of the five phenols increased with an increase of their hydrophobicities but there was no certain correlation with the hydrophobicities of the ILs. The authors suggested that the driven force for the extraction of resorcinol was mainly hydrogen bonding and the extraction of the other four phenols from water into IL phases was due to hydrophobic interactions.

Pilli et al. [62] reported the judicious screening of various ILs for the extraction of pentachlorophenol (PCP) and dichlorodiphenyltrichloroethane (DDT) from aqueous solutions. The authors used COSMO-RS for the prediction of selectivity of these compounds in aqueous medium at infinite dilution. About 1015 possible ILs were used to determine the best extractant for the removal of these two endocrine disruptor chemicals (EDCs) from aqueous solution. The authors used five different cations namely imidazolium, pyridinium, pyrrolidinium, ammonium and phosphonium that were combined with 29 different anions and were then investigated using COSMO-RS for screening. It turned out that among the studied cations, phosphonium based cations gave the highest selectivity for both PCP and DDT extraction. Fan et al. [63] and Blanchard et al. [64] described the hydrophobic nature of ILs, which are suitable for liquid–liquid extraction of organic chemicals from their aqueous solutions. The hydrophobic nature of any component depends on the activity coefficient of the extractant. Even though selectivity values were very high, their corresponding activity coefficients at infinite dilution (γ_i^∞) were very less and in most of the cases these were less than one indicating high miscibility with water. This may cause the solubility of the IL to increase in aqueous phase. The authors' selection of the proper IL was based on higher values of selectivity as well as high values of γ_i^∞ water as the requirements [63].

Kermanioryani et al. [65] investigated three newly synthesized task specific ionic liquids (TSILs) incorporating different types of aromatic group on imidazolium ring namely, 1-butyl-3-benzoimidazolium bis(trifluoromethylsulfonyl)imide [bzbIm][Tf$_2$N]; 1-butyl-2-phenyl-imidazolium bis(trifluoromethylsulfonyl)imide [bphIm][Tf$_2$N]; and 1-benzyl-3-butyl-imidazolum bis(trifluoromethylsulfonyl)imide [bnbIm][Tf$_2$N]. These TSILs have been employed to remove swiftly heavy loads (up to 1,500 ppm) of cationic methylene blue dye (MB) from aqueous solution without pH adjustment. The results showed that the presence of additional aromatic groups has a significant effect on the extraction efficiency of MB removal through enhancement of the π–π interactions between ILs and the aromatic structure of MB. In addition, the authors used COSMO-RS to analyze the IL-MB interaction in the studied systems. Moreover, the authors designed an original microbiological method not only to assess the toxicity of the aqueous solution before and after the extraction process to reveal the industrial potential of this methodology but also to highlight the high hydrophobicity of these new TSILs.

Vidal et al. [66] evaluated a series of [C$_n$mIm][PF$_6$] and [CnmIm][BF$_4$] ILs for the extraction of phenol, tyrosol and p-hydroxybenzoic acid from aqueous solutions. A near quantitative extraction of the three phenols was obtained using [omIm][BF$_4$]. The results were similar to those observed using n-octanol. Fan et al. [63] reported similar results for the extraction of phenol, bisphenol A, pentachlorophenol, 4-octylphenol, and 4-

nonylphenol for the same series of ILs. The increase in extraction efficiency for ILs as the length of the alkyl chain attached to the cation increased was assigned to the considerable importance of solute hydrophobicity on the extraction mechanism. The higher distribution constants observed for ILs containing the [BF$_4$] anion in comparison to the [PF$_6$] anion were assigned to the stronger hydrogen-bonding interactions of the phenols with the [BF$_4$] anion. The ionic liquids were 10-fold more efficient for extracting the phenols from water than dichloromethane.

Ferreiraa et al. [67] proposed a novel approach to remove dyes from aqueous discharges by the use of IL-based aqueous two-phase systems (ATPS). A detailed study on the partition coefficients and extraction efficiencies of a set of textile dyes (chloranilic acid, indigo blue and sudan III) using ATPS composed of ILs (phosphonium- and imidazolium-based) and an inorganic (aluminium sulphate) or organic salt (potassium citrate) were addressed. The gathered data allowed the evaluation of the effect of chemical structure of IL, the nature of the salt, as well as the pH of the aqueous medium. The results obtained revealed that a proper selection of the IL and salt could lead to complete extraction of the three dyes into the IL-rich phase in a single-step procedure. Moreover, the authors described the dyes recovery to allow the reuse of the IL-rich phase. This new approach, using hydrophilic ILs, is consequently more environmentally friendly and could be envisaged as a promising process for reducing the pollution of wastewaters.

Lu and Pei [68] studied the use of support liquid membrane separation from phenolic wastewater in hydrated environment. The process of separation was conducted with hydrophobic IL, 1-butyl-3-methylimidazole hexafluorophosphate [bmIm][PF$_6$] and applied to the inner coupling liquid membrane system. The authors investigated the effect of temperature, stirring speed, feed liquid phase acidity, and NaOH concentration in the stripping phase on the phenol migration. The authors reported that under optimum conditions of the process phenol migration rate is approximately 100% and the IL may be recycled for several times. Table 5 shows the most commonly used ILs in water-contaminant membrane separations.

Ma and Hong [78] reviewed the applications of ILs in organic pollutants control. An overview of separation, recycling and control of organic pollutants with ILs was presented, for example, phenolic compounds, polycyclic aromatic hydrocarbon (PAHs) and dyes from wastewater, dioxins from waste gas, polyamides from solid wastes, chlorophenothane (DDT) and dieldrin from contaminated soils. Furthermore, the problems and challenges of ILs in organic pollutants control were discussed.

Table 5: Most commonly used ionic liquids in water-contaminant membrane separations, adapted from [69].

IL name	Abbreviations	Contaminants	Ref.
1,3-Dialkylimidazolium Tetrafluoroborate	$[dC_nim][BF_4]$	Tetrahydrofurane	[70]
1-Ethyl-3-Methylimidazolium Acetate	[emim][Ace]	Ethanol	[71]
1-Ethyl-3-Methylimidazolium Tetrafluoroborate	$[emim][BF_4]$	Ethanol	[72]
1-(2-Hydroxyethyl)-3-Methylimidazolium Chloride	[eOHmim][Cl]	Tetrahydrofurane	[73]
1-Ethyl-3-Methylimidazolium Acetate	[emim][Ace]	2-Propanol	[74]
1-Ethyl-3-Methyl-Imidazolium Chloride	[emim][Cl]	Tert-Butanol	[75]
1-Ethenyl-3-Ethylimidazolium Hexafluorophosphate	$[eeim][PF_6]$	Acetone And Butanol	[76]
1-Ethenyl-3-Ethylimidazolium Bromide	[eeim][Br]	Acetone And Butanol	[77]

Lakshmi et al. [79] reported the reduction of the volume requirement of tributyl phosphate by mixing with imidazolium based ILs for the solvent extraction of phenol, p-chlorophenol, 2,4-dichlorophenol, 2,4,6-trichlorophenol, and pentachlorophenol from aqueous solutions. The authors also investigated and optimized the effects of aqueous phase pH, solute concentration in feed, temperature (303–333 K), treat ratio (aqueous to solvent phase ratio), and [Bmim][BF$_4$] volume in tributyl phosphate (TBP); 0–0.7% (v/v) on extraction of phenols. It has been found that 0.5% (v/v) of ionic liquid [Bmim][BF$_4$] in solvent TBP extracted more than 97.5% of phenol and chlorophenols from aqueous solutions with a treat ratio of 5. Transport mechanism for extraction and stripping of phenol and chlorophenols using ionic liquid [Bmim][BF$_4$] have been discussed. The results show that by appropriate selection of extraction and stripping conditions, it was possible to remove nearly all phenols with a treat ratio of 5.

Bekou et al. [80] investigated the extraction of chlorinated phenols from aqueous solutions using two room temperature water immiscible ionic liquids, 1-ethyl-3-methylimidazolium bis(perfluoroethylsulfonyl)imide and 1-butyl-3-methylimidazolium hexafluorophosphate, [bmim][PF$_6$]. Partitioning of phenol, 2-chlorophenol, 2,4-dichlorophenol, 2,4,6-trichlorophenol, 2,3,4,5-tetrachlorophenol, and pentachlorophenol were measured in both aqueous and IL phases using HPLC. Extraction efficiency was found to be greater when [bmim][PF$_6$] was used and when the pH of the aqueous solution was at least one unit below the value of the dissociation constant (pKa). Partitioning, for both ILs, increased as the number of chlorine atoms in the chlorophenol increased. The IL-water distribution ratio for the extraction efficiency of chlorinated phenols using these two ILs was one order of magnitude lower than the corresponding 1-octanol-water partition coefficient. It was found that the ionic strength of the aqueous phase had no significant effect on the IL-water distribution ratios of chlorinated phenols but had a dramatic effect on the solubility of ILs in water. In addition, the IL-water distribution ratios of chlorinated phenols were not influenced significantly by the concentration of chlorophenols in the aqueous phase.

4.   Degradation of HHCs in ILs

Martin et al. [81] reviewed the electrochemical reduction of halogenated organic compounds as a strategy for the remediation of environmental pollutants. The authors discussed the effect of main components of the process, e.g. cells, electrodes, solvents, and electrolytes on the design of a procedure for degrading a targeted pollutant. In addition, they described and contrasted some different experimental techniques in order to explore and characterize the electrochemical behavior of that pollutant. Moreover, the authors proposed different mechanisms for the dehalogenation. The authors outlined techniques, instrumentation, and cell designs involved in scaling up a benchtop experiment to an industrial-scale system. The authors also surveyed the electrochemical studies of various categories of halogenated pollutants.

Subramanian et al. [82] investigated the photolytic degradation of chlorinated aromatics namely pentachlorophenol (PCP) in 1-butyl-3-methylimidazolium hexafluorophosphate [BMIM][PF$_6$] IL using UV radiation of 253.7 nm. It was found that at low concentrations (less than 1.57 mM) PCP could be degraded in this IL following pseudo-first-order kinetics. Quenching studies suggested the generation of hydroxyl radicals in the IL. The authors also reported the *in situ* degradation of PCP extracted from soils using the same IL. Results showed enhanced reaction rates compared to those of the samples prepared by dissolving the compounds directly in the IL.

In another study Yang and Dionysiou [83] reported the photolytic degradation of 2-chlorophenol (2-CP), 4-chlorophenol (4-CP), 2,4-dichlorophenol (2,4-DCP), 2,4,6-trichlorophenol (2,4,6-TCP), 2,3,4,5-tetrachlorophenol (2,3,4,5-TeCP) and PCP in [bmIm][PF$_6$] and 1-ethyl-3-methylimidazolium bis(perfluoroethylsulfonyl)imide [emim][beti] ILs using UV radiation of 253.7 nm. Stability study revealed that these ILs were relatively resistant to phototransformation when used as pure phases. At low concentrations (less than 1.57 mM), chlorinated phenols could be degraded in these ILs following pseudo-first-order kinetics. Intermediate product identification using electrospray TOFMS, GC–MS and HPLC revealed the formation of phenol among the stable phototransformation intermediates of 2-CP. The increase of the chlorine atoms in the phenolic ring generally resulted in a decrease in the phototransformation rates, with exception of 2,4,6-TCP. Recyclability study indicated that the recycling of ILs was hindered to some extent by the presence of organic impurities that were resistant to photolysis or by the generation of byproducts that were resistant to photolytic degradation and compete with the chlorophenols for photon absorption.

Levec and Pintar [2] reviewed the developments in the field of catalytic wet-air oxidation (CWAO). Catalysts awere reviewed first, followed by mechanistic speculations and kinetics that have been proposed for the CWAO process. The process was discussed more in detail only in those cases where it has been already commercialised or at least foreseen to be in the near future. Particular attention was given to the heterogeneously catalyzed wet-air oxidation of real industrial wastewaters (such as Kraft bleach plant effluents) in batch and continuous-flow oxidation reactors. Finally, the considerable potential of the CWAO process to ultimately destroy organic pollutants in industrial effluents and detoxify them by using novel titania-supported Ru catalysts was presented.

Luan et al. [1] reviewed the published literature related to the treatment of refractory organic pollutants in industrial wastewaters, such as dyes. Phenolics were taken as model pollutants in most cases. Reports on the effect of treatment for the WAO of refractory organic pollutants in industrial wastewaters are reviewed, such as emulsified wastewater, TNT red water, etc. Discussions are also made on the mechanism and kinetics of WAO and main technical parameters influencing WAO. Finally, development direction of WAO is summed up.

Wet air oxidation has been a promising technology for the treatment of refractory organic pollutants in the effluent from various industries in view of high pollutant removal efficiency at under mild experimental conditions, and environmental friendliness not involving any harmful chemical reagent. Studies on the WAO of model pollutants and industrial effluents illustrate the potential of WAO as a treatment technology for industrial wastewater. Most of the studies have usually employed phenolics as a model pollutant. And phenol is mainly taken as a model compound because it is widely used in industries and is generated as an intermediate during the oxidation of higher molecular weight aromatic compounds. Generally, the complete oxidation of organic pollutants to CO_2 and H_2O is to difficult achieve due to the formation of more refractory intermediates like short-chain carboxylic acids. WAO integrated with the biological process can be more attractive for the treatment of industrial wastewater containing toxic pollutants.

Babuponnusami and Muthukumar [84] reviewed advanced oxidation processes (AOPs) as a promising technology for the treatment of wastewaters. The investigated processes were developed to decolorize and/or degrade organic pollutants. The authors discussed fundamentals and main applications of typical methods such as Fenton, electro-Fenton, photo-Fenton, sono-Fenton, sono-photo-Fenton, sono-electro-Fenton and photo-electro-Fenton. In addition, the application of nano-zero valent iron in treating refractory compounds was discussed.

AlNashef et al. [56] reported the use of chemically or electrochemically generated superoxide ion for the degradation of several chlorinated hydrocarbons in different ILs. It was found that the degradation efficiency can reach up to 99%. Using GC-MS it was confirmed that the formation of degradation products depended on the HHC, the IL, the molar ratio of superoxide to HHC, and the number of chlorine atoms in the HHC. However, in all cases no hazardous by-products were formed. It was also found that the superoxide ion reacted with imidazolium cation to give the corresponding 2-imidazolone compound. Table 4 shows the physical properties of $O_2^{\bullet-}$.

Hayyan et al. [85] investigated the chemical generation and long term stability of superoxide ion in three bis(trifluoromethylsulfonyl) imide anion based ILs with cations of 1-(3-methoxypropyl)-1-methylpiperidinium, 1-hexyl-1-methyl-pyrrolidinium and trihexyl (tetradecyl) phosphonium. The chemical generation of superoxide ion in ILs was carried out by dissolving potassium superoxide, KO_2, in the corresponding IL, while electrochemical generation was carried out by reduction of O_2 to superoxide ion. It was found that the solubility of KO_2 in the studied ILs depends on the structure of ILs in addition to temperature. However, in all cases the solubility was high enough for many applications. The long-term stability of the generated superoxide ion was carried out by the application of KO_2 in aprotic solvent (dimethyl sulfoxide) in the presence of the corresponding IL. UV–visible spectrophotometry in the absorbance range of 190 – 400 nm was used to determine the stability of superoxide ion. It was found that ILs containing methylpiperidinium and 1-hexyl-1-methyl-pyrrolidinium cations offer a promising long term stability of superoxide ion for various reactions and applications, while the superoxide ion was unstable in the phosphonium based IL.The chemically generated superoxide ion was then used for the destruction of chlorobenzenes under ambient experimental conditions. Table 6 shows

Table 6 Summary of studied ILs for $O_2^{\bullet-}$ generation, Adapted from [23].

Ionic Liquid	Ref.
[hmPyr] [Tf$_2$N]	[76]
1-(3-methoxypropyl)- 1-methylpiperidinium [Tf$_2$N]	[76]
trihexyl (tetradecyl) phosphonium [Tf$_2$N]	[76]
[dmpIm][Tf$_2$N]	[97]
[bmPyrr][Tf$_2$N]	[86,87,88,95,97,98]
[bdmIm][Tf$_2$N]	[89, 90, 95]
[mbIm][Tf$_2$N]	[85,89,91]
[bdmIm][PF$_6$]	[23,85,89,93,94]
[emIm][Tf$_2$N]	[86,97]
[emIm]Cl/AlCl$_3$	[94]
[emIm][BF$_4$]	[90,93]
[bmIm][BF$_4$]	[90, 93]
[pmIm][BF$_4$]	[93]
[tmbAm][Tf$_2$N]	[91]
[C5H5N+C8F18][Tf$_2$N]	[91]
[C5H5N+C18H38][Tf$_2$N]	[91]
[htmAm][Tf$_2$N]	[97]
[hteAm][Tf$_2$N]	[89,92,95,96]
[tebAm][Tf$_2$N]	[87, 85,99,100]
Tris(N-Hexyl)Tetradecylphosphonium Trifluoro tris (Pentafluoroethyl) Phosphate	[95]
Tris(N-Hexyl)Tetradecylphosphonium Bis (Trifluoromethylsulfonyl) Imide	[95]
[tbAm][PF$_6$]/AcN	[101]
[Li][PF$_6$]/AcN	[101]
[tbAm][ClO$_4$]/AcN	[101]
[K][PF$_6$]/AcN	[101]
[Na][PF$_6$]/AcN	[101]

Conclusions and future directions

The use of ILs for wastewater treatment is gaining momentum globally. The unique properties of ILs and their convergence with current treatment technologies present great opportunities to improve wastewater treatment. Although all applications of ILs discussed in this chapter are still in the laboratory research stage and none has made their way to pilot testing, there is an ample chance for some of these applications to reach even full-scale application.

The most important challenges that faces the transfer of the use of ILs from lab scale into commercial applications include cost, commercial availability, toxicity, miscibility with water, and compatibility with the existing infrastructure.

The challenges faced by wastewater treatment using ILs are important, but many of these challenges are temporary, including technical hurdles, high cost, and potential environmental and human risk.

To overcome these barriers, collaboration between research institutions, industry, government, and other stakeholders is imprtant. It is believed that advancing applications of ILs by carefully steering its direction while avoiding unintended consequences can continuously provide robust solutions to the wastewater treatment challenges.

References

[1] M. Luan, G. Jing, Y. Piao, D. Liu, L. Jin, Treatment of refractory organic pollutants in industrial wastewater by wet air oxidation, Arab. J. Chem. 10 (2017) S769-S776. https://doi.org/10.1016/j.arabjc.2012.12.003

[2] J. Levec, A. Pintar, Catalytic wet-air oxidation processes: A review, Catal. Today 124 (2007) 172-184. https://doi.org/10.1016/j.cattod.2007.03.035

[3] H. Debellefontaine, M. Chakchouk, J. Foussard, D. Tissot, P. Striolo, Treatment of organic aqueous wastes: wet air oxidation and wet peroxide oxidation®, Environ. Pollut. 92 (1996) 155-164. https://doi.org/10.1016/0269-7491(95)00100-X

[4] T.G. Danis, T.A. Albanis, D.E. Petrakis, P.J. Pomonis, Removal of chlorinated phenols from aqueous solutions by adsorption on alumina pillared clays and mesoporous alumina aluminum phosphates, Water Res. 32 (1998) 295-302. https://doi.org/10.1016/S0043-1354(97)00206-6

[5] F. Zimmermann, New waste disposal process, Chem. Eng. 65 (1958) 117-121.

[6] V.S. Mishra, V.V. Mahajani, J.B. Joshi, Wet air oxidation, Ind. Eng. Chem. Res 34 (1995) 2-48. https://doi.org/10.1021/ie00040a001

[7] Metcalf & Eddy Inc., T. Asano, F. Burton, H. Leverenz, R. Tsuchihashi, G. Tchobanoglous, Water Reuse: Issues, Technologies, and Applications, Mc-Graw Hill, NewYork, USA, 2007.

[8]   Metcalf & Eddy Inc., G. Tchobanoglous, H.D. Stensel, R. Tsuchihashi, F.L. Burton, Wastewater engineering: treatment and reuse, 5th Edition, McGraw Hill, 2013.

[9]   B. Tawabini, N. Fayad, and M. Morsy, The impact of groundwater quality on the removal of methyl tertiary-butyl ether (MTBE) using advanced oxidation technology, Water Sci. Technol. 60 (2009) 2161-2165. https://doi.org/10.2166/wst.2009.586

[10]   R. J. Watts, Hazardous wastes: sources, pathways, receptors, John-Wiley & Sons, Inc. New Jersy, USA (1998).

[11]   P.M. Kutty, A.A. Nomani, T. Thankachan, Analyses of water samples from Jeddah seawater RO/MSF plants for organic pollutants,Technical Report No. SWCC (RDC)-14, Citeseer, 1991.

[12]   I.M. AlNashef, M.L. Leonard, M.C. Kittle, M.A. Matthews, J.W. Weidner, Electrochemical generation of superoxide in room-temperature ionic liquids, Electrochem. Solid State Lett. 4 (2001) D16-D18. https://doi.org/10.1149/1.1406997

[13]   M.S. Callahan, B. Green, Hazardous solvent source reduction, McGraw-Hill, Texas, USA (1995).

[14]   E. Kalu, R. E. White, In situ degradation of polyhalogenated aromatic hydrocarbons by electrochemically generated superoxide ions, J. Electrochem. Soc. 138 (1991) 3656-3660. https://doi.org/10.1149/1.2085475

[15]   J. Jensen, Chlorophenols in the Terrestrial Environment, in: G. W. Ware, ed. Reviews of Environmental Contamination and Toxicology: Continuation of Residue Reviews, Springe, New York, 1996, pp. 25-51. https://doi.org/10.1007/978-1-4613-8478-6_2

[16]   M.H. El-Naas, H.A. Mousa, M.E. Gamal, Microbial Degradation of Chlorophenols, in: S. Singh (Ed.) Microbe-Induced Degradation of Pesticides. Environmental Science and Engineering, Springer, New York, 2017, p.p. 23-58. https://doi.org/10.1007/978-3-319-45156-5_2

[17]   D. Hale, W. Reineke, and J. Wiegel, Chlorophenol degradation, in: GR. Chaudhry (Ed.), Biological degradation and bioremediation of toxic compounds, Diosorides Press, Portland, OR, USA, 1994, p.p. 74-91.

[18]   W. Berson, "Ionic Liquids in Synthesis", Edited by P. Wasserscheid, T. Welton, Wiley-VCH, (2002).

[19]   L. Zaijun, L. Junkang, S. Xiulan, Ionic liquid as novel solvent for extraction and separation in analytical chemistry, in: A. Kokorin (Ed.), Ionic Liquids: Applications and Perspectives, 2011, p.p.181-206. https://doi.org/10.5772/14250

[20]   Y. Deng, Physico-chemical properties and environmental impact of ionic liquids, Ph. D. thesis, Université Blaise Pascal-Clermont-Ferrand II, 2011.

[21]   E. Aguilera-Herrador, R. Lucena, S. Cárdenas, M. Valcárcel, Sample treatments based on ionic liquids, in: A. Kokorin, Ionic Liquids: Applications and Perspectives, 2011, p.p. 181-206. https://doi.org/10.5772/14337

[22]   A. Chapeaux, Extraction of alcohols from water using ionic liquids, Ph. D. thesis, University of Notre Dame, 2009.

[23]   M. Hayyan, F.S. Mjalli, M.A. Hashim, I.M. AlNashef, Generation of superoxide ion in pyridinium, morpholinium, ammonium, and sulfonium-based ionic liquids and the application in the destruction of toxic chlorinated phenols, Ind. Eng. Chem. Res. 51 (2012) 10546-10556. https://doi.org/10.1021/ie3006879

[24]   J. W. Lee, J. Y. Shin, Y. S. Chun, H. B. Jang, C. E. Song, S.-g. Lee, Toward understanding the origin of positive effects of ionic liquids on catalysis: formation of more reactive catalysts and stabilization of reactive intermediates and transition states in ionic liquids, Accounts Chem. Res. 43 (2010) 985-994. https://doi.org/10.1021/ar9002202

[25]   E. Siedlecka, M. Czerwicka, J. Neumann, P. Stepnowski, J. Fernández, J. Thöming, Ionic liquids: methods of degradation and recovery, in: A. Kokorin, Ionic Liquids: Theory, Properties, New Approaches, InTech, 2011, p.p. 701-721. https://doi.org/10.5772/15463

[26]   A.S. Barnes, E.I. Rogers, I. Streeter, L. Aldous, C. Hardacre, G.G. Wildgoose, R.G. Compton, Unusual voltammetry of the reduction of O_2 in [C4dmim][N (Tf)2] reveals a strong interaction of $O^{2\cdot-}$ with the [C4dmim]+ cation, J. Phys. Chem. C 112 (2008)13709-13715. https://doi.org/10.1021/jp803349z

[27]   SIGMA; SAFC; SIGMA-ALDRICH; ISOTEC; ALDRICH; FLUKA; and SUPELCO, Enabling Technologies: Ionic Liquids, Sigma-Aldrich Co., 5 (2005) 1-23.

[28]   K. Vijayaraghavan, T. Padmesh, K. Palanivelu, M. Velan, Biosorption of nickel (II) ions onto Sargassum wightii: application of two-parameter and three-parameter isotherm models, J. Hazard. Mater. 133 (2006) 304-308. https://doi.org/10.1016/j.jhazmat.2005.10.016

[29] D. Han, K. H. Row, Recent applications of ionic liquids in separation technology, Molecules 15 (2010) 2405-2426. https://doi.org/10.3390/molecules15042405

[30] D.-x. Chen, X.-k. OuYang, Y.-g. Wang, L.-y. Yang, C.-h. He, Liquid–liquid extraction of caprolactam from water using room temperature ionic liquids, Separ. Purif. Tech. 104 (2013) 263-267. https://doi.org/10.1016/j.seppur.2012.11.035

[31] S.S. AlSaleem, Using Potassium Superoxide for Destruction of Extracted Halogenated Hydrocarbons from Water by Ionic Liquid, Ph. D. Thesis, King Saud University, Riyadh, Saudi Arabia, 2014 .

[32] E.W. Neuman, Potassium superoxide and the three-electron bond, Chem. Phys. 2 (1934) 31-33. https://doi.org/10.1063/1.1749353

[33] G.F. Carter, D.H. Templeton, Polymorphism of sodium superoxide, J. Am. Chem. Soc. 75 (1953) 5247-5249. https://doi.org/10.1021/ja01117a031

[34] B.H.J. Bielski, D.E. Cabelli, R.L. Arudi, A.B. Ross, Reactivity of HO_2/O_2 radicals in aqueous solution, J. Phys. Chem. Ref. Data 14 (1985) 1041-1100. https://doi.org/10.1063/1.555739

[35] T. Ozawa, A. Hanaki, H. Yamamoto. On a spectrally well-defined and stable source of superoxide ion, O^{-2}, FEBS Lett. 74 (1977) 99-102. https://doi.org/10.1016/0014-5793(77)80762-X

[36] I. Dzidic, D.I. Carroll, R.N. Stillwell, E.C. Horning, Gas phase reactions. Ionization by proton transfer to superoxide anions, J. Am. Chem. Soc. 96 (1974) 5258-5259. https://doi.org/10.1021/ja00823a045

[37] L. Andrews, Matrix Infrared spectrum and bonding in the lithium superoxide molecule, LiO_2, J. Am. Chem. Soc. 90 (1968) 7368-7370. https://doi.org/10.1021/ja01028a048

[38] D. Vasudevan, H. Wendt, Electroreduction of oxygen in aprotic media, J. Electroanal. Chem. 392 (1995) 69-74. https://doi.org/10.1016/0022-0728(95)04044-O

[39] X.J. Huang, E.I. Rogers, C. Hardacreand, R.G. Compton, The reduction of oxygen in various room temperature ionic liquids in the temperature range 293-318 K: Exploring the applicability of the stokes-einstein relationship in room temperature ionic liquids, J. Phys. Chem. B 113 (2009) 8953-8959. https://doi.org/10.1021/jp903148w

[40] P.H. Krupenie, The spectrum of molecular oxygen, J. Phys. Chem. Ref. Data 1 (1972) 423-534. https://doi.org/10.1063/1.3253101

[41]   K.M., Ervin, I. Anusiewicz, P. Skurski, J. Simons, W.C. Lineberger, The only stable state of O_2^- is the X $^2\prod$g ground state and it (still!) has an adiabatic electron detachment energy of 0.45 eV, J. Phys. Chem. A 107 (2003) 8521-8529. https://doi.org/10.1021/jp0357323

[42]   M.R. Green, H. Allen, O. Hill, D.R. Turner, The nature of the superoxide ion in dipolar aprotic solvents: The electron paramagnetic resonance spectra of the superoxide ion in N,N-dimethylformamide–evidence for hydrated forms, FEBS Lett. 103 (1979) 176–180. https://doi.org/10.1016/0014-5793(79)81276-4

[43]   P.S. Jain, S. Lal, Electrolytic reduction of oxygen at solid, electrodes in aprotic solvents-the superoxide ion, Electrochim. Acta 27 (1982) 759–763. https://doi.org/10.1016/0013-4686(82)85071-8

[44]   M.E. Peover, B.S. White, Electrolytic reduction of oxygen in aprotic solvents: the superoxide ion, Electrochim. Acta 11 (1966) 1061–1067. https://doi.org/10.1016/0013-4686(66)80043-9

[45]   J.D. Wadhawan, P.J. Welford, H.B. McPeak, C.E.W. Hahn, R.G. Compton, The simultaneous voltammetric determination and detection of oxygen and carbon dioxide: a study of the kinetics of the reaction between superoxide and carbon dioxide in non-aqueous media using membrane-free gold disc microelectrodes, Sens. Actuators B 88 (2003) 40–52. https://doi.org/10.1016/S0925-4005(02)00307-6

[46]   Y. Wei, K. Wu, Y Wu, S. Hu, Electrochemical characterization of a new system for detection of superoxide ion in alkaline solution, Electrochem. Commun. 5 (2003) 819–824. https://doi.org/10.1016/j.elecom.2003.08.001

[47]   M.M. Islam, M.S. Saha, T. Okajima, T. Ohsaka, Current oscillatory phenomena based on electrogenerated superoxide ion at the HMDE in dimethylsulfoxide, J. Electroanal. Chem. 577 (2005) 145–154. https://doi.org/10.1016/j.jelechem.2004.12.003

[48]   T. Okajima, T. Ohsaka, Chemiluminescence of indole and its derivatives induced by electrogenerated superoxide ion in acetonitrile solutions, Electrochim. Acta 47 (2002) 1561–1565. https://doi.org/10.1016/S0013-4686(01)00888-X

[49]   G.J. Hills, L.M. Peter, Electrode kinetics in aprotic media, J. Electroanal. Chem. Interfacial Electrochem. 50 (1974) 175–185. https://doi.org/10.1016/S0022-0728(74)80149-X

[50]   R.A. Johnson, E.G. Nidy, M.V. Merritt, Superoxide chemistry. Reactions of superoxide with alkyl halides and alkyl sulfonate esters, J. Am. Chem. Soc. 100 (1978) 7960–7966. https://doi.org/10.1021/ja00493a028

[51]   E.J. Corey, K.C.Nicolaou, M. Shibasaki, Y. Machida, C.S. Shiner, Superoxide ion as a synthetically useful oxygen nucleophile, Tetrahedron Lett. 16 (1975) 3183–3186. https://doi.org/10.1016/S0040-4039(00)91450-3

[52]   S. Randström, G.B. Appetecchi, C. Lagergren, A. Moreno, S. Passerini, The influence of air and its components on the cathodic stability of N-butyl-n-methylpyrrolidinium bis (Trifluoromethanesulfonyl) imide, Electrochim. Acta 53 (2007) 1837–1842. https://doi.org/10.1016/j.electacta.2007.08.029

[53]   C. Villagrán, L. Aldous, M.C. Lagunas, R.G. Compton, C. Hardacre, Electrochemistry of phenol in bis {(Trifluoromethyl) sulfonyl} amide ([NTf$_2$]−) based ionic liquids, J. Electroanal. Chem. 588 (2006) 27–31. https://doi.org/10.1016/j.jelechem.2005.11.023

[54]   Silvester, L. Aldous, C. Hardacre, R.G. Compton, An electrochemical study of the oxidation of hydrogen at platinum electrodes in several room temperature ionic liquids, J. Phys. Chem. B 111 (2007) 5000–5007. https://doi.org/10.1021/jp067236v

[55]   S.S. AlSaleem, W. M. Zahid, I.M. AlNashef, M.K. Hadj-Kali, Solubility of halogenated hydrocarbons in hydrophobic ionic liquids: Experimental study and COSMO-RS prediction, J. Chem. Eng. Data 60 (2015) 2926-2936. https://doi.org/10.1021/acs.jced.5b00310

[56]   I.M. AlNashef, M.A. Hashim, F.S. Mjalli, M. Hayyan, Benign degradation of chlorinated benzene in ionic liquids, IJCEBS 1 (2013) 2320–4087.

[57]   A. Makowska, A. Siporska, K. Kobierska, J. Szydłowski, Phase behavior of 1-alkyl-3-methylimidazolium bis (trifluoromethylsulfonyl) imide with chloroform and/or chloroform/carbon tetrachloride mixed solvent, J. Mol. Liq. 199 (2014) 364-366. https://doi.org/10.1016/j.molliq.2014.09.036

[58]   A. Siporska, J. Szydłowski, Phase behavior of ionic liquids 1-alkyl-3-methylimidazolium bis(trifluoromethylsulfonyl)imides with halogenated benzenes, J. Chem. Therm. 88 (2015) 22-29. https://doi.org/10.1016/j.jct.2015.04.012

[59]   S.S. AlSaleem, W.M. Zahid, I.M. AlNashef, H. Haider, Extraction of halogenated hydrocarbons using hydrophobic ionic liquids, Separ. Purif. Tech. 184 (2017) 231-239. https://doi.org/10.1016/j.seppur.2017.04.047

[60]   N. Deng, M. Li, L. Zhao, C. Lu, S.L. de Rooy, I. M. Warner, Highly efficient extraction of phenolic compounds by use of magnetic room temperature ionic liquids for environmental remediation, J. of Hazard. Mater. 192 (2011) 1350-1357. https://doi.org/10.1016/j.jhazmat.2011.06.053

[61]   Y. Fan, Y. Li, X. Dong, G. Hu, S. Hua, J. Miao, D. Zhou, Extraction of phenols from water with functionalized ionic liquids, Ind. Eng. Chem. Res. 53 (2014) 20024-20031. https://doi.org/10.1021/ie503432n

[62]   S.R. Pilli, T. Banerjee, K. Mohanty, Extraction of pentachlorophenol and dichlorodiphenyltrichloroethane from aqueous solutions using ionic liquids, J. of Ind. Eng. Chem. 18 (2012) 1983-1996. https://doi.org/10.1016/j.jiec.2012.05.017

[63]   J. Fan, Y. Fan, Y. Pei, K. Wu, J. Wang, M. Fan, Solvent extraction of selected endocrine-disrupting phenols using ionic liquids, Separ. Purif. Tech. 61 (2008) 324-331. https://doi.org/10.1016/j.seppur.2007.11.005

[64]   L. A. Blanchard, D. Hancu, E. J. Beckman, J. F. Brennecke, Green processing using ionic liquids and CO_2, Nature 399 (1999) 28-29. https://doi.org/10.1038/19887

[65]   M. Kermanioryani, M. I. A. Mutalib, K. A. Kurnia, K. C. Lethesh, S. Krishnan, J.-M. Leveque, Enhancement of π–π aromatic interactions between hydrophobic ionic liquids and methylene blue for an optimum removal efficiency and assessment of toxicity by microbiological method, J. Clean Prod. 137 (2016) 1149-1157. https://doi.org/10.1016/j.jclepro.2016.07.193

[66]   S.T.M. Vidal, M.J. Neiva Correia, M.M. Marques, M.R. Ismael, M.T. Angelino Reis, Studies on the use of ionic liquids as potential extractants of phenolic compounds and metal ions, Separ. Sci. Tech. 39 (2005) 2155-2169. https://doi.org/10.1081/SS-120039311

[67]   A.M. Ferreira, J.A.P. Coutinho, A.M. Fernandes, M.G. Freire, Complete removal of textile dyes from aqueous media using ionic-liquid-based aqueous two-phase systems, Separ. Purif. Tech. 128 (2014) 58-66. https://doi.org/10.1016/j.seppur.2014.02.036

[68]   S. Lu, L. Pei, A study on phenol migration by coupling the liquid membrane in the ionic liquid, Int. J. Hydrogen Energ. 41 (2016) 15724-15732. https://doi.org/10.1016/j.ijhydene.2016.05.008

[69]   C. Petra, B. Katalin, Application of Ionic Liquids in Membrane Separation
       Processes, in: A. Kokorin (Ed.), Ionic Liquids: Applications and Perspectives,
       InTech, 2011, pp. 561-586. https://doi.org/10.5772/14862

[70]   D. Han, K.H. Row, Recent applications of ionic liquids in separation technology
       (Review), Molecule 15 (2010) 2405-2426.
       https://doi.org/10.3390/molecules15042405

[71]   M Seiler, J. Jork, K. Asimina, A. Wolfgang, H. Rolf, Separation of azeotropic
       mixtures using hyperbranched polymers or ionic liquids, AIChE J. 50 (2004)
       2439-2454. https://doi.org/10.1002/aic.10249

[72]   Y. Ge, L. Zhang, X. Yuan, W. Geng, J. Ji, Selection of ionic liquids as entrainers
       for separation of (water + ethanol), J. Chem. Thermodyn. 40 (2008) 1248-1252.
       https://doi.org/10.1016/j.jct.2008.03.016

[73]   X. Hu, J. Yu, H. Liu, Separation of THF and water by room temperature ionic
       liquids, Water Sci. Technol. 53 (2006) 245-249.
       https://doi.org/10.2166/wst.2006.359

[74]   Q. Zhang, Z. Li, J. Zhang, S. Zhang, L. Zhu, J. Yang, X. Zhang, Y. Deng,
       Physicochemical properties of nitrile-functionalized ionic liquids, J. Phys. Chem.
       B 111 (2007) 2864-2872. https://doi.org/10.1021/jp067327s

[75]   L. Zhang,B. Qiao, Y. Ge, D. Deng, J. Ji, Effect of ionic liquids on (vapor + liquid)
       equilibrium behavior of (water + 2-methyl-2-propanol), J. Chem. Thermodyn. 41
       (2009) 138-143. https://doi.org/10.1016/j.jct.2008.07.004

[76]   P. Izák, W. Ruth, Z. Fei, P.J. Dyson, U. Kragl, Selective removal of acetone and
       butan-1-ol from water with supported ionic liquid-polydimethylsiloxane
       membrane by pervaporation, Chem. Eng. J. 139 (2008) 318-321.
       https://doi.org/10.1016/j.cej.2007.08.001

[77]   X.J. Huang, L. Aldous, A.M. O'Mahony, F.J. Del Campo, R.G. Compton, Toward
       membrane-free amperometric gas sensors: A microelectrode array approach, Anal.
       Chem. 82 (2010) 5238−5245. https://doi.org/10.1021/ac1006359

[78]   J. Ma, X. Hong, Application of ionic liquids in organic pollutants control, J.
       Environ. Manage. 99 (2012) 104-109.
       https://doi.org/10.1016/j.jenvman.2012.01.013

[79]   A. Brinda Lakshmi, A. Balasubramanian, S. Venkatesan, Extraction of phenol and
       chlorophenols using ionic liquid [Bmim]+[BF4]−dissolved in tributyl phosphate,

Clean - Soil, Air, Water 41 (2013) 349-355.
https://doi.org/10.1002/clen.201100632

[80]   E. Bekou, D.D. Dionysiou, R.Y. Qian, and G.D. Botsaris, Extraction of Chlorophenols from Water Using Room Temperature Ionic Liquids, ACS Symposium Series, 856 (2003) 544-560. https://doi.org/10.1021/bk-2003-0856.ch042

[81]   E.T. Martin, C.M. McGuire, M.S. Mubarak, D.G. Peters, Electroreductive remediation of halogenated environmental pollutants, Chem. Rev. 116 (2016) 15198-15234. https://doi.org/10.1021/acs.chemrev.6b00531

[82]   B. Subramanian, Q. Yang, Q. Yang, A.P. Khodadoust, D.D. Dionysiou, Photodegradation of pentachlorophenol in room temperature ionic liquids, J. Photochem. Photobiol. Chem. 192 (2007) 114-121. https://doi.org/10.1016/j.jphotochem.2007.05.012

[83]   Q. Yang, D. D. Dionysiou, Photolytic degradation of chlorinated phenols in room temperature ionic liquids, J. Photochem. Photobiol. Chem. 165 (2004) 229-240. https://doi.org/10.1016/j.jphotochem.2004.03.022

[84]   A. Babuponnusami, K. Muthukumar, A review on Fenton and improvements to the Fenton process for wastewater treatment, Int. J. Chem. Environ. Eng. 2 (2014) 557-572. https://doi.org/10.1016/j.jece.2013.10.011

[85]   M. Hayyan, F. S. Mjalli, M. A. Hashim, I. M. AlNashef, S. M. Al-Zahrani, K. L. Chooi, Long term stability of superoxide ion in piperidinium, pyrrolidinium and phosphonium cations-based ionic liquids and its utilization in the destruction of chlorobenzenes, J. Electroanal. Chem. 664 (2012) 26-32. https://doi.org/10.1016/j.jelechem.2011.10.008

[86]   J. Ghilane, C. Lagrost, P. Hapiot, Scanning electrochemical microscopy in nonusual solvents: inequality of diffusion coefficients problem, Anal. Chem. 79 (2007) 7383−7391. https://doi.org/10.1021/ac071195x

[87]   C. Villagrán, L. Aldous, M.C. Lagunas, R.G. Compton, C. Hardacre, Electrochemistry of phenol in bis {(trifluoromethyl) sulfonyl} amide ([NTf2]−) based ionic liquids, J. Electroanal. Chem. 588 (2006) 27−31. https://doi.org/10.1016/j.jelechem.2005.11.023

[88]   D. Zhang, T. Okajima, F. Matsumoto, T. Ohsaka, Electroreduction of dioxygen in 1-n-alkyl-3-methylimidazolium tetrafluoroborate room-temperature ionic liquids. J. Electrochem. Soc. 151 (2004) D31−D37. https://doi.org/10.1149/1.1649748

[89] A. Rene, D. Hauchard, C. Lagrost, P. Hapiot, Superoxide Protonation by Weak Acids in Imidazolium Based Ionic Liquids. J. Phys. Chem. B 113 (2009) 2826–2831. https://doi.org/10.1021/jp810249p

[90] M.C. Buzzeo, O.V. Klymenko, J.D. Wadhawan, C. Hardacre, K.R. Seddon, R.G. Compton, Kinetic analysis of the reaction between electrogenerated superoxide and carbon dioxide in the room temperature ionic liquids 1-ethyl-3-methylimidazolium bis-(trifluoromethylsulfonyl)imide and hexyltriethylammonium bis-(trifluoromethylsulfonyl)imide, J. Phys. Chem. B 108 (2004) 3947–3954. https://doi.org/10.1021/jp031121z

[91] M.M. Islam, T. Ohsaka, Two-electron quasi-reversible reduction of dioxygen at hmde in ionic liquids: Observation of cathodic maximum and inverted peak, J. Electroanal. Chem. 623 (2008) 147–154. https://doi.org/10.1016/j.jelechem.2008.07.004

[92] B. Martiz, R. Keyrouz, S. Gmouh, M. Vaultier, V. Jouikov, Superoxide-stable ionic liquids: New and efficient media for electrosynthesis of functional siloxanes, Chem. Commun. 2004 (2004) 674–675. https://doi.org/10.1039/b313832a

[93] M.C. Buzzeo, O.V. Klymenko, J.D. Wadhawan, C. Hardacre, K.R. Seddon, R.G. Compton, Voltammetry of oxygen in the room-temperature ionic liquids 1-ethyl-3-methylimidazolium bis((trifluoromethyl) sulfonyl) imide and hexyltriethylammonium bis((trifluoromethyl) sulfonyl) imide: One-electron reduction to form superoxide. steady-state and transient behavior in the same cyclic voltammogram resulting from widely different diffusion coefficients of oxygen and superoxide. J. Phys. Chem. A 107 (2003) 8872–8878. https://doi.org/10.1021/jp0304834

[94] K. Ding, The electrocatalysis of multi-walled carbon nanotubes (MWCNTs) for oxygen reduction reaction (ORR) in room temperature ionic liquids (RTILs) Por, Electrochim. Acta 27 (2009) 165–175. https://doi.org/10.4152/pea.200902165

[95] M.T. Carter, C.L. Hussey, S.K.D. Strubinger, R.A. Osteryoung, Electrochemical reduction of dioxygen in room-temperature imidazolium chloride-aluminum chloride molten salts, Inorg. Chem. 30 (1991) 1149–1151. https://doi.org/10.1021/ic00005a051

[96] R.G. Evans, O.V. Klymenko, S.A Saddoughi, C. Hardacre, R.G. Compton, Electroreduction of oxygen in a series of room temperature ionic liquids composed of group 15-centered cations and anions, J. Phys. Chem. B 108 (2004) 7878–7886. https://doi.org/10.1021/jp031309i

[97] E.I. Rogers, X.J. Huang, E.J.F. Dickinson, C. Hardacre, R.G. Compton, Investigating the mechanism and electrode kinetics of the oxygen| superoxide $(O_2|O_2^{\cdot-})$ couple in various room-temperature ionic liquids at gold and platinum electrodes in the temperature range 298−318 K, J. Phys. Chem. C 113 (2009) 17811−17823. https://doi.org/10.1021/jp9064054

[98] Y. Katayama, H. Onodera, M. Yamagata, T Miura, Electrochemical reduction of oxygen in some hydrophobic room-temperature molten salt systems, J. Electrochem. Soc. 151 (2004) A59−A63. https://doi.org/10.1149/1.1626669

[99] Y. Katayama, K. Sekiguchi, M. Yamagata, T. Miura, Electrochemical behavior of oxygen/superoxide ion couple in 1-butyl-1-methylpyrrolidinium bis (trifluoromethylsulfonyl) imide room-temperature molten salt, J. Electrochem. Soc. 152 (2005) E247−E250. https://doi.org/10.1149/1.1946530

[100] D. Zigah, A. Wang, C. Lagrost, P. Hapiot, diffusion of molecules in ionic liquids/organic solvent mixtures. Example of the reversible reduction of O_2 to superoxide. J. Phys. Chem. B 113 (2009) 2019−2023. https://doi.org/10.1021/jp8095314

[101] C.O. Laoire, S. Mukerjee, K.M. Abraham, E.J. Plichta, M.A. Hendrickson, Elucidating the mechanism of oxygen reduction for lithium-air battery applications, J. Phys. Chem. C 113 (2009) 20127−20134. https://doi.org/10.1021/jp908090s

Chapter 4

Biohydrogen and Bioethanol Production from Agro-Industrial Wastewater

Eduardo Lucena Cavalcante de Amorim*, Norma Candida dos Santos Amorim, Williane Vieira Macêdo[1], Eric Avilino Batista

Technology Center – Federal University of Alagoas. Av. Lourival Melo Mota, s/n – Cidade Universitária – CEP 57072-900 – Maceió/AL – Brazil

eduardo.lucena@ctec.ufal.br

Abstract

The co-digestion of cassava wastewater with swine residues was evaluated for hydrogen and methane production in separate phases. Four different dilution rates were used in the acidogenic phase. The highest hydrogen yield (HY) (1.13 mol H_2/mol glucose) was achieved in the 25-50-25% dilution (percentage of swine, water and cassava wastewater). For the methanogenic phase, the effluent from the acidogenic process was used as substrate. *Sururu* shells (*Mytella falcata*) were used as support material and pH buffer to control the alkalinity of the reaction medium. The best yield obtained was 1.73 ± 0.13 mL CH_4/gCOD in the hydraulic retention time (HRT) of 12 h.

Keywords

Co-digestion, Anaerobic Fluidized Bed Reactor, Anaerobic Fixed-bed Reactor, *Sururu* Shells

Contents

1. Introduction

Anaerobic digestion processes of solid and liquid residues allow the production of renewable energy and, at the same time, promote environmental control through the reduction and treatment of organic waste. The production of biohydrogen and biomethane from biomass has attracted much attention due to emerging environmental problems particularly related to global warming [1]. Large amounts and variety of biomasses are available especially in the form of organic waste. However, for organic matters to be potentially useful as substrates for sustainable energy production, they must be abundant and readily available as well as inexpensive and highly biodegradable. In general, agro-industrial waste meets these requirements [2].

Among the agroindustrial residues, those of animal origin are considered most polluting due to high content of organic matter, nitrogen and phosphorus. The manure (feces and urine) generated by animals kept in confinement (poultry, swine and cattle farming) can cause soil and water degradation [3].

Swine farming has economic and social relevance in the Brazilian agricultural complex for enabling food production, employment generation, and the increase of many families' income. It has been a profitable practice as it produces large amounts of meat in a shorter period of time and in reduced physical space. In the last 15 years, Brazil has increased exports by more than 600% and swine production by 40%. It ranks fourth in the world ranking of producing and exporting countries. On export, it is behind the US, the European Union (27 countries) and Canada, and in production Brazil comes after China, the European Union and the United States [4].

In addition to the swine activity, the cassava crops are noteworthy in the Brazilian north east. Cassava is a typical Brazilian agricultural product characterized by being tolerant to drought conditions and low soil fertility. Brazil is one of the world's largest producers of cassava, which is processed to produce flour and starch. During its processing about 7,000 L of wastewater per kg of processed cassava root is generated. The cassava wastewater is rich in carbohydrates and organic matter and has high chemical oxygen demand (COD) as well as biochemical oxygen demand (BOD) [5]. Despite its potential pollution, the characteristics of this wastewater make it attractive for hydrogen production.

Recently, a co-digestion of different organic wastes has been highlighted. Some of the benefits of anaerobic co-fermentation include: better pH conditions and better carbon/nitrogen balance [6], simultaneous treatment of wastewaters, dilution of toxic or inhibitory compounds, the increase of hydrogen production and adjustment of the carbohydrate/protein ratio [7]. Wang et al. [6] evaluated the co-digestion of cassava wastewater in batch reactors with four different substrates (cassava wastewater, swine manure, cattle manure, and activated sludge waste). The authors observed about 46% increase in hydrogen yield of the cassava wastewater with cassava sludge excess in relation to the cassava wastewater by itself. Hernandez et al. [8] studied the co-digestion of coffee mycilage with swine manure; the authors obtained a maximum of 39% hydrogen content in the biogas. Marrone et al. [9] studied the co-digestion of three different substrates (buffalo slurry, cheese whey, and crude glycerol) in order to evaluate the best composition for hydrogen production. The authors concluded that hydrogen production was better when cheese whey was predominant in the substrate composition. Rosa et al. [10] evaluated the inoculum and hydraulic retention time (HRT) effects on the hydrogen production in anaerobic fluidized bed reactor using a mixture of cassava wastewater and glucose as substrate. The authors obtained better yields for hydrogen production when it was applied 6 h HRT. Because in fermentation complex residues of agroindustrial cultures can be used as substrates, there are several opportunities to develop anaerobic digestion of two or more substrates, which could have complementary characteristics, such as carbon and nitrogen contents, pH, alkalinity, and microorganisms. Residues from swine farming have high concentration of protein and lipids and therefore, have been widely used in anaerobic digestion as substrate support while treating other wastes [8]. The cassava wastewater has high carbohydrate content and the co-digestion of both wastes could be feasible if they are produced in the same geographic regions. In this scenario, swine and cassava wastewaters were used to produce hydrogen and methane gases in fluidized and fixed bed anaerobic reactors, respectively.

2. Materials and methods

2.1 Anaerobic reactors configuration and operational conditions

The experiment consisted in the use of two anaerobic reactors: a fluidized bed reactor (AFBR) for the acidogenic phase and a fixed bed reactor (FBR) for the methanogenic phase. The operational scheme of the reactors is illustrated in Fig. 1.

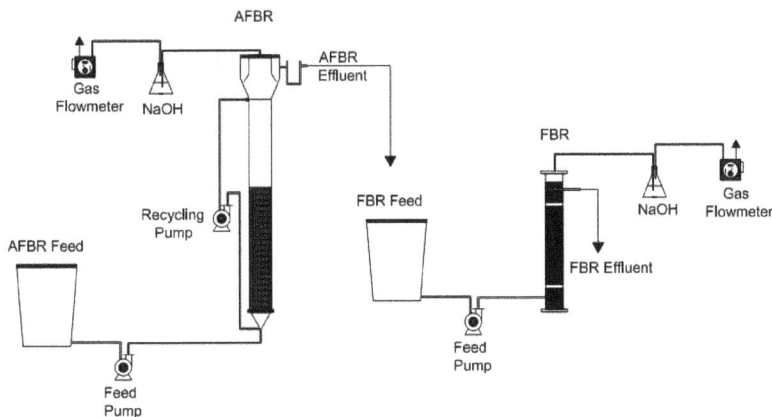

Figure 1. Operational scheme.

Table 1: Substrate composition (%v/v) and reactors HRT.

Phase	Substrate composition (%v/v)			HRT (h)	
	Swine residues	Water	Cassava wastewater	AFBR	FBR
1	25	50	25	2	24
2	50	25	25	2	12
3	75	10	15	2	12
4	85	0	15	2	12

The AFBR was built in transparent acrylic as a base material with a thickness of 5 mm, height of approximately 90 cm, and diameter of 5.3 cm. The fixed bed anaerobic reactor was constructed using acrylic tubes with internal diameter of 80 mm, external diameter of 88 mm, and 750 mm of length. The reactors were operated at room temperature (26 to 34°C). Table 1 lists the composition of the substrate in each phase as well as the hydraulic retention time (HRT) applied in each reactor. This composition corresponds to the AFBR feed and the fixed bed reactor was fed with the effluent from the AFBR.

2.2 Reactors' start-up

In the AFBR, the microorganisms' (biomass) adaptation occurred inside the reactor itself, which contained expanded clay particles (diameter of 2.8-3.35 mm) [11] as support material for microbial adhesion. The inoculum used was sludge from the swine farming residues. Initially, the inoculum diluted in water remained for 48 hours in an adaptation period to enable the adhesion of microorganisms to the support material and maintenance of the microbial activity. The appropriate microorganisms to the process were previously selected by the pre-heat treatment method at 90°C for 10 min [12]. In the fixed bed anaerobic reactor, the effluent from the AFBR was used for inoculation. The adaptation of the microbial population took place in the reactor itself using *sururu* shells for microbial adherence.

2.3 Hydrogen and methane production

A MilliGas-counter meter from the manufacturer Ritter model MGC-1 V3 1 AMMA was used to quantify the volumetric production of hydrogen and methane. The monitoring of pH and chemical oxygen demand (COD) were performed using the methodology of standard methods [13]. The carbohydrate concentrations were measured using the methodology developed by Dubois et al. [14]. The concentration of organic acids and alcohols were determined by gas chromatography according to the methodology of Maintinguer et al. [12].

3. Results and discussion

The obtained results will be discussed in the following two sections, the first corresponding to the operation of the acidogenic reactor used for hydrogen production and the second corresponding to the methanogenic reactor fed with the effluent from the acidogenic reactor.

3.1 Anaerobic fluidized bed reactor – hydrogen production

Fig. 2 demonstrates the variation of the hydrogen production yield and the volumetric hydrogen production in each experimental phase. It can be observed that the volumetric hydrogen production presented an increase with the reduction of cassava wastewater concentration (from 25 to 15% of the substrate), increasing from 0.18 to 0.45 L/h/L when going from phase 1 to phase 4, respectively. However, the yield of hydrogen behaved inversely, decreasing along the phases from 1.13 in phase 1 to 0.14 molH$_2$/mol glucose in phase 4.

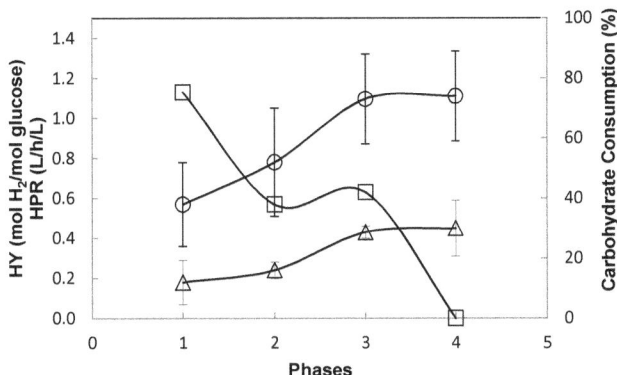

Figure 2. Hydrogen production yield (□), hydrogen volumetric production (Δ), and carbohydrate consumption (○) for AFBR in each experimental phase.

The decrease in the hydrogen yield through the phases can be justified by the reduction of the cassava wastewater concentration in the substrate since it is the main carbohydrate source in this study. Fig. 3 reprents the soluble metabolites produced during the experiment and the pH variation in each phase. The pH variation in the range between 5.5 and 6 was considered satisfactory for hydrogen production according to the literature [15]. An increase in the pH (from 4.91 in phase 1 to 5.66 in phase 4) was observed with an increase in the percentage of swine manure in the substrate (from 25 to 85% in Phases 1 and 4, respectively). This increase can be attributed to the fact that the swine waste used has a higher pH than the cassava wastewater (7.41 and 4.31, respectively). The hydrogen yield (HY) in phase 1 was maximum in the experiment (1.13 mol H_2/mol glucose) although a lower pH of 4.91 was observed in this phase, which is below the optimum range 5.5-6.0 considered by Chen et al. [15]. However, Amorim et al. [11], also using AFBR in their experiments, obtained satisfactory results in the production of hydrogen from glucose (HY between 1.19-2.49 mol H_2/mol glucose) on applying low pH (between 3.7 and 6.8), indicating the feasibility of hydrogen production at lower pH values.

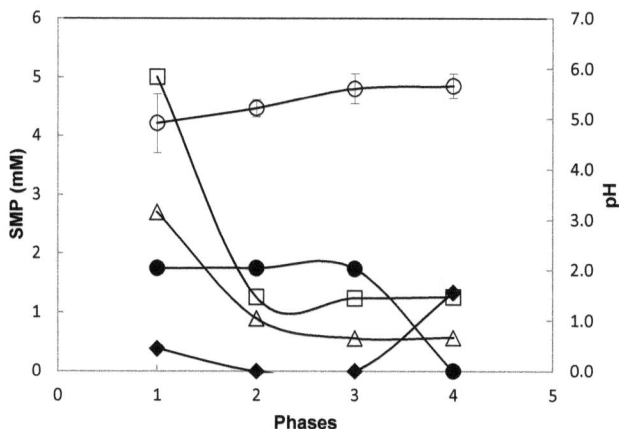

Figure 3. pH variation and the distribution of the soluble metabolites produced (SMP) during the AFBR operation. pH (○), Acetic acid: HAc (□), Butyric acid: HBu (Δ), Propionic acid: HPr (♦), Ethanol: EtOH (●).

The following metabolites were detected: acetic acid, butyric acid, propionic acid, and ethanol (Fig. 3). Table 2 presents the distribution of each metabolite in each applied condition. Acetic acid (HAc) and butyric acid (HBu) were the only metabolites present in all phases of the reactor operation and acetic acid was the predominant one with a general average of 39% of the detected metabolites. Concentrations ranged from 1.26-5.00 mM for acetic acid, 0.55-2.70 mM for butyric acid, 0-1.33 mM for propionic acid (HPr), and 0-1.75 mM for ethanol.

Table 2: Soluble metabolites production during AFBR operation.

Phase	HAc/SMP (%)	HBu/SMP (%)	HPr/SMP (%)	EtOH/SMP (%)	TVFA (mM)	SMP (mM)	HAc/HBu
1	51	27	4	18	8.09	9.84	1.85
2	32	23	0	45	2.15	3.90	1.40
3	35	16	0	49	1.80	3.54	2.22
4	40	18	42	0	3.17	3.17	2.21

The production of acetic acid is related to the production of hydrogen (Eq. 1). Likewise, the production of butyric acid also contributes to the formation of hydrogen (Eq. 2). Differently, the propionic acid route must be avoided due to the consumption of two moles of hydrogen in the reaction (Eq. 3). The formation of ethanol does not consume hydrogen even though it consumes the substrate that could have been used in the hydrogen formation as can be seen in Eq. 4.

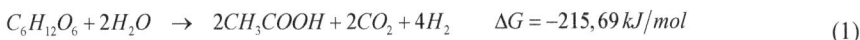

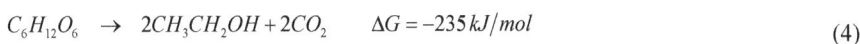

$$C_6H_{12}O_6 + 2H_2O \rightarrow 2CH_3COOH + 2CO_2 + 4H_2 \qquad \Delta G = -215,69\,kJ/mol \tag{1}$$

$$C_6H_{12}O_6 + 2H_2O \rightarrow CH_3CH_2CH_2COOH + 2CO_2 + 2H_2 \quad \Delta G = -257,1\,kJ/mol \tag{2}$$

$$C_6H_{12}O_6 + 2H_2 \rightarrow 2CH_3CH_2COOH + 2H_2O \qquad \Delta G = -358\,kJ/mol \tag{3}$$

$$C_6H_{12}O_6 \rightarrow 2CH_3CH_2OH + 2CO_2 \qquad \Delta G = -235\,kJ/mol \tag{4}$$

According to Cardozo [16], a high volumetric production does not imply necessarily a high hydrogen yield because the fermentative route taken in the process will directly influence the HY. In this experiment, for example, the volumetric production of hydrogen increased (0.18-0.45 L/h/L) throughout the operational phases (from phase 1 to phase 4) while the yield decreased (1.13-0.14 mol H_2/mol glucose). Phase 1, in which the highest HY (1.13 mol H_2/mol glucose) was observed there was a predominance of acetic acid (51% of SMP), confirming the relationship between this metabolite and hydrogen production. In the second phase, the HY reduced about 50% in relation to the previous phase reaching 0.57 mol H_2/mol glucose. This may be attributed to the reduction of the acetic acid (32% of SMP) concentration while ethanol production increased from 18% (in phase 1) to 45% in SMP (in phase 2). It can be concluded that the formation of ethanol may have competed with the production of hydrogen. The experiment trend in the third phase was very similar to phase 2. The percentages of acetic acid and ethanol were 35 and 49% of SMP, respectively, and HY was 0.63 mol H_2/ mol glucose. In Phase 4, there was a reduction in the HY (0.14 mol H_2/mol glucose). This can be justified by the content of propionic acid in this phase (42% of SMP). The predominance of the propionic acid fermentative route may have occurred because during this phase most of the substrate consisted of swine manure (85%). This hypothesis is reinforced by the work of Santos [17] which observed a similar behavior while using swine manure with 2 g of sucrose as a supplement or none to produce hydrogen in an AFBR. Similar to this study, the authors observed slight predominance of propionic acid (27 and 29% of SMP) in relation to acetic acid (21 and 28% of SMP). The production of ethanol in Phases 2 and 3 (45 and 49% of SMP) exhibits the possibility of formation of this metabolite simultaneously with the production of hydrogen and thus, enhancing the economical feasibility of the process. Santos [17] obtained averages of ethanol and HY of 44% of SMP (65 mM) and 2.59 mol H_2/mol sucrose, respectively. Also using AFBR in the treatment of dairy effluents, Macário [18] also observed ethanol production (42% of SMP and 7.37 mM) simultaneous to the production of hydrogen (0.44 mol H_2/mol glucose). Another fact that may influence the hydrogen production according to Reis et al. [19] is the acetic acid/butyric acid (HAc/HBu) ratio, the higher is the value of this ratio, the greater is the hydrogen production. It is because of the fact that acetic acid production route yields four moles of hydrogen instead of two moles generated by the butyric route (Eqs 1 and 2, respectively). In this study, the ratio of acetic and butyric acids (Table 2) was higher in Phases 3 and 4 (2.22 and 2.21, respectively), which did not affect the HY behavior (Fig. 2). Thus, it can be assumed that the ethanol competition in Phase 3 and the inhibitory effect of the propionic acid production in Phase 4 had more effect on the HY than the HAc/HBu ratio. The results of the detected soluble metabolites in this study showed that the concentrations of the produced acids and ethanol were influenced by the composition of the substrate fed throughout the operational phases. In addition, the results obtained and referenced in this study demonstrate the need to control the maintenance of acidogenic populations and to prevent the contamination with other non-hydrogen producing organisms such as yeasts which leads to substrate competition in the system [20].

3.1.1 COD Removal and carbon balance

Fig. 4 shows the COD variation in each operational phase. COD removal efficiency was mostly constant throughout all experimental phases (29% in average), which corroborated with the expected COD removal during acidogenic processes in anaerobic digestion according to the literature [21].

The influent COD varied throughout the experiment due to the different proportions of substrate components in the mixture. As the cassava wastewater had higher COD than the swine manure (27.3 and 3.0 g/L, respectively), it had a greater influence on the COD variation in the substrate fed. In Phases 1 and 2, the cassava wastewater was responsible for 25% of the substrate composition while in Phases 3 and 4 it was only 15%, causing a decrease in the COD influent in these phases. Table 3 shows the carbon balance for each metabolite, the total theoretical COD, the measured COD in the effluent, and the difference between the total theoretical COD and the measured in the effluent. The consistency between the measured COD effluent and total theoretical COD were 93, 93, 75 and, 46% for phases 1, 2, 3, and 4, respectively. The results showed that there was not significant interference of non-quantified metabolites in Phases 1, 2 and 3. However, in Phase 4 there was a considerable difference between the measured COD effluent and the theoretical COD (1629.6 mg/L). It can be attributed to the presence of metabolites which have contributed to the lower HY in this phase (0.14 mol H_2/mol glucose) and were not quantified.

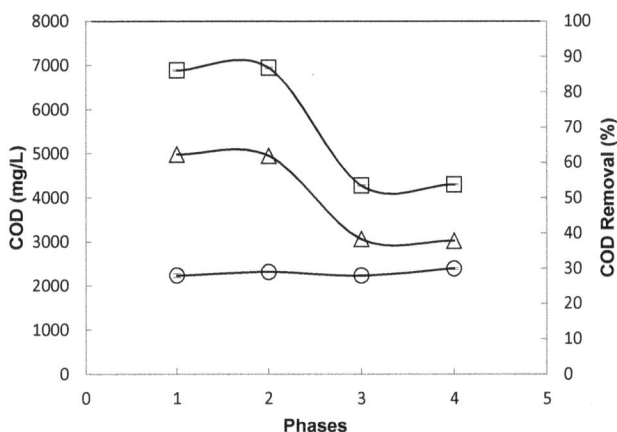

Figure 4. COD influent (□), COD effluent (Δ), and removal efficiency (○) in each experimental phase.

Table 3: Theoretical COD, Biomass COD and measured effluent COD.

Phase	COD_t, HAc (mg/L)	COD_t, HBu (mg/L)	COD_t, HPr (mg/L)	COD_t, EtOH (mg/L)	COD_t, Gulose (mg/L)	COD_t, Total (mg/L)	COD_t, Measured (mg/L)	COD_{Others} (mg/L)
1	320.33	433.22	43.03	168.35	3671.97	4636.9	4978.6	341.7
2	80.34	143.43	0	168.39	4229.09	4621.2	4947.9	326.7
3	79.30	89.42	0	167.43	1959.96	2296.1	3064.6	768.5
4	80.70	91.30	149.82	0	1075.11	1396.9	3026.5	1629.6

Shida et al. [21] observed a minimum difference of 248.31 mg/L and maximum of 764.20 mg/L using glucose as a substrate at the concentration of 2000 mg/L. Amorim et al. [11] quantified a difference ranging from 25 to 1259 mg/L when using the glucose concentration of 4000 mg/L. However, Reis and Silva [22] obtained a maximum difference of 1602 mg/L when applying similar fluidodynamic conditions to this work (Vmf = 1.24 cm/s) and using glucose as a substrate at 5000 mg/L concentration. The results observed in the literature as well as those obtained in this research pointed to the possible production of other undetected metabolites such as lactic acid and formic acid, among others.

3.2   Anaerobic fixed bed reactor – methane production

The fixed bed reactor was fed with the effluent generated in the AFBR. Table 4 presents the data of influent and effluent COD as well as the removal of organic matter in the FBR. The pH variation and the total volatile acids present in the influent and effluent are also included in Table 4.

Table 4: Influent and effluent COD, removal efficiency and pH.

HRT (h)	COD (mg/L) Influent	COD (mg/L) Effluent	E (%)	pH	VFA Influent	VFA Effluent
24	4963.25	3750.53	25	7.38 ± 0.43	1112.60±508.02	900.22±466.66
12	3045.62	2224.90	27	7.42 ± 0.39	1148.60±270.34	1145.91±269.79

The pH was varied from 7.38 (HRT of 24 h) to 7.42 (HRT of 12 h) because methanogenic archaeas have been reported to develop best at a neutral pH [23, 24]. The FBR had an average COD removal of 26%. Combining the two steps (acidogenic and methanogenic), the total COD removal was about 47%. Angonese et al. [25] carried out an experiment with swine manure without pre-treatment and observed a 77% COD removal, which was greater than the one found in this study. Buitron et al. [26] in their experiment used tequila vinasse as substrate and achieved COD removal of 56, 65, and 67% for the initial concentrations of 400, 1085 and 1636 mg COD/L, respectively. According to Faria [27], the low removal of organic matter occures in the presence of compounds resistant to degradation. In this study, the average volatile acid values of the influent and the effluent were 1112.60 and 900.22 mg/L, for the 24 h HRT, and 1148.60 and 1145.91 mg/L for the 12 h HRT, respectively (Table 4). The low conversion of the total volatile acids (19.1 and 0.2%, for the HRT of 24 and 12 h, respectively) probably has influenced the removal of COD. According to Onodera et al. [28], high concentrations of acetic, propionic and butyric acids in the influent can lead to an unstable reactor performance. Kongjan et al. [29] noted that even with the maintenance of favorable pH, methane production may be inhibited by a significant increase of volatile fatty acids, especially acetic acid and butyric acid with concentrations close to 3000 and 1600 mg/L, respectively. In this study, the chromatography of the acids present in the effluent of the FBR was not performed. However, it is known that the influent fed (generated in the acidogenic reactor) was composed mainly of acetic acid. Therefore, the production of methane was affected by the accumulation of this metabolite.

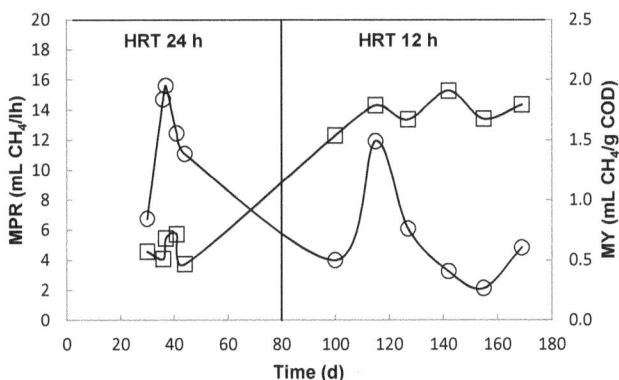

Figure 5. COD influent (□), effluent (Δ) and removal efficiency (○) in each experimental phase.

The average of MPR and MY were 12.12 ± 3.50 mL CH_4/L.h and 0.59 ± 0.11 mL CH_4/g COD for the 24 h HRT, and 5.38 ± 3.48 mL CH_4/L.h and 1.73 ± 0.13 mL CH_4/g COD for the 12 h HRT. Despite the highest volume production occurred in the 24 hour HRT, the highest yield was observed in the HRT of 12 h. The methane production rate in this study was much lower than the MPR observed by Amorim et al. [24] which used only cassava wastewater as substrate under similar experimental conditions, achieving in the HRT of 12 h a MY of 0.430 ± 0.150 L/g COD and an MPR of 42.463 L/h/L. According to Moller et al. [30], animal manures are highly diluted and have resistant fractions of organic compounds for anaerobic digestion and consequently, for methane production if compared to the easily degradable substrates of organic wastes from agro industrial productions. Apparently, the co-digestion of the cassava wastewater with swine manure may have inhibited the methane production in this study.

4. Conclusions

Two agroindustrial residues were used to produce hydrogen and methane gases as renewable energy sources in separate processes. The cassava wastewater is a residue rich in carbohydrates which is essential for the production of hydrogen while the swine waste has microorganisms favorable to acidogenesis. Thus, hydrogen production in the anaerobic fluidized bed reactor was evaluated from the co-digestion of these two residues. Throughout this process, the cassava wastewater was reduced (from 25 to 15%) in the fed mixture, which resulted in the decrease of the hydrogen yield while an increase in the volumetric hydrogen production was observed throughout the operational phases. The main metabolites produced were acetic, butyric and propionic acids, and ethanol. The presence of ethanol in the first three phases indicated the possibility of producing hydrogen and ethanol simultaneously. The influent composition of 85% of swine manure and 15% of cassava wastewater resulted in a lower hydrogen yield and the inhibition of ethanol production and the production of propionic acid was predominant. In addition, the carbon balance showed the possibility of the generation of undetected soluble metabolites that may have affected the process. The chemical parameters analyzed in the Anaerobic Fixed Bed Reactor presented conditions in accordance with the available literature. *Sururu* shells were efficient in the buffering process for the initial acidic pH, bringing it close to neutrality and avoiding the presence of acidogenic microorganisms. The COD removal reached efficiencies around 26% due to the presence of recalcitrant compounds and the greatest methane production was obtained in the 24 h HRT even though when the HRT was reduced to 12 h there was an optimization in the yield.

References

[1]   B. Demirel, P. Scherer, O. Yenigun, T.T. Onay, Production of methane and hydrogen from biomass through conventional and high-rate anaerobic digestion processes, Crit. Rev. Environ. Sci. Tech. 40 (2010) 116-46. https://doi.org/10.1080/10643380802013415

[2]   X.M. Guo, E. Trably, E. Latrille, H. Carrère, J-P. Steyer, Hydrogen production from agricultural waste by darkfermentation: A review. Int. J. Hydrogen Energ. 35 (2010) 10660-10673. https://doi.org/10.1016/j.ijhydene.2010.03.008

[3]   R.K. Hubbard, R.R. Lowrance, Dairy cattle manure management. In: Agricultural
        utilization of municipal, animal and industrial wastes. USDA, Agricultural
        Research Service, Conservation Res. Rep. 44 (1998) 91-100.

[4]   ABIPECS. Em 15 anos, Brasil se tornou o 4° maior produtor e exportadormundial
        de carne suína. Disponível em <http://www.abipecs.org.br/news/710/134/Em-15-
        anos-Brasil-se-tornou-o-4-maior-produtor-e-exportador-mundial-de-carne-
        suina.html>. Acesso em: 10 de fevereiro, 2014.

[5]   B.M. Cappelletti, V. Reginatto, E.R. Amante, R.V. Antonio, Fermentative
        production of hydrogen from cassava processing wastewater by Clostridium
        acetobutylicum. Renew Energy. 36 (2011) 3367-3372.
        https://doi.org/10.1016/j.renene.2011.05.015

[6]   W. Wang, L. Xie, G. Luo, Q. Zhou, Enhanced fermentativehydrogen production
        from cassava stillage by co-digestion: the effects of different co-substrates. Int. J.
        Hydrogen Energ. 38 (2013) 6980-6988.
        https://doi.org/10.1016/j.ijhydene.2013.04.004

[7]   G. De Gioannis, A. Muntoni, A. Polettini, R. Pomi, A review of dark fermentative
        hydrogen production from biodegradable municipal waste fractions. Waste
        Management. 33 (2013) 1345-1361. https://doi.org/10.1016/j.wasman.2013.02.019

[8]   M.A. Hernández, M.R. Susa, Y. Andres, Use of coffee mucilage as a new
        substrate for hydrogen production in anaerobic co-digestion with swine manure.
        Bioresource Technology. 168 (2014) 112-118.
        https://doi.org/10.1016/j.biortech.2014.02.101

[9]   A. Marone, C. Varrone, F. Fiocchetti, B. Giussani, G. Izzo, L. Mentuccia, S. Rosa,
        A. Signorini. Optimization of substrate composition for biohydrogen production
        from buffalo slurryco-fermented with cheese whey and crude glycerol, using
        microbial mixed culture. Int. J. Hydrogen Energ. 40 (2015) 209-218.
        https://doi.org/10.1016/j.ijhydene.2014.11.008

[10]   P.R.F. Rosa, S.C. Santos, I.K. Sakamoto, M.B.A. Varesche, E.L. Silva , The
        effects of seed sludge and hydraulic retention time on the production of hydrogen
        from a cassava processing wastewater and glucose mixture in an anaerobic
        fluidized bed reactor. Int. J. Hydrogen Energ. 39 (2014) 13318-13127.
        https://doi.org/10.1016/j.ijhydene.2014.06.152

[11]   E.L.C. Amorim, A.R. Barros, M.H.R.Z. Damianovic, E.L.Silva, Anaerobic
        fluidized bed reactor with expanded clay as support for hydrogen production
        through dark fermentation of glucose. Int. J. Hydrogen Energ. 34 (2009) 783-790.
        https://doi.org/10.1016/j.ijhydene.2008.11.007

[12] S.I. Maintinguer, B.S. Fernandes, I.C.S. Duarte, N.C. Saavedra, M.A.T. Adorno, M.B. Varesche, Fermentative hydrogen production by microbial consortium. Int. J. Hydrogen Energ. 33 (2008) 4309-4317. https://doi.org/10.1016/j.ijhydene.2008.06.053

[13] APHA, Standard Methods for the examination for water and wastewater. 20th Ed. Washington, DC, American Public Health Association/American Water Works Association/Water Environmental Federation (1998).

[14] S.M. Dubois, K.A. Gilles, J.L. Hamilton, P.A. Rebers, F. Smith, Colorimetric methods for determination of sugar and related substance. Anal. Chem. 28 (1956) 350-356. https://doi.org/10.1021/ac60111a017

[15] C.C. Chen, C.Y. Lin, M.C. Lin. Acid-base enrichment enhances anaerobic hydrogen production process, Appl. Microbiol. Biotechnol. 58 (2002) 224-228. https://doi.org/10.1007/s002530100814

[16] P.H.G. Cardoso, Produção de hidrogênio a partir da manipueira em reator anaeróbio de leito fluidificado: Efeito do pH. Dissertação (Mestrado em Recursos Hídricos e Saneamento) – Universidade Federal de Alagoas, Maceió, 2013.

[17] E.C.L. Santos, Produção biológica de hidrogênio utilizando resíduo de suinocultura suplementado com sacarose. Dissertação (Mestrado em Recursos Hídricos e Saneamento) - Universidade Federal de Alagoas, 2014.

[18] S.C.A. Macário, Influência do inóculo na produção de hidrogênio a partir do soro de queijo em pó, da lactose isolada e do efluente da indústria de laticínios em reator anaeróbio de leito fluidificado. Dissertação (Mestrado em Recursos Hdricos e Saneamento) - Universidade Federal de Alagoas, 2016.

[19] C.M. Reis, E. L. Silva, Effect of upflow velocity and hydraulic retention time in anaerobic fluidized-bed reactors used for hydrogen production. Chem. Eng. J. 172 (2011) 28-36. https://doi.org/10.1016/j.cej.2011.05.009

[20] N.C.S Amorim, I. Alves, J.S. Martins, E.L.C. Amorim, Biohydrogen production from cassava wastewater in an anaerobic fluidized bed reactor. Braz. J. Chem. Eng. 31 (2014) 603-612. https://doi.org/10.1590/0104-6632.20140313s00002458

[21] G.M. Shida, A.R. Barros, C.M. Reis, E.L.C. Amorim, M.H.R.Z. Damianovic, E.L.Silva, Long-term stability of hydrogen and organic acids production in an anaerobic fluidized-bed reactor using heat treated anaerobic sludge inoculum. Int. J. Hydrogen Energ. 34 (2009) 3679–3688. https://doi.org/10.1016/j.ijhydene.2009.02.076

[22] C.M. Reis, E.L. Silva, Effect of upflow velocity and hydraulic retention time in anaerobic fluidized-bed reactors used for hydrogen production. Chem. Eng. J. 172 (2011) 28-36. https://doi.org/10.1016/j.cej.2011.05.009

[23] G. Peixoto, J.L.R. Pantoja-Filho, J.A.B. Agnelli, M. Barboza , M. Zaiat, Hydrogen and Methane Production, Energy Recovery, and Organic Matter Removal from Effluents in a Two-Stage Fermentative Process. Appl. Biochem. Biotech. 168 (2012) 651-671. https://doi.org/10.1007/s12010-012-9807-4

[24] N.C.S. Amorim,A.M. Olivera, K.R.Solomon,E.L.C. Amorim, Produção de hidrogênio e metano a partir da manipueira em fases separadas. 8° Congresso Internacional de Bioenergia. São Paulo (2013).

[25] A.R. Angonese, A.T. Campos, C.E. Zacarkim, M.S. Matsuo, F. Cunha, Eficiência Energética de Sistema de Produção de Suínos com Tratamento dos Resíduos em Biodigestor. Revista Brasileira de Engenharia Agrícola e Ambiental. Campina Grande. 10(3) (2006).

[26] G. Buitrón, G. Kumar, A. Martinez-Arce, G. Moreno, Hydrogen and methane production via a two-stage processes (H2-SBR + CH4-UASB) using tequila vinasses. Int. J. Hydrogen Energ. 39 (2014) 19249-19255. https://doi.org/10.1016/j.ijhydene.2014.04.139

[27] A.A.A. Faria, fermentação metanogênica mesofílica de Melaço e termofílica de vinhaça em reatores Uasb. Dissertação (Mestrado em Microbiologia Agropecuária) - Universidade Estadual Paulista, 2014.

[28] T. Onodera, S. Sase, P. Choeisai, W. Yoochatchaval, H. Sumino, T. Yamaguchi, Y. Ebie, K. Xu, N. Tomioka, K. Syutsubo, High-rate treatment of molasses wastewater by combination of an acidification reactor and a USSB reactor. Journal of Environmental Science and Health. Part A, Toxic/hazardous Substances & Environmental Engineering. 46 (2011) 1721-1731. https://doi.org/10.1080/10934529.2011.623975

[29] P. Kongjan, S. O-Thong, I. Angelidaki, Hydrogen and methane production from desugared molasses using a two-stage thermophilic anaerobic process. Engineering in Life Sciences. 13 (2013) 118–125. https://doi.org/10.1002/elsc.201100191

[30] H.B. Moller, S.G. Sommer, B.K. Ahring, Methane productivity of manure, straw and solid fractions of manure. Biomass Bioenerg. 26 (2004) 485-495. https://doi.org/10.1016/j.biombioe.2003.08.008

Chapter 5

Methods for the Treatment of Dairy Wastewater

P. Stanchev[1], A. Mousavi[2], V. Vasilaki[1], E. Katsou[1*]

[1] Department of Mechanical, Aerospace and Civil Engineering; Institute of Environment, Health and Societies, Uxbridge Campus, Middlesex, UB8 3PH, Uxbridge, Brunel University, UK

[2] Department of Electronic and Computer Engineering, College of Engineering, Design and Physical Sciences, Uxbridge Campus, Middlesex, UB8 3PH, Uxbridge, Brunel University, UK

*evina.katsou@brunel.ac.uk

Abstract

This chapter aims to investigate the main treatment methods for dairy wastewater treatment along with the key aspects for improvement of the efficiency and sustainability of the treatment plants. In this regard, the anaerobic treatment process has been found most useful as a potential approach for sustainable and circular dairy wastewater management.

Keywords

Dairy Wastewater, Treatment Processes, Environmental Footprint, Sustainability, Resource Recovery

Contents

1. Introduction

Dairy products form an essential staple of the world's food consumption and thus have economic and strategic nutritional importance, [1]. Dairy industries are one of the most important components of Europe's food sector [2]. Wastewater originating from dairy processes contains complex and slowly biodegradable organic compounds [3], and hence the appropriate treatment of industrial wastewater is important in order to avoid adverse environmental impacts, such as eutrophication of surface waters, hypoxia and algal bloom of the scarce clean water resources [4–7]. This chapter reviews and assesses the key factors that affect dairy wastewater treatment with the view to increase the sustainability of the sector. The specific characteristics of dairy process effluents, such as high Chemical Oxygen Demand (COD) concentration, variable pH, origin, and salinity; production process stages and materials used poses a considerable challenge to the treatment of dairy wastewater. [8,9]. In order to achieve sustainable performance, dairy wastewater treatment plants are expected to: (i) produce high-quality effluents satisfying the increasingly strict discharge legislation, (ii) expand wastewater reuse and energy recovery potential in accordance with the concept of circular economy, (iii) have the capacity for upgrading and retrofitting energy-efficient and cost-effective technologies, (iv) decrease the investment costs and (v) exhibit low environmental impact [4,5], [10–

12]. The implementation of environmentally sustainable methods for dairy wastewater treatment should be supported by cleaner production strategies for reclamation, recycling, and reuse of the generated wastewater [13].

Life cycle assessment (LCA) is one of the most widely applied and standardized methodologies for quantifying the environmental impact of a system (ISO 14040, 2006). To date, LCA studies in dairy sector have been limited to the dairy farm activities [14,15] with little emphasis on the processes [16]. One of the major gaps in the literature regarding wastewater treatment of dairy processing is the LCA based environmental impact assessment of full-scale anaerobic processes. According to the study of Korsström and Lampi [17], the environmental impact of dairy wastewater treatment mainly depends on the high organic load of the influent and energy requirements of the applied process.

An orderly improvement in the production process and improvement of the quality of final products via an integrated plant and inter-plant monitoring as well as performance measurement system seems desirable. As our understanding and control complex biological and chemical processes evolve and improve, the methods and techniques to acquire the essential data (sensors), measure (data analytics and performance modelling), and control (actuation-optimisation) needs to adjust.

2. Dairy processes and characteristics of dairy wastewater

The wastewater from dairy factories is generated through various processing steps, such as milk receiving/storage, pasteurisation, homogenisation, separation/clarification, cheese/butter/milk powder/cream making, packaging and during cleaning, heating, cooling or floor washing (Fig. 1).

The dairy wastewater contains milk components, as well as acid and alkaline detergents used in the equipment cleaning. It has been estimated that about 2% of the total milk processed is wasted and discharged into drains. Due to the high pollution load of dairy wastewater, a set of regulations for the effluent discharge has been imposed to protect the environment [18]. The use of detergents in the cleaning process causes high variation of pH between 2 to 12 [17]. The European Commission Directorate has reported variable volumes for wastewater generation from the milk, milk powder and cheese processing [19]. In the case of cheese processing, the levels and type of wastewater depend on the production processes, the materials used, and the type of final product. The amount of wastewater generated varies from 0.4 to 60 l/kg of the processed milk. The wide range of dairy effluent characteristics reported in literature are depicted in Table 1.

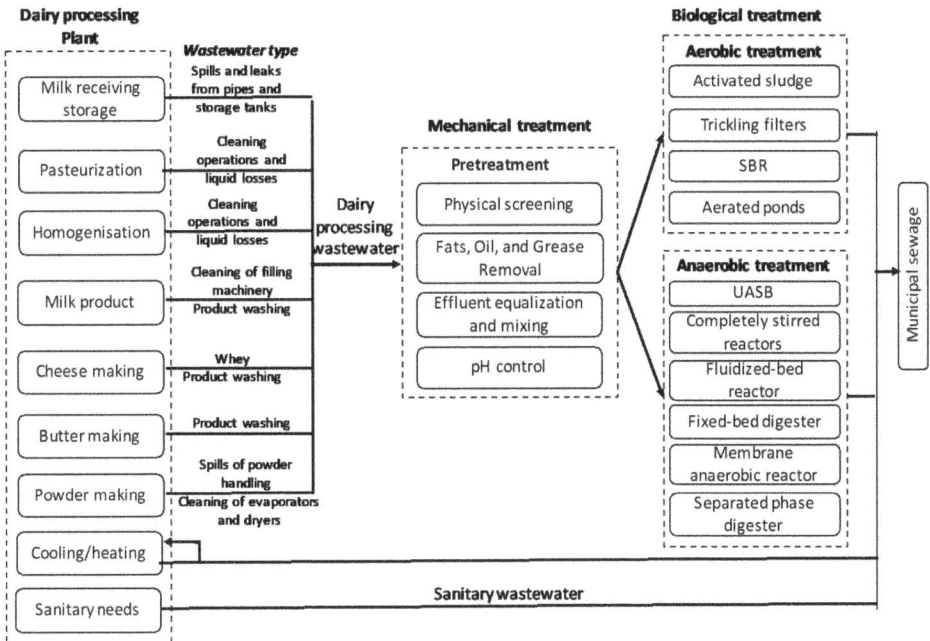

Figure 1. Sources of dairy processing wastewater and common treatment methods.

Typically, the COD of dairy wastewater ranges from 0.1-100 kg/m^3 (Table 1), mainly due to milk carbohydrates and proteins, along with the presence of fats (0.07-2.9 kg/m^3), nutrients (nitrogen and phosphorous) and suspended solids (0.2-5.1 kg/m^3) [34]. The variable composition of dairy effluents affects their biodegradability [35]. The hourly, daily and seasonal fluctuations make the prediction of wastewater flow rates and pollution loads difficult [36]. In addition, high volumes of water are required for the cleaning and cooling processes resulting in high volumes of contaminated water containing detergents and large quantities of unused milk [17].

Table 1: Dairy effluent characteristics reported in literature.

Product portfolio	BOD (mg/L)	COD (mg/L)	pH	TS (mg/L)	TSS (mg/L)	Ref.
Milk/cream	—	4656	6.92	—	—	[20]
Milk/cream	500-1300	950-2400	5.0-9.5	—	90-450	[21]
Milk/cream	—	713-1410	7.1-8.1	900-1470	360-920	[22]
Milk/cream	1200-4000	2000-6000	8-11	—	350-1000	[23]
Cheese/whey	71526	20000	4	56782	22050	[24]
Cheese/whey	80000	120000	6	—	8000	[25]
Cheese/whey	—	68814	—	—		[26]
Cheese/whey	588-5000	1000-7500	5.5-9.5	—	500-2500	[27]
Cheese/whey	—	61000		—	1780	[28]
Cheese/whey	565-5722	785-7619	6.2-11.3	1837-14205	326-3560	[29]
Cheese/whey	377-2214	189-6219	5	—	188-2330	[30]
Butter/milk powder	—	1908	5.8	—	—	[20]
Butter/milk powder	1500	—	10-11	—	—	[31]
Mixed dairy processing	2100	1040	7-8	2500	1200	[32]
Mixed dairy processing	—	1150-9200	6-11	2705-3715	340-1730	[33]

3. Dairy wastewater treatment processes

Figure 1 summarizes the main sources of dairy processing wastewater and methods that have been tested at various scales of application for the treatment of the dairy effluents. A detailed description of main processes for dairy wastewater treatment and target parameters that affect their performance is provided in this section. The pre-treatment options can be broken down into, screening, pH flow control, and separation of Fat, Oil, and Grease (FOG).

3.1 Screening

Screening is an important step applied to protect pumps and remove large particles or debris that may cause damage to downstream machines and equipment. [37]. Screening further removes solids from wastewater and large quantities of non-emulsified FOG. Screening is accompanied by other technical and operational measures to avoid clogging [19].

3.2 Flow and pH balancing and control

Detergents are used in the dairy process for cleaning purposes. They have an additional impact on the pH and discharge flow pattern of the generated effluents. The effluent from the different processes is normally mixed in equalization tanks in order to reach a near-neutral pH required for the subsequent biological treatment [17,38]. However, in cases of large variations, the pH should be adjusted with the addition of chemicals, such as sulphuric or nitric acid, sodium hydroxide, carbon dioxide or lime.

3.3 Separation of fat, oil and grease (FOG)

The presence of fat in the dairy effluents remains an issue since it can exert significant inhibition problems in biological processes. Dairies manufacturing butter usually separate the fat from the fatty rinsing water during the production process and reprocess the recovered fat. The removal of fat residues by the produced wastewater can be partly performed in special gravity fat traps or by passing the effluent through a flotation unit. The fat traps consist of a concrete tank divided into cells and the fat floats on the surface. The fat can be removed by manual or mechanical skimming. However, the traps require frequent cleaning and maintenance; increasing the cost and environmental impact of the process. Wastewater is aerated in the flotation plant by applying finely dispersed air, which transfers the fat to the surface, from where it is automatically strained off. The fatty sludge can be further treated in an anaerobic digestion unit [17,38].

4. Biological processes

Direct application of biological processes for dairy wastewater treatment is challenging due to the high flow and pollutant concentration variations of the dairy effluents [39].

4.1 Aerobic treatment

Aerobic processes are widely used for the treatment of wastewater in milk-producing industries [40]. Trickling filters and activated sludge process have been extensively used for the treatment of wastewater in dairy processing [41–43]. The activated sludge (AS) process requires the use of chemicals and involves high capital, operational and maintenance cost, while suitable pre-treatment is often required in order to avoid inhibition problems [7]. The sequencing batch reactors (SBRs) have been used for industrial wastewater treatment as an improved version of the conventional activated sludge (CAS) systems. SBRs are a single tank 5 phase process, filling, reaction, settling, idle and decantation. The SBR tanks act as equalization and clarifier. SBRs can produce high-quality effluent with flexible operation and low energy consumption under optimized industrial wastewater treatment conditions [44–47]. The SBRs can be more economical since equalization, aeration, and clarification can all be performed within a single reactor, making clarifiers redundant.

A common operational problem in treating dairy effluents with SBR technology is the overgrowth of filamentous bacteria [48], [49]. Meunier et al. [49] developed a method to convert the bulking activated sludge into a well-settling biomass at low oxygen levels. The results indicated that with the application of the aerobic granular sludge (AGS) technology, the high level of filamentous bacteria could be reduced [49].

A study by [50] reported that the quantity of sludge produced from the aerobic treatment of the dairy wastewater was higher than the respective one in municipal wastewater treatment plants (WWTPs). The processing scheme usually includes the energy-intensive phases of stabilization and dewatering of the excess sludge (57.6–187.3 m^3/d). The energy requirements ranged between 0.94 and 1.5 kWh/m^3 [51]. Based on energy modeling of the aerobic treatment process, Dąbrowski et al. [41] concluded that the aerobic sewage sludge stabilization and dewatering were responsible for 3-25% of the total energy consumption of a dairy WWTP plant with a capacity of 13,500 population equivalent (PE).

4.2   Anaerobic treatment

The anaerobic process involves the degradation of organic matter by bacteria in the absence of oxygen. It has been adopted in the treatment of various industrial effluents (e.g. aqueous extractions of winery wastes, biodiesel industry wastewater, soluble fraction of food industry wastes etc.) [13], [52–54]. Dairy wastewater treatment technologies are being continuously improved with the aim to reduce the environmental burdens. Attention has been focused mainly on anaerobic processes, due to the absence of energy requirements for aeration, small sludge production and low footprint [43,55]. Dairy effluents have been attractive feedstock for anaerobic digestion [34,56,57]. However, data from the application of anaerobic treatment of dairy processing wastewater are still limited due to an inadequate number of full-scale applications [58]. The anaerobic wastewater treatment processes are preferred over aerobic treatment methods due to certain benefits. One of the major advantages in anaerobic wastewater treatment is the recovery of biogas, which usually has the following composition: 55-75% methane (CH_4), 25-45% carbon dioxide (CO_2) and 1-5% nitrogen (N_2) [13]. The operating cost of the anaerobic process can be offset by the benefits obtained from biogas recovery. To avoid further treatment before its utilization for digester heating, electricity generation and fuel production, the level of methane in the biogas should be more than 60% [13,59]. Lower methane yields can be explained by the high methane solubility especially at low temperatures; i.e. 15°C [13], [60]; Xing et al. [61] observed the low methanogenic activity during the study, lab-scale EGSB treating brewery wastewater. Higher temperatures were beneficial for generation of methane [13]. Kothari et al. [62] explored the opportunity for energy recovery from anaerobic wastewater treatment of dairy effluents by applying a bacterial strain of Enterobacter aerogens and methanogenic bacteria of cow dung to produce simultaneously hydrogen and methane. The results showed that treatment targeting at simultaneous bioenergy production was economically viable, while higher efficiency was achieved for phosphates, sulfates and nitrates pollutants [62]. The anaerobic process has been implemented in various configurations, such as upflow anaerobic sludge blankets (UASB), anaerobic membrane bioreactors (AnMBR), anaerobic hybrid (AH) reactors, expanded granular sludge bed reactors (EGSB) and inverse fluidized bed reactors (IFBR).

4.2.1 Up-flow anaerobic sludge blanket reactor (UASB)

UASB is a well-established process for anaerobic wastewater treatment representing a suspended-growth system, where sludge granules grow in a tubular reactor [63]. The degradation occurs as the water flows up the reactor and contaminants are in contact with the sludge granules. It is used for anaerobic wastewater treatment mainly in warm climates as high temperatures offer the appropriate conditions for anaerobic degradation. The latter along with simple operation, efficient pathogen removal, limited land requirements and the ability to treat high organic loads justify the wide UASB application in developing tropical countries [64–66]. Arnaiz et al. [67] reported COD removal efficiency higher than 90% was obtained for dairy wastewater treatment (influent COD=30,000 mg/L). Gavala et al. [40] achieved stable operation in UASB reactor at organic loading rate (OLR) of 6.2 g COD/L, while the maximum OLR was 7.5 g COD/L.d. However, the authors concluded that the application of the UASB reactors might not be the most cost-effective option due to the high retention times necessary to treat the non-diluted wastewater of the dairy effluents [40].

4.2.3 Anaerobic membrane bioreactors (AnMBR)

Anaerobic digestion (AD) has been widely applied for wastewater treatment during the last 30 years. Nonetheless, its application to high-rate and/or low-strength wastewater treatment (e.g. urban wastewater) was limited by the difficulty in retaining slow-growth-rate anaerobic microorganisms when operating at short hydraulic retention times (HRTs). The complexity of the anaerobic process can be balanced by the energy production and the reduction of carbon footprint. In this regard, AnMBR offers the potential to operate the system in energy neutral or even positive net energy balance due to biogas generation. Membrane bioreactor (MBR) technology combines activated sludge process with membrane filtration. It is considered a well-established, mature technology with many full-scale MBR plants treating municipal and industrial wastewaters. The use of membranes in the anaerobic reactor allows the separation of components taking advantage of the ability of membranes to control the rate of permeation of different species. The suspended or dissolved pollutants are reduced or removed, while the membrane allows the purified water to pass through the membrane.

AnMBRs exhibit lowers energy requirements compared to the aerobic MBRs; however anaerobic conditions are less favorable for filtration and more prone to fouling. The operational cost for the membrane cleaning, replacement of membrane and the energy required for the gas recirculation have put restrictions on extensive AnMBR applications. Operation at low temperature, nutrients removal limitations, and optimum configuration have been the other identified challenges. This is why AnMBR has only very few full-scale applications (Technology Readiness Level – TRL = 8). Given the early stage of development and uncertainties around AnMBR performance, it is unclear how detailed design and operational decisions influence the economic and environmental performance of AnMBRs.

5.   Factors affecting the performance of anaerobic processes

The anaerobic technologies for the treatment and the main factors such as OLR, pH, temperature, HRT affecting the performance of dairy wastewater in terms of contaminant removal is discussed in this section. The key parameters.

5.1   pH value

Anaerobic digestion strongly depends on medium pH [68]. The anaerobic process of higher than 9 pH with membranes results in lesser biogas production and poor membrane performance. In [13,69–71] optimum pH range between 7 and 8 for anaerobic digestion was reported. The optimum pH for the methanogenic bacteria is 6.8-7.2; when the pH drops below 6.8, the acidogenic bacteria prevail over the methanogens. As a consequence, acid zones are formed inside the reactor and methane production is reduced [52]. Moreover, pH shocks lead to dispersion of the sludge flocs. Small-sized particles (e.g. colloids) exist in suspended sludge and result in increased fouling when membranes are applied in the anaerobic reactor [70]. Membrane fouling issues are more prevalent in AnMBRs than in aerobic MBRs. Thus, AnMBRs operate at lower transmembrane fluxes. The improvement of the membrane performance and the minimisation of relevant costs remains a challenge that should be addressed for the full-scale expansion of AnMBR technology. Moreover, AnMBRs are more prone to inorganic fouling caused by calcium, phosphorus and sulfur precipitates; struvite is the dominant inorganic foulant in the treatment of industrial wastewater streams. The need for pH neutralization increases the overall operational cost, as well as the environmental footprint of the applied process.

It is important to understand the role of a combination of factors such as OLR, pH etc. in the sustainable operation of the anaerobic processes in dairy wastewater treatment. The use of chemicals for the adjustment of the influent pH at ~7 must be optimized in order to reduce the environmental impact and the cost of the process. High COD removal can be achieved when the process operates under OLRs ranging from 1 to 8 kgCOD/m^3.d [58]. Yu and Fang [72] report that with pH increase from 4.0 to 5.5, the degradation of pollutants present in dairy wastewater also increased and almost half of the COD was converted into volatile fatty acids (VFAs) and alcohols. The further increase of pH to 6.5 resulted in a slight increase of degradation [72].

5.2   Temperature

Higher temperatures are favorable for methane production but disadvantageous for the immobilization of anaerobic biomass.

The effect of temperature on the performance of the process has been investigated in several studies. Bialek et al. [73] applied psychrophilic conditions in a lab-scale IFBR for dairy wastewater treatment in order to investigate the efficiency of anaerobic treatment in northern countries, where the yearly average temperature remains lower than 15°C. At 10°C, the system exhibited low average removal efficiency (~69%) and unstable operation with hydrolysis being the rate-limiting step. Biofilm overgrowth caused the decrease in the population of hydrogenotrophic methanogens.

5.3   Hydraulic retention times (HRT)

Long HRTs are usually applied during anaerobic treatment of industrial effluents in order to provide sufficient time for substrate hydrolysis and the methanogens to occur [13]. This is in accordance with the study of Wang et al. [54] who observed a decrease of the COD removal from 93 to 80% with the decrease of HRT from 10 to 2 days applying AnMBR for industrial wastewater treatment. Shorter HRTs lead to insufficient contact time between the sludge and the substrate and, consequently, to a poorer system performance. As a result, part of biomass could not settled and was washed out without appropriate treatment [54]. On the other hand, the need for decreasing the overall cost and to operate at smaller reactor volumes, the application of shorter HRTs is required [74]. Thus, it is important to identify the optimum HRT for each configuration applying anaerobic process for industrial wastewater treatment. The decision on optimal HRT should be based on efficient substrate degradation and limited accumulation of soluble substances onto the anaerobic reactor, and cost optimization of the process in terms of reactor volumes.

However, the application of higher OLR and lower HRT (OLR=27.65 kg COD/m^3.d and HRT=2 days) is accompanied by smaller reactor volumes, satisfactory biogas methane content (57.4%) and COD removal efficiency of 91.8%. Thus, process optimization should take into consideration technical cost and environmental indicators. Methane generation under the optimal combination of parameters, such as OLR and HRT can reduce operational cost (e.g. reactor volume) resulting in energy recovery. Demirel and Yenigun reported that the variations in HRT can affect the volatile fatty acids (VFAs) production as the decrease in HRT resulted in an increase of acid production, proportionally to the organic loading rate [75].

6. Sustainable dairy wastewater treatment

Dairy products are irreplaceable components of human diet constituting the most important sub-sector of the food industry in Europe [76]. However, their production is associated with severe environmental impacts due to the high water use, energy demand as well as direct and indirect greenhouse gas (GHG) emissions [1]. In 2007, the dairy sector globally contributed by 4% to the GHG emissions [77]. Specifications connected with hygiene and cleanliness in the dairy sector lead to the production of vast amounts of wastewater that has to be treated before being discharged back to water sources. Therefore, dairy waste streams are associated with high environmental impacts due to their significant organic and microbiological load and the high salinity of the wastewater produced [78]. The dairy sector is constantly innovating new strategies to decrease environmental burdens and improve social and ethical issues in the supply chain. Benchmarking is an important tool for continuous improvement of organizational performance, total quality management and competitive advantage [79]. The environmental benchmarking of the dairy industry initiated by dairy UK has led to 15% reduction of water consumption during milk production processes. The efforts have been now focused on reducing the COD loads of dairy processing discharges by 20%, mainly by the introduction of on-site wastewater treatment systems [80].

The selection of appropriate layouts for wastewater treatment should take into account not only economic parameters (i.e. investment, operation, and maintenance) but also environmental issues (e.g. eutrophication, global warming potential (GWP) and marine ecotoxicity etc.). Thus, several studies have evaluated the techno-economic viability of dairy effluents treatment by applying anaerobic digestion [43,56,57, 81–83]. Different configurations have been used, including UASB, CST reactors and anaerobic filters [84,85]. However, the instability of these processes together with the slow reaction rates have led to a limited number of full-scale applications [86]. The lack of appropriate measuring and control technologies have been the main constraints of implementing any known optimization techniques for the anaerobic digestion (AD) process in the dairy industry [57]. The environmental footprint of the process should be quantified and optimized in order to estimate and increase the environmental performance of the anaerobic technologies [13,44,87]. Options for improvement of the anaerobic digestion efficiency have been studied. Co-digestion of dairy effluents (mainly cheese whey) with livestock manure has been also tested at pilot scale with better results for variable dairy effluents [81,82,88–90]. Co-digestion of these waste streams is advantageous since it results in higher CH_4 rates and acidification problems can be partially neutralized [81,82,89]. However, feeding rates of the substrates can significantly affect the stability of the process [81,82,89].

Concerning the environmental impact of AD processes for dairy wastewater treatment, there is still a gap from the life-cycle perspective. Life cycle assessment (LCA) can be used as a decision-making aid; it has been used for the environmental impact assessment of various food waste management schemes that include anaerobic digestion [91–93].

To date, little has been reported on the utilization of LCA for the assessment of the environmental performance of AnMBR for dairy treatment. This is mainly attributed to the lack of full-scale data. Research regarding the water footprint of dairy products has until now almost exclusively focused on the farm stage. Recently, the water footprint for the production of conventional and organic milk of Brazilian dairy farming has been assessed following the WFN methodology [94]. Aydener et al. [95] demonstrated that the application of membrane-based solutions for recovery of water and whey powder could improve the techno-economic and environmental of the dairy wastewater treatment.

On the other hand, the anaerobic technology is promising in terms of greenhouse gases emissions mitigation. Methane produced during anaerobic digestion can be used for heating purposes, electricity, fuel production and other, instead of being emitted into the atmosphere in order to make the AnMBR technology as a potential candidate to convert the dairy WWTPs into water resource recovery facilities [96].

7. Resource recovery in line with the circular economy approach

Dairy wastewater can be a source of recovered products moving towards the circular concept. The adoption of AnMBRs can facilitate the WWTPs transition towards the water, energy and resource recovery facilities and close loops by enhanced biogas production, resource recovery and recycling as well as reuse of reclaimed water in agriculture (Fig. 2). The latter is in line with the new green concept, which considers wastewater as a renewable source of energy, reclaimed water and valuable nutrients [97]. Moreover, it contributes to the reduction of GHG emissions and biosolids production. Fig. 2 illustrates the integration of anaerobic treatment in the dairy supply chain in order to increase sustainability and achieve the transition to the circular economy concept.

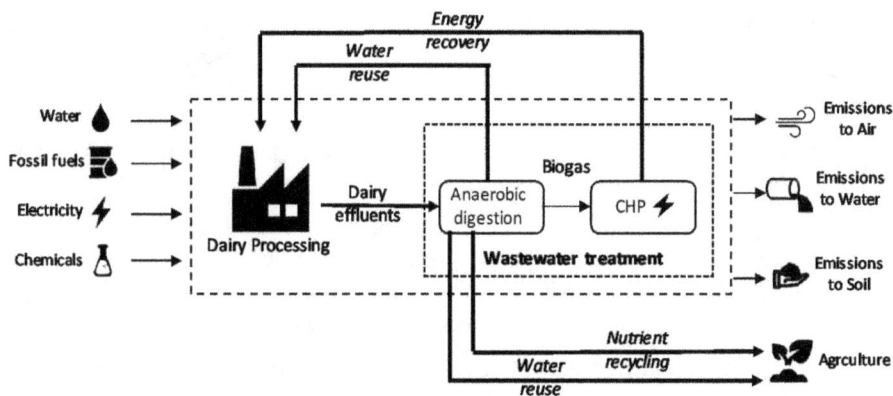

Figure 2. Dairy wastewater treatment towards a circular economy.

Potential options that can reduce the water footprint of the process through water recovery include reuse of secondary water, e.g. use of reclaimed water after reverse osmosis treatment for cleaning of less critical equipment; closed-circuit cooling systems for reuse of cooling water and reuse of cooling water for cleaning purposes. However, the quality of reclaimed water should comply with the hygienic safety and quality requirements in order to be reused [17]. Furthermore, uncertainties in the regulatory framework for the reuse of the treated effluent hinder the wider adoption and market uptake of the AnMBR technology. Thus, legislative barriers should be tackled in order to make use of the value of wastewater and promote valorization potential. The current legislative framework, legal barriers, and obstacles for the reuse of the treated effluent hinder the wider adoption and market uptake of the AnMBR technology. Sustainable wastewater treatment combining anaerobic membrane technology and water reuse was selected as one of the two innovation deals within the concept of the circular economy (European Commission). The main target was to address legislative barriers in order to make use of the value of wastewater and promote a paradigm shift from the conventional wastewater treatment plants to water resource recovery facilities. [98]. Thus, the end-users of the recovered resources should not be considered as potential consumers, but as active participants of the wastewater treatment process and its valorization potential.

8.   Management of dairy wastewater treatment: Real-time control and optimization through ICT technologies

The outline of a generic but customisable, data acquisition, data analytics, performance monitoring, and control architecture is proposed. The suggested architecture has based the nature and principles of the feedback loop control system where the system input, output and control parameters coupled with the process (plant) definition and model are designed for the dairy plant processes, the anaerobic treatment and the combined heat and power plant (CHP) (Fig. 3).

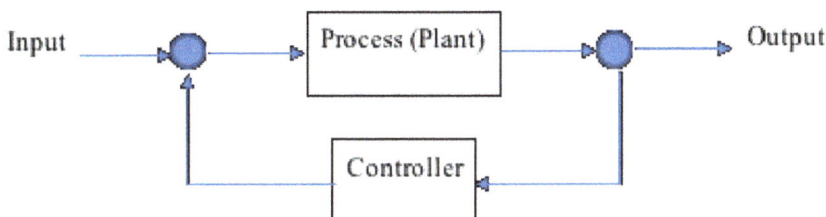

Figure 3. Feedback Loop control system.

Fig. 4 provides an overview of the processes and the system parameters in a typical dairy industry with anaerobic wastewater treatment reactor attached to it.

The input data of the system originates from sensors, counters, actuators, soft sensors (latent parameters), and other sources of information that may come from different sources and databases. In the case, in hand, the main input data will be extracted from the output of the Dairy process plant. For example, the effluent characteristics released from the process provide the information about the properties of the raw material that feeds the anaerobic reactor. The levels of energy consumed to maintain or treat the biological and chemical processes (e.g. fuel or electricity used) could be considered the key input parameters. In addition, the data that assists in measuring the performance of the anaerobic reactor, such as volume wastewater fed, hydraulic retention time (HRT), organic loading rate (OLR), the pH, and the current temperature of the reactor are measurable input variables. We will add to that the environmental data such as ambient temperature, humidity as well as the state of the machines and equipment as the input parameters of the system. The output parameters will be production output, greenhouse gas emissions and overall productivity of the reactors. Figure 5 explains the proposed data acquisition and process modeling structure for monitoring and control, under the auspices of the industrial control network.

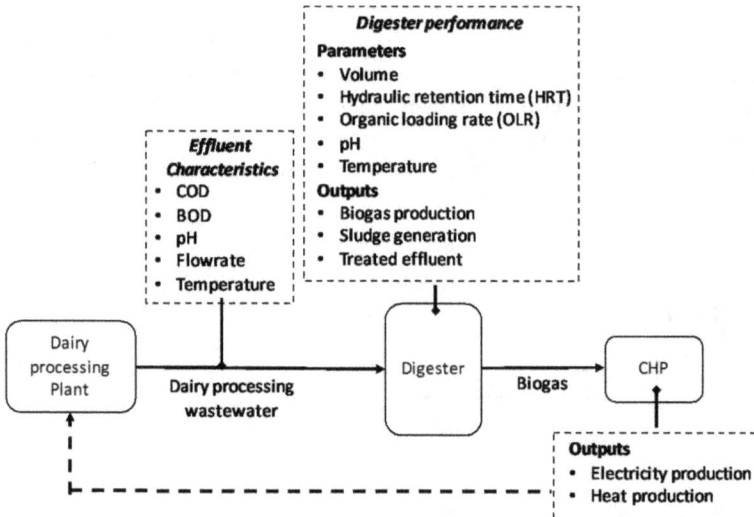

Figure 4. Overview of the main parameters in the anaerobic wastewater treatment system.

The data analytics layer collects real-time data and with respect to the historical (known information and knowledge), implements the data interpretation and cross relationships between system parameters (i.e. input, output, and control). Machine learning, data mining, clustering techniques pending the nature and application of the data gathered will provide the necessary knowledge to confirm the existing transfer/inferential models or where necessary to modify or create new ones [99]. The three performance models of Product, GHG emissions, and Productivity provides accurate indicators of overall performance of the system (i.e. the outputs), that can be shared between various stakeholders.

Further, the measured metrics can be compared with regulations, customer requirements and production performance targets to create an overall satisfaction function (unifying metric) [100,101].

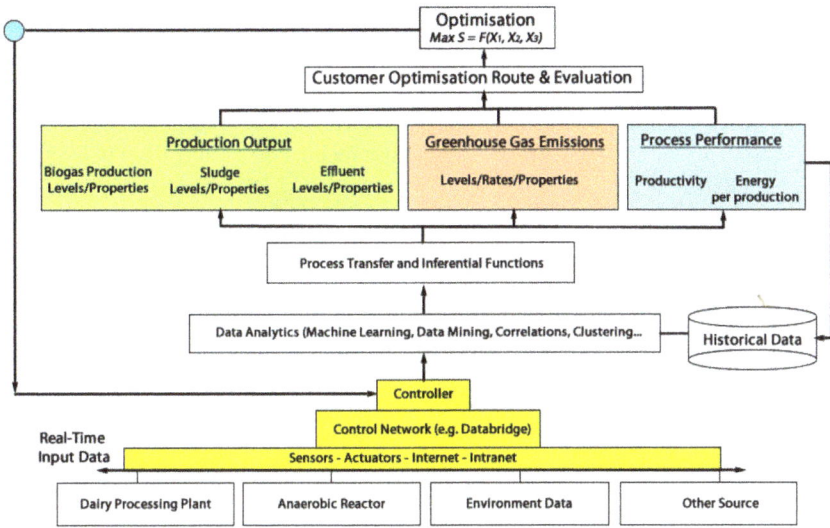

Figure 5. Data acquisition and process modeling structure for monitoring and control.

Conclusions

Dairy products are irreplaceable components of human diet and one of the largest and most important sub-sectors of the food industry in Europe. In this chapter, the authors attempted to highlight the key parameters that affect the performance of dairy plants waste water treatment. It later went on to describe a monitoring and control technology that will be capable of measuring the key performance parameters of the treatment processes and provided a strategy for control and optimisation of those performances against sustainability and Life Cycle Assessment targets.

Acknowledgments

The authors would like to acknowledge the Royal Society for the funding of the current research: Ad-Bio: Advanced Biological Wastewater Treatment Processes, Newton Advanced Fellowship-2015/R2.

References

[1]   F.X. Milani, D. Nutter, and G. Thoma, Invited review: Environmental impacts of dairy processing and products: a review, J. Dairy Sci. 94 (2011) 4243–4254. https://doi.org/10.3168/jds.2010-3955

[2]   M.R. Kosseva, "Management and processing of food wastes,"Comprehensive Biotechnology, 2011, pp. 557–593. https://doi.org/10.1016/B978-0-08-088504-9.00393-7

[3]   A. Baban, "Biodegradability assessment and treatability of high strength complex industrial park wastewater," Clean no. 41 (10), pp. 976–983, 2013.

[4]   B. Lesjean and E.H. Huisjes, "Survey of the European MBR market: trends and perspectives," DES, vol. 231, no. 231, pp. 71–81, 2008. https://doi.org/10.1016/j.desal.2007.10.022

[5]   E. Huisjes, K. Colombel, and B. Lesjean, "The European MBR market: specificities and future trends," Final MBR-Network Work., 2009.

[6]   M. Shakerkhatibi, P. Monajemi, M.T. Jafarzadeh, S.A. Mokhtari, and M.R. Farshchian, "Feasibility study on EO/EG wastewater treatment using pilot scale SBR," Anaerob. Membr. Bioreact. mini Rev. with Emphas. Ind. wastewater Treat. Appl. limitations Perspect., vol. 7, no. 1, pp. 195–204, 2013.

[7]   V.G. Gude, "Wastewater treatment in microbial fuel cells - an overview," pp. 287–307, 2016.

[8]   H. Ozgun, R.K. Dereli, M.E. Ersahin, C. Kinaci, H. Spanjers, and J.B. Van Lier, "A review of anaerobic membrane bioreactors for municipal wastewater treatment: Integration options, limitations and expectations," Sep. Purif. Technol., vol. 118, pp. 89–104, 2013. https://doi.org/10.1016/j.seppur.2013.06.036

[9]   J.D. & A.Č. Lukáš Dvořák, Marcel Gómez, "Anaerobic membrane bioreactors-a mini review with emphasis on industrial wastewater treatment: applications, limitations and perspectives," pp. 1–15, 2017.

[10]   J. Desloover et al., "Floc-based sequential partial nitritation and anammox at full scale with contrasting N_2O emissions," Water Res., vol. 45, pp. 2811–2821, 2011. https://doi.org/10.1016/j.watres.2011.02.028

[11]   A. Santos, W. Ma, and S.J. Judd, "Membrane bioreactors: Two decades of research and implementation," Desalination, vol. 273, pp. 148–154, 2011. https://doi.org/10.1016/j.desal.2010.07.063

[12]   P. Krzeminski, J.H.J.M. van der Graaf, and J.B. van Lier, "Specific energy consumption of membrane bioreactor (MBR) for sewage treatment," Water Sci. Technol., vol. 65, no. 2, p. 380, Jan. 2012. https://doi.org/10.2166/wst.2012.861

[13]   H. Lin et al., "A review on anaerobic membrane bioreactors: Applications, membrane fouling and future perspectives," DES, vol. 314, pp. 169–188, 2013. https://doi.org/10.1016/j.desal.2013.01.019

[14]   J.C.P. Palhares and J.R.M. Pezzopane, "Water footprint of conventional and organic dairy production system," IV Symp. Agric. Agroindustrial waste Manag. May, vol. 93, pp. 5–8, 2015.

[15]   M.A. Zonderland-Thomassen and S.F. Ledgard, "Water footprinting–a comparison of methods using New Zealand dairy farming as a case study," Agric. Syst., vol. 110, pp. 30–40, 2012. https://doi.org/10.1016/j.agsy.2012.03.006

[16]   V. Vasilaki, E. Katsou, S. Pons, and J. Colon, "Water and carbon footprint of selected dairy products: a case study in Catalonia," J Clean Prod, 2016. https://doi.org/10.1016/j.jclepro.2016.08.032

[17]   E. Korsström and M. Lampi, "Best available techniques (BAT) for the Nordic dairy industry."

[18]   W.B. Group, "Environmental, Health, and Safety Guidelines, Dairy Processing."

[19]   "European Commission Directorate: General JRC Institute for Prospective Technological Studies Integrated Pollution Prevention and Control. Reference Document on Best available techniques in the food, drink and milk industries."

[20]   J.P. Strydom, JT.J. Britz, J.F. Mostert, "Two-phase anaerobic digestion of three different effluents using a hybrid bioreactor," Water Salination, vol. 23, pp. 151–6, 1997.

[21]   I. Ozturk, V. Eroglu, G. Ubay, I.Demir "Hybrid upflowanaerobic sludge blanket reactor (HUASBR) treatment of dairy effluents" Water Sci Technol, no. 28, pp. 77–85, 1993.

[22]   P.J.Samkutty, R.H. Gough, "Filtration treatment of dairy processing wastewater" J. Environ. Sci. Heal., vol. A37 (2), pp. 195–199, 2002. https://doi.org/10.1081/ESE-120002582

[23]   O. Ince, "Performance of a two-phase anaerobic digestion system when treating dairy wastewater," Water Res., vol. 32, no. 9, pp. 2707–2713, 1998. https://doi.org/10.1016/S0043-1354(98)00036-0

[24] U.B. Deshannavar, R.K. Basavaraj, and N.M. Naik, "High rate digestion of dairy industry effluent by upflow anaerobic fixed-bed reactor," J. Chem. Pharm. Res., vol. 4, no. 6, pp. 2895–2899, 2012.

[25] J.I. Qazi, M. Nadeem, S.S. Baig, S. Baig, and Q. Syed, "Anaerobic fixed film biotreatment of dairy wastewater," Middle-East J. Sci. Res., vol. 8, no. 3, pp. 590–593, 2011.

[26] T.A. Malaspina F, Stante L, Cellamare CM, "Cheese whey and cheese factory wastewater treatment with a biological anaerobic– aerobic process," Water Sci Technol, pp. 59–72, 1995.

[27] G.J. Monroy OH, Vazquez FM, Derramadero JC, "Anaerobic– aerobic treatment of cheese wastewater with national technology in Mexico: the case of 'El Sauz,'" Water Sci Technol, no. 32, pp. 149–56, 1995.

[28] K.K. van den Berg L, "Dairy waste treatment with anaerobic stationary fixed film reactors," Des. Anaerob. Process. Treat. Ind. Munic. wastes. Pennsylvania Technomic Publ. Co., p. p.89-96., 1992.

[29] L. Danalewich, J.R Papagiannis, T.G Belyea, R.L Tumbleson, M.E Raskin, "Characterization of dairy waste streams, current treatment practices, and potential for biological nutrient removal.," Water Res., vol. 32 (12), pp. 3555–3568, 1998. https://doi.org/10.1016/S0043-1354(98)00160-2

[30] R. Andreottola, G., Foladori, P., Ragazzi, M Villa, "Dairy wastewater treatment in a moving bed biofilm reactor," Wat. Sci. Technol., vol. 45 (12), pp. 321–328, 2002.

[31] J. Donkin, "Bulking in aerobic biological systems treating dairy processing wastewaters.," Int. J. Dairy Tech., vol. 50, pp. 67–72, 1997. https://doi.org/10.1111/j.1471-0307.1997.tb01740.x

[32] K.V.R. and V.V.N.K.T. Kusum Lata, Arun Kansal, Malini Balakrishnan, "Evaluation of Biomethanation Potential of Selected Industrial Organic Effluents in India," Energy Res. Inst. , Biogas Users Surv. 1998/99 Nepal.

[33] D.B., "Acidogenesis in two-phase anaerobic treatment of dairy wastewater. Ph.D. Thesis," Bogazici Univ. Istanbul,Turkey, 2003.

[34] F. Carvalho, A.R. Prazeres, and J. Rivas, "Cheese whey wastewater: characterization and treatment," Sci. Total Environ., vol. 445–446, pp. 385–396, 2013. https://doi.org/10.1016/j.scitotenv.2012.12.038

[35] W. Janczukowicz, M. Zieli Ski, and M. Debowski, "Biodegradability evaluation of dairy effluents originated in selected sections of dairy production," 2007.

[36] J.R. Danalewich, T.G. Papagiannis, R.L. Belyea, M. E. Tumbleson, and L. Raskin, "Characterization of dairy waste streams, current treatment practices, and potential for biological nutrient removal", Wat. Res., 1998. https://doi.org/10.1016/S0043-1354(98)00160-2

[37] R.L. Droste, "Theory and practice of water and wastewater treatment," John Wiley &Sons Inc New York, USA, no. 28, 1997.

[38] T.J. Britz and C. van Schalkwyk, "Treatment of dairy processing wastewaters," Waste Treat. Food Process. Ind., no. c, pp. 1–24, 2006.

[39] J.R. Danalewich, T.G. Papagiannis, R.L. Belyea, M.E. Tumbleson, and L. Raskin, "Characterization of dairy waste streams, current treatment practices, and potential for biological nutrient removal" , Wat. Res., 1998." https://doi.org/10.1016/S0043-1354(98)00160-2

[40] H.N. Gavala, H. Kopsinis, I.V Skiadas, K. Stamatelatou, and G. Lyberatos, "Treatment of dairy wastewater using an upflow anaerobic sludge blanket reactor," J. Agric. Eng. Res., vol. 73, no. 1, pp. 59–63, 1999. https://doi.org/10.1006/jaer.1998.0391

[41] S. Tchamango, C.P. Nanseu-Njiki, E. Ngameni, D. Hadjiev, and A. Darchen, "Treatment of dairy effluents by electrocoagulation using aluminium electrodes," Sci. Total Environ., vol. 408, pp. 947–952, 2009. https://doi.org/10.1016/j.scitotenv.2009.10.026

[42] M. Amini, H. Younesi, A. Akbar, Z. Lorestani, and G. Najafpour, "Determination of optimum conditions for dairy wastewater treatment in UAASB reactor for removal of nutrients," Bioresour. Technol., vol. 145, pp. 71–79, 2013. https://doi.org/10.1016/j.biortech.2013.01.111

[43] D. Traversi et al., "Environmental advances due to the integration of food industries and anaerobic digestion for biogas production: perspectives of the italian milk and dairy product sector," Bioenergy Res., vol. 6, no. 3, pp. 851–863, 2013. https://doi.org/10.1007/s12155-013-9341-4

[44] J.A. Gil, L. Túa, A. Rueda, B. Montaño, M. Rodríguez, and D. Prats, "Monitoring and analysis of the energy cost of an MBR," DES, vol. 250, pp. 997–1001.

[45]   A. Mahvi, "Sequencing batch reactor: a promising technology in wastewater treatment," Environ. Technol. Rev., vol. 5, no. Environ. Health Sci. Eng, pp. 79–90, 2008.

[46]   Y.J. Chan, M. F. Chong, and C.L. Law, "Biological treatment of anaerobically digested palm oil mill effluent (POME) using a Lab-Scale Sequencing Batch Reactor (SBR)," J. Environ. Manage., vol. 91, pp. 1738–1746, 2010. https://doi.org/10.1016/j.jenvman.2010.03.021

[47]   S. Sirianuntapiboon and K. Chairattanawan, "Comparison of sequencing batch reactor (SBR) and granular activated carbon-SBR (GAC-SBR) systems on treatment textile wastewater containing basic dye," Desalin. Water Treat., vol. 57, no. 56, pp. 27096–27112, Dec. 2016. https://doi.org/10.1080/19443994.2016.1167629

[48]   A.M. Martins, K. Pagilla, J.J. Heijnen, and M.C. van Loosdrecht, "Filamentous bulking sludge—a critical review," Water Res., vol. 38, no. 4, pp. 793–817, 2004. https://doi.org/10.1016/j.watres.2003.11.005

[49]   C. Meunier, O. Henriet, B. Schoonbroodt, J.M. Boeur, J. Mahillon, and P. Henry, "Influence of feeding pattern and hydraulic selection pressure to control filamentous bulking in biological treatment of dairy wastewaters," Bioresour. Technol., vol. 221, pp. 300–309, 2016. https://doi.org/10.1016/j.biortech.2016.09.052

[50]   M.L. Christensen,. K. Keiding, P.H. Nielsen, M.K. Jørgensen, "Dewatering in biological wastewater treatment: a review," Water Res., vol. 82, pp. 14–24, 2015. https://doi.org/10.1016/j.watres.2015.04.019

[51]   W. Dąbrowski, R. Żyłka, and P. Malinowski, "Evaluation of energy consumption during aerobic sewage sludge treatment in dairy wastewater treatment plant," Environ. Res., vol. 153, pp. 135–139, 2017. https://doi.org/10.1016/j.envres.2016.12.001

[52]   R.M. Osman, "Anaerobic Fermentation of Industrial Wastewater (Review Article).," OJCES, pp. 50–78, 2014.

[53]   K. Rajeshwari, M. Balakrishnan, A. Kansal, K. Lata, and V.V. Kishore, "State-of-the-art of anaerobic digestion technology for industrial wastewater treatment," Renew. Sustain. Energy Rev., vol. 4, no. 2, pp. 135–156, 2000. https://doi.org/10.1016/S1364-0321(99)00014-3

[54] W. Wang, Q. Yang, S. Zheng, and D. Wu, "Anaerobic membrane bioreactor (AnMBR) for bamboo industry wastewater treatment," Bioresour. Technol., vol. 149, pp. 292–300, 2013. https://doi.org/10.1016/j.biortech.2013.09.068

[55] B. Demirel and O. Yenigü, "Changes in microbial ecology in an anaerobic reactor," 2005.

[56] D. Karadag, O.E. Köroğlu, B. Ozkaya, and M. Cakmakci, "A review on anaerobic biofilm reactors for the treatment of dairy industry wastewater," Process Biochem., vol. 50, no. 2, pp. 262–271, 2015. https://doi.org/10.1016/j.procbio.2014.11.005

[57] A.R. Prazeres, F. Carvalho, and J. Rivas, "Cheese whey management: a review," J. Environ. Manage., vol. 110, pp. 48–68, 2012. https://doi.org/10.1016/j.jenvman.2012.05.018

[58] B. Demirel, O. Yenigun, and T. T. Onay, "Anaerobic treatment of dairy wastewaters : a review.", 2005.

[59] P. Poh, M. Chong, and B. Sunway, "Upflow anaerobic sludge blanket-hollow centered packed bed (UASB-HCPB) reactor for thermophilic palm oil mill effluent (POME) treatment," Biomass Bioenergy, vol. 67, pp. 231–242, 2014. https://doi.org/10.1016/j.biombioe.2014.05.007

[60] N. Brown, "Methane dissolved in wastewater exiting UASB reactors: concentration measurement and methods for neutralization," Dep. Energy Technol. R. Inst. Technol., 2006.

[61] J. Xing, W. Zuo, J-e. Dai, N. Cheng, J. Li, "Reactor performance and microbial community of an EGSB reactor operated at 20 and 15^0C.," J. Appl. Microbiol., pp. 848–857, 2009. https://doi.org/10.1111/j.1365-2672.2009.04260.x

[62] R. Kothari, V. Kumar, V.V. Pathak, and V.V. Tyagi, "Sequential hydrogen and methane production with simultaneous treatment of dairy industry wastewater: Bioenergy profit approach," Int. J. Hydrogen Energy, vol. 42, no. 8, pp. 4870–4879, 2017. https://doi.org/10.1016/j.ijhydene.2016.11.163

[63] A.B.T. Abdullah Yasar, "Anaerobic treatment of industrial wastewater by UASB reactor integrated with chemical oxidation processes," Polish J. Environ. Stud., vol. 19, no. No. 5, pp. 1051–1061, 2010.

[64] B. Lew, S. Tarre, M. Belavski, and M. Green, "UASB reactor for domestic wastewater treatment at low temperatures: a comparison between a classical UASB and hybrid UASB-filter reactor", 2004.

[65]   A. Sá Nchez, D. Buntner, and J. M. Garrido, "Impact of methanogenic pre-treatment on the performance of an aerobic MBR system," Water Res., vol. 47, pp. 1229–1236, 2013. https://doi.org/10.1016/j.watres.2012.11.042

[66]   D.F.C. Dias, T. E. Possmoser-Nascimento, V.A.J. Rodrigues, and M. von Sperling, "Overall performance evaluation of shallow maturation ponds in series treating UASB reactor effluent: ten years of intensive monitoring of a system in Brazil," Ecol. Eng., vol. 71, pp. 206–214, 2014. https://doi.org/10.1016/j.ecoleng.2014.07.044

[67]   C. Arnaiz, P. Buffiere, S. Elmaleh, J. Lebrato, and R. Moletta, "Anaerobic digestion of dairy wastewater by inverse fluidization: the inverse fluidized bed and the inverse turbulent bed reactors," Environ. Technol., vol. 24, no. 11, pp. 1431–1443, Nov. 2003. https://doi.org/10.1080/09593330309385687

[68]   M.M.A. Saleh and U. F. Mahmood, "Anaerobic digestion technology for industrial wastewater treatment," 2004.

[69]   P. Weiland, "Biogas production: current state and perspectives," Appl. Microbiol. Biotechnol., vol. 85, no. 4, pp. 849–860, Jan. 2010. https://doi.org/10.1007/s00253-009-2246-7

[70]   W.J.J. Gao, H.J. Lin, K.T. Leung, and B.Q. Liao, "Influence of elevated pH shocks on the performance of a submerged anaerobic membrane bioreactor," Process Biochem., vol. 45, pp. 1279–1287, 2010. https://doi.org/10.1016/j.procbio.2010.04.018

[71]   G. Skouteris, D. Hermosilla, P. López, C. Negro, and Á. Blanco, "Anaerobic membrane bioreactors for wastewater treatment: A review," Chem. Eng. J., vol. 198–199, pp. 138–148, 2012. https://doi.org/10.1016/j.cej.2012.05.070

[72]   H.Q. Yu and H.H.P. Fang, "Acidogenesis of dairy wastewater at various pH levels", Water Sci Technol., 2002.

[73]   K. Bialek, D. Cysneiros, and V. O'Flaherty, "Hydrolysis, acidification and methanogenesis during low-temperature anaerobic digestion of dilute dairy wastewater in an inverted fluidised bioreactor," Appl. Microbiol. Biotechnol., vol. 98, no. 20, pp. 8737–8750, Oct. 2014. https://doi.org/10.1007/s00253-014-5864-7

[74]   D.C. Stuckey, "Recent developments in anaerobic membrane reactors," Bioresour. Technol., vol. 122, pp. 137–148, 2012. https://doi.org/10.1016/j.biortech.2012.05.138

[75]   O. Demirel, B. and Yenigun, "Anaerobic acidogenesis of dairy wastewater: the effects of variations in hydraulic retention time with no pH control.," J Chem. Technol. Biotechnol. 79, vol. 79, pp. 755–760, 2004.

[76]   K.J. Wijnands, J.H., van der Meulen, H.A.B., and Poppe, "Competitiveness of the European Food Industry: an economic and legal assessment 2007 (Office for Official Publications of teh European Communities).," 2007.

[77]   "Greenhouse gas emissions from the dairy sector: a Life Cycle Assessment," FAO, p. Available at: www.fao.org/docrep/012/k7930e/k7930e, 2010.

[78]   K.A.S. and O.U.B. Yavuz Y, Ocal E, "Treatment of dairy industry wastewater by EC and EF processes using hybrid Fe-Al plate electrodes.," J. Chem. Technol. Biotechnol., vol. 86, pp. 964–969, 2011. https://doi.org/10.1002/jctb.2607

[79]   S. Manning, L., Baines, R., and Chadd, "Benchmarking the poultry meat supply chain.," Benchmarking an Int. J., vol. 15(2), pp. 148–165, 2008.

[80]   "The dairy roadmap," no. Available at: http://www.dairyuk.org/images/documents/publications/2015_Dairy_Roadmap.pd f (accessed: 23/4/2017), 2015.

[81]   J. Gelegenis, D. Georgakakis, I. Angelidaki, and V. Mavris, "Optimization of biogas production by co-digesting whey with diluted poultry manure ARTICLE IN PRESS," Renew. Energy, vol. 32, pp. 2147–2160, 2007. https://doi.org/10.1016/j.renene.2006.11.015

[82]   M. Carlini, S. Castellucci, and M. Moneti, "Biogas production from poultry manure and cheese whey wastewater under mesophilic conditions in batch reactor," Energy Procedia, vol. 82, no. 82, pp. 811–818, 2015. https://doi.org/10.1016/j.egypro.2015.11.817

[83]   J. Zhong, D.K. Stevens, and C.L. Hansen, "Optimization of anaerobic hydrogen and methane production from dairy processing waste using a two-stage digestion in induced bed reactors (IBR)," Int. J. Hydrogen Energy, vol. 40, pp. 15470–15476, 2015. https://doi.org/10.1016/j.ijhydene.2015.09.085

[84]   M. Rodgers, X.M. Zhan, and B. Dolan, "Mixing characteristics and whey wastewater treatment of a novel moving anaerobic biofilm reactor.," J. Environ. Sci. Health. A. Tox. Hazard. Subst. Environ. Eng., vol. 39, no. 8, pp. 2183–93, 2004. https://doi.org/10.1081/ESE-120039383

[85] V. Blonskaja and T. Vaalu, "Investigation of different schemes for anaerobic treatment of food industry wastes in Estonia," Proc. Est. Acad. Sci. Chem, vol. 55, no. 1, pp. 14–28, 2006.

[86] H. Gannoun, E. Khelifi, H. Bouallagui, Y. Touhami, and M. Hamdi, "Ecological clarification of cheese whey prior to anaerobic digestion in upflow anaerobic filter," 2008.

[87] A.Y. Hu and D.C. Stuckey, "Treatment of Dilute Wastewaters Using a Novel Submerged Anaerobic Membrane Bioreactor," J. Environ. Eng., vol. 132, no. 2, pp. 190–198, Feb. 2006. https://doi.org/10.1061/(ASCE)0733-9372(2006)132:2(190)

[88] G. Antonopoulou, K. Stamatelatou, N. Venetsaneas, M. Kornaros, and G. Lyberatos, "Biohydrogen and methane production from cheese whey in a two-stage anaerobic process," Ind. Eng. Chem. Res., vol. 47, no. 15, pp. 5227–5233, Aug. 2008. https://doi.org/10.1021/ie071622x

[89] B. Kavacik and B. Topaloglu, "Biogas production from co-digestion of a mixture of cheese whey and dairy manure," Biomass and Bioenergy, vol. 34, pp. 1321–1329, 2010. https://doi.org/10.1016/j.biombioe.2010.04.006

[90] D. Hidalgo and J.M. Martín-Marroquín, "Biochemical methane potential of livestock and agri-food waste streams in the Castilla y León Region (Spain)," FRIN, vol. 73, pp. 226–233, 2015. https://doi.org/10.1016/j.foodres.2014.12.044

[91] M. Franchetti, "Economic and environmental analysis of four different configurations of anaerobic digestion for food waste to energy conversion using LCA for: A food service provider case study," J. Environ. Manage., vol. 123, pp. 42–48, 2013. https://doi.org/10.1016/j.jenvman.2013.03.003

[92] O. Eriksson, M. Bisaillon, M. Haraldsson, and J. Sundberg, "Enhancement of biogas production from food waste and sewage sludge e Environmental and economic life cycle performance," 2016.

[93] K. Sin Woon, I.M. Lo, S.L. Chiu, and D.Y. Yan, "Environmental assessment of food waste valorization in producing biogas for various types of energy use based on LCA approach," 2016.

[94] A. Hoekstra, "How sustainable is Europe's water footprint?," Water Wastewater Int., vol. 26, no. 2, pp. 24–26, 2011.

[95] C. Aydiner, U. Sen, D.Y. Koseoglu-Imer, and E.C. Dogan, "Hierarchical prioritization of innovative treatment systems for sustainable dairy wastewater

management," J. Clean. Prod., vol. 112, pp. 4605–4617, 2016. https://doi.org/10.1016/j.jclepro.2015.08.107

[96]   Z. Chen, J. Luo, Y. Wang, W. Cao, B. Qi, and Y. Wan, "A novel membrane-based integrated process for fractionation and reclamation of dairy wastewater," Chem. Eng. J., vol. 313, pp. 1061–1070, 2017. https://doi.org/10.1016/j.cej.2016.10.134

[97]   E. Romgens, B. Kruizinga, "Wastewater management roadmap towards 2030," Roadmap Wastewater Manag., 2013.

[98]   IWA, "Water Utility Pathways in a Circular Economy," 2016.

[99]   Y. LeCun, Y. Bengio, and G. Hinton, "Deep learning," Nature, vol. 521, no. 7553, pp. 436–444, May 2015. https://doi.org/10.1038/nature14539

[100] A. Mousavi, P. Adl, R.T. Rakowski, A. Gunasekaran, and N. Mirnezami, "Customer optimization route and evaluation (CORE) for product design," Int. J. Comput. Integr. Manuf., vol. 14, no. 2, pp. 236–243, Jan. 2001. https://doi.org/10.1080/09511920150216350

[101] S. Tavakoli, A. Mousavi, and S. Poslad, "Input variable selection in time-critical knowledge integration applications: A review, analysis, and recommendation paper," Adv. Eng. Informatics, vol. 27, no. 4, pp. 519–536, 2013. https://doi.org/10.1016/j.aei.2013.06.002

Chapter 6

Occurrence of Pesticides and Hormones in Municipal Wastewater and their Removal by Membrane Bioreactors

V. Naddeo[1]*, L. Borea[1], S.W. Hasan[2], V. Belgiorno[1]

[1] Sanitary Environmental Engineering Division (SEED), Department of Civil Engineering, University of Salerno, Via Giovanni Paolo II 132, 84084 Fisciano, SA, Italy

[2] Department of Chemical Engineering, Khalifa University of Science and Technology, Masdar City Campus, P.O. Box 54224, Abu Dhabi, United Arab Emirates

vnaddeo@unisa.it*

Abstract

The presence of emerging contaminants (ECs), including pharmaceutically active compounds (PhACs), steroid hormones and pesticides, in municipal wastewater as well as in the aquatic environment has been an issue of concern from an environmental and human health point of view. These ECs have been found in municipal wastewater in the concentration range of several µg/L. Considering the importance assumed by many of ECs, it is not possible to ban their usage. Therefore, it is essential to effectively remove these contaminants to protect the environment and drinking water resources. There is evidence that conventional treatment methods are not able to completely remove many ECs, therefore, advanced processes are demanded in order to degrade these ECs and protect both water environment and human health. Membrane bioreactors (MBRs), coupling biological degradation with membrane filtration, are one possible option due to the advantages that characterize this technology. The present chapter discusses the occurrence and fate of pesticides and hormones in wastewater and proposes a comparison between removal efficiency of conventional activated sludge treatments and MBRs, also defining common concentrations at which these ECs are found in wastewater.

Keywords

Emerging Contaminants, MBR, Trace Organic Contaminants, Wastewater

Contents

1.   Introduction

The removal of emerging contaminants (ECs) such as endocrine disrupting chemicals (EDCs), pesticides and flame retardants is an important consideration for the production of safe drinking water and the environmentally responsible release of wastewater [1]. Their presence in the environment is currently rising [2–6], because of increased consumption of pharmaceuticals and personal care products, as well as pesticides. For most ECs, there is little information regarding their potential toxicological significance in ecosystems, particularly effects from long-term, low-level environmental exposures, even if there is already some evidence of significant impacts on the environment [7]. The main pathways by which ECs may end up in the environment are shown in Fig. 1 [8].

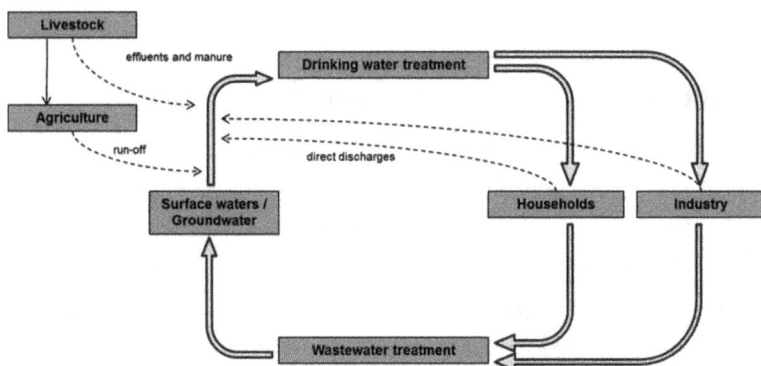

Figure 1. Pathways by which ECs may end up in the environment.

Rather than removing ECs from groundwater and surface waters during potabilization treatments, the best option is to remove the contaminants at an early stage, during the wastewater treatment, in order to avoid any further contamination of the environment.

However, conventional wastewater treatment processes do not provide an effective barrier against many ECs [9–11]. Therefore, efficient treatment technologies to achieve effective ECs removal from wastewater need to be identified.

While advanced oxidation processes (AOPs) can be effective for the removal of ECs, these processes lead to the formation of oxidation intermediates that may be even more toxic than the original compounds [12–14]. In addition, unwanted oxidation byproducts, such as halogenated organic compounds or bromate, are produced in oxidation processes involving the use of chlorine and ozone as oxidants.

Membrane bioreactors (MBRs), combining membrane filtration with biological degradation, are a full-grown and reliable system for wastewater treatment and later on its reuse [15,16]. Therefore, the application of this technology can be potentially useful. In contrast, there are certain reports [17–19] indicating that conventional activated sludge (CAS) reactors have a low removal efficiency, regarding many ECs, which are present at μg/L or ng/L concentration [20]. Therefore, a better performing system is required to handle the ECs problem.

It is well known that trace amount of hormones and pesticides are nowadays often detected in the environment because of the improvement in analyzing techniques [21]. In addition, some of these chemicals have proved to be bioaccumulative and/or ecotoxicological [22]. Moreover, many ECs have been found to be persistent in the environment for many years [23], and their consumption is increasing [21,24,25].

Pesticides are regarded as ECs and have been included in the list of the broad range of persistent organic pollutants (POPs) which are carbon-based chemical substances that have ecological impacts, persist in the environment, bioaccumulate through food, are capable of adverse effects to human health and to the environment [26,27]. However, effects on living organisms depend on the category of the pesticide, therefore generalization is not possible. Water contaminated by pesticide runoff is the principal pathway that causes ecological impacts, even though terrestrial impacts by pesticides could happen; farm workers must manage risks associated with inhalation and skin contact [28]. Pesticide occurrence in surface water has two main human health impacts. The first is related to human consumption of fish and shellfish which are contaminated by pesticides; the second is the direct ingestion of pesticide-contaminated water [29]. WHO [30] has established drinking water guidelines for 33 pesticides. "Acceptable daily intake" (ADI) values have been set by several agencies for the protection of human health and environment which indicate the maximum allowable daily ingestion over a person's lifetime without appreciable risk to the individual [31]. Among substituted phenols, tetrachlorohydroquinone – a toxic metabolite of the biocide pentachlorophenol has been found to produce significant and dose-dependent DNA damage [32,33].

Hormones have been discharged in the environment for a long time. However, only in the early 1990s researchers began reporting several environmental threats including feminization of male fish, which was found especially where wastewater effluents were discharged [34] The 17β-estradiol, ethinylestradiol and estrone played a key role in the feminization of fish, and it must be noted that current increase in use of oral contraceptives (OCs) has to be taken into account as a new input of hormones into the environment [35,36].

Because of the complexity of the ECs related to their mechanisms of action, standard methods for monitoring and detection of these compounds are lacking. Furthermore, many of these contaminants have been reported to be persistent in the environment [23,37,38].

The aim of this chapter is to analyze both fates of ECs in the environment and their occurrence in wastewater, as well as to discuss the application of MBRs to wastewater in order to remove ECs, focusing on pesticides and hormones and trying to evaluate the performance of different MBR configurations.

2. MBR configurations and performance

The advantages offered by MBR technology have been extensively recognized [39]. MBRs have been made by coupling CAS process with membrane separation, with the aim of retaining the biomass and concentrate it up reducing in turn the necessary tank size, also increasing the efficiency of the biotreatment process [15,40–42]. Advantages of MBRs include the avoidance of secondary clarifiers and tertiary filtration processes, thereby reducing plant footprint [43]. In addition, they produce a quality effluent suitable for reuse applications [44,45], providing a barrier to certain chlorine-resistant pathogens such as Cryptosporidium and Giardia [46–49].

The configuration of the membrane, basically its geometry and the way it is mounted and oriented in relation to the flow of water, is crucial in determining the overall process performance. Other practical considerations concern the way in which the membrane elements, are housed in "shells" to produce modules and the complete vessels through which the water flows. Ideally, the membrane should be configured to facilitate the following features [39]:

- high membrane area to module bulk volume ratio;
- high degree of turbulence for mass transfer promotion on the feed side;
- low energy expenditure per unit product water volume;
- low cost per unit membrane area;
- the design which facilitates cleaning and permits modularization.

Membranes can be directly immersed into the biological tank (iMBR) or located in a different container hydraulically linked with the tank (sMBR), and their porosity can be in the microfiltration (10 to 0.1 µm) or ultrafiltration (0.1 to 0.005 µm) range. Such membranes have been produced from ceramic or polymeric materials [15,39].

Membrane fouling, the main issue related to MBR performance, results from the interaction between the membrane material and the components in the activated sludge liquor [50]. The latter include substrate components, cells, cell debris, and microbial metabolites such as extracellular polymeric substances (EPS) [50]. Thus, though many investigations of membrane fouling have been published [42,51], the diverse range of operating conditions and feed water matrices employed and the limited information reported in most of the studies on the suspended biomass composition have made it difficult to establish any generic behavior pertaining to membrane fouling in MBRs specifically. Given that membrane fouling represents the main limitation to membrane process operation [51], it is not surprising that the majority of membrane materials and processes in connection of researches conducted have focused on characterization and amelioration [39]. Fouling can take place through a number of physicochemical and biological mechanisms which all related to increased deposition of solid material onto the membrane surface and within the membrane structure [52].

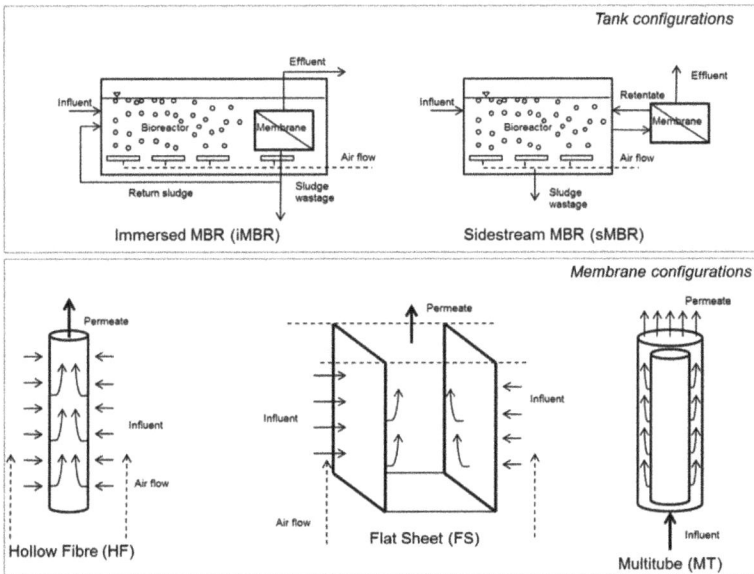

Figure 2. MBRs configurations: membrane modules immersed into the tank (iMBR) or not (sMBR), and membrane configurations.

Physical and chemical cleaning are applied in order to suppress fouling. Ultimately, fouling can be reduced by two methods: promoting turbulence and reducing flux [53]. In side-stream MBRs, turbulence can be promoted simply by increasing the crossflow velocity, whereas for an immersed system this can only reasonably be achieved by increasing the membrane aeration [39]. Turbulence promotion can also arise through passing either the feed water or an air/water mixture along with the surface of the membrane in order to aid the passage of permeate through it [43]. Physical cleaning is most simply affected by reversing the flow, at a rate 2-3 times higher than the forward flow [54] back through the membrane to remove some of the fouling layers on the retentate side. For this to be feasible, the membrane must have sufficient integrity to withstand the hydraulic stress imparted [55]. In other words, the membrane must be strong enough not to break or buckle when the flow is reversed. Recently, relaxation processes were developed in order to restrict fouling formation [56,57]; relaxation means that after a fixed amount of time the forced transmembrane pressure goes down, leading to a substantial reduction of the stress field over the membrane [39].

Regarding flux, there are three principal configurations currently employed in MBR processes, which all have various practical benefits and limitations. The configurations are based on either a planar or cylindrical geometry and comprise flat sheet (FS), hollow fibre (HF) and multitubular (MT), the functional scheme is demonstrated in Fig. 2.

The rejection of contaminants ultimately places a fundamental constraint on every membrane process. Rejected constituents in the retentate tend to accumulate at the membrane surface, producing various phenomena which lead to a reduction in the flow of water through the membrane at a given transmembrane pressure (TMP), or conversely an increase in the TMP for a given flux [58]. On one hand, wastewater treatment plants managers are interested in higher removal efficiencies regarding macropollutants; on the other hand, the more is blocked by the membrane, the more that membrane is fouled.

Regarding macropollutants removal, MBR systems can remove COD efficiently. According to literature reports [59–62], the overall organic degradation efficiencies in MBR have been higher than 95% of municipal wastewater. This high removal efficiency implies that in the membrane bioreactor system, organic matter can be degraded at high rates, because of high concentration of biomass. This overall organic degradation efficiency can be attributed to both biological degradation and membrane filtration. Membrane filtration played a significant role in maintaining high and stable organic removal efficiency [63].

The total Kjeldahl nitrogen removal efficiencies have been high [64], which showed that the removal of nitrogen could be attributed mainly to the action of simultaneous nitrification and denitrification happening in the MBR [65]. In addition, nitrogen compounds in the effluent appeared mostly in the form of nitrate indicating the complete nitrification [66].

3.   Emerging contaminants

In the following sections, information regarding local, national and international regulations about ECs, fate, and occurrence of each group of compounds in wastewaters and the removal efficiencies of ECs by conventional activated sludge (CAS) reactors will be presented and adequately discussed.

3.1   Policy and regulations

At the present time, there is no general agreement among states about enforceable regulations for ECs in wastewater treatment plant effluents. Because pesticides have been in use for a long time, the European Union sets limits for them in drinking water: 100 ng/l for individual compounds and 500 ng/l for the total pesticides. The presence of pesticides in European surface waters has intensively investigated and monitored [19,67–71]. In the last decades, there has been enormous efforts by companies and authorities to study and collect all possible information concerning the safe introduction of pesticides in the market and their release to the environment [72]. Regarding wastewater reclamation and reuse, many countries have already developed detailed laws in which some of the ECs have restricted in reclaimed wastewater effluents.

3.2   Pesticides

Pesticides, such as herbicides, fungicides, insecticides, and bactericides have been a matter of concern as regards to surface water quality. Widespread use in agricultural practice and household consumption are important sources of pesticides and their residues in the aquatic environment [8]. After the second world war, the introduction of relatively non-polar and very persistent pesticides, such as chlordane, aldrin and DDT, enabled impressive increases in food production and crops security [8]. After application on the field, the pesticides contaminated surface waters by drift, runoff, and leaching; for some compounds, degradation products can be present at greater levels in the environment than the parent pesticide [19].

Currently registered pesticides include hundreds of different compounds, such as glyphosate, triazines, organophosphorus herbicides, thiocarbamates, and chlorophenoxy acetic acids, of which some have been used extensively worldwide [19].

Fig. 3 represents the connection between pesticide consumption and human targets from where it is clear that conventional wastewater treatments may have some problems in removing persistent compounds at low concentration levels.

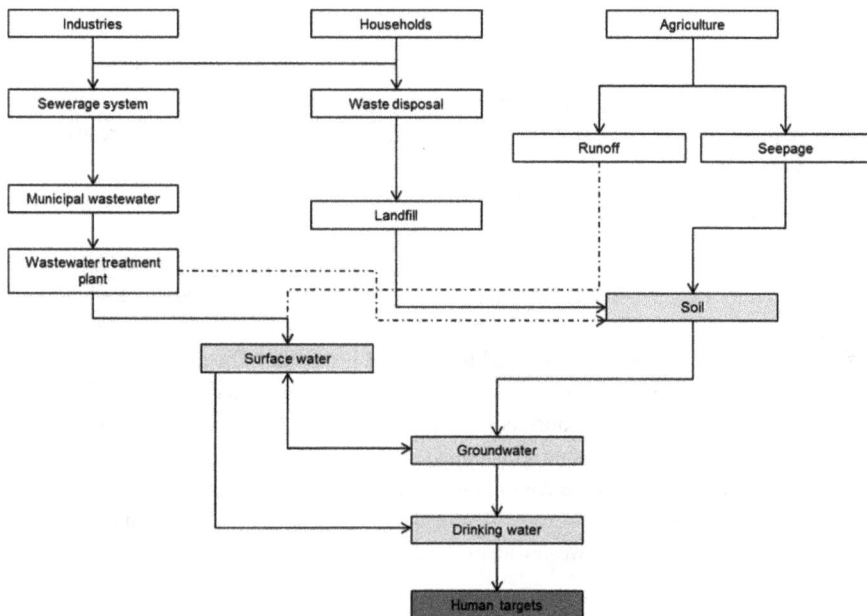

Figure 3. Fate and disposal of ECs in the environment in function of human activities.

Once applied, pesticides can evaporate and, transported by wind or may precipitate and seep. On the other hand, runoff may lead to pollution of surface water, and direct seepage links soils and groundwater. For these reasons, pesticides are gaining increasing attention as harmful substances.

In Table 1 chemical properties of pesticides, along with their occurrence in wastewater and related removal efficiencies by CAS reactors are shown.

3.3   Hormones in wastewater

The report on the first detection of human hormones in water published in 1965, pointed out that steroids were not completely eliminated during wastewater treatment [84]. However, the presence of hormones in the aquatic environment was not given much attention until the 1990s when a link was recognized between a synthetic birth-control pharmaceutical (ethinylestradiol) and its impact on fish when research showed that hormones could affect fish at very low concentrations (1.0 ng/L) [85]. At these low concentrations, there were alterations in the endocrine system leading to growth, development, or reproduction alterations in exposed animals. These changes may be expressed later in the life cycle or even in future generations [86].

Table 1: Chemical properties of pesticides and their occurrence in wastewaters.

Common Name	Use	Formula	Molecular Weight	CAS number	Concentration in WWs, [μg/L]	CAS efficiency [%]	Ref.
2,4-D	Herbicide	$C_8H_6Cl_2O_3$	221,04	94-75-7	20 - 30	30 - 60	[73,74]
2,4-DP	Herbicide	$C_9H_8Cl_2O_3$	235,064	120-36-5	0,001 - 0,02	90 - 99	[74,75]
4-chlorophenol	Herbicide metabolite	C_6H_5ClO	128,56	95-57-8	0,1 - 0,2	65 - 75	[74]
Aldicarb	Insecticide	$C_7H_{14}N_2O_2S$	190,26	116-06-3	-	-	
Atrazine	Herbicide	$C_8H_{14}ClN_5$	215,68	1912-24-9	0,03 - 9	20 - 50	[76,77]
Ametryn	Herbicide	$C_9H_{17}N_5S$	227,33	834-12-8	-	-	[78]
Bayrepel acid	Insect repellent metabolite	$C_{12}H_{23}NO_3$	229,3	119515-38-7	0,1 - 0,8	20 - 28	[79]
Bentazone	Herbicide	$C_{10}H_{12}N_2O_3S$	240,279	25057-89-0	13,2 - 24,5	10 - 55	[80]
Clofibric acid	Herbicide	$C_{10}H_{11}ClO_3$	214.64	882-09-7	-	-	[78]
DEET	Insect repellent	$C_{12}H_{17}NO$	191,27	134-62-3	0,2 - 0,3	5 - 10	[77]
Fenoprop	Herbicide and plant growth regulator	$(C_9H_7Cl_3O_3)$	269.51	93-72-1	-	-	[20]
Isoproturon	Herbicide	$C_{12}H_{18}N_2O$	206,28	34123-59-6	0,001 - 0,2	33 -53	[81]
Linuron	Herbicide	$C_9H_{10}Cl_2N_2O_2$	249,11	330-55-2	-	1 - 5	[82]
MCPA	Herbicide	$C_9H_9ClO_3$	200,62	94-74-6	0,03 - 1,3	90-99	[68,74]
MCPP	Herbicide	$C_{10}H_{11}ClO_3$	214,646	93-65-2	0,1 - 0,2	10 - 16	[79]
Pentachlorophenol	Anfifungal agent	C_6HCl_5O	266,34	87-86-5	4 - 6	-	[83]
Propoxur	Insecticide	$C_{11}H_{15}NO_3$	209.24	114-26-1	-	-	[78]

In Fig. 3 it is exposed how hormones may come up to human beings, via both drinking water or fruit and vegetables; even if new treatment units are capable of removing estrogens, runoffs from fields may end up into surface waters [87]. As stated, agricultural products contained a significant amount of estrogens [88].

Steroidal estrogens identified in domestic sewage effluent are, therefore, present in sufficient quantity to cause synthesis of vitellogenin in fish in vivo, and their effects are additive.

In some rivers where the effluent contributes to a large volume of the flow, it is possible that aquatic organisms may be exposed to concentrations of estrogenic chemicals sufficient to produce biological responses [89–91].

At the present level of knowledge, any review of estrogens effects is an attempt into unquestionable biological facts and epidemiologic observations and, at the same time, an indulgence in hypothetical claims and still unvalidated theories [92]. The progressive impairment over the last 50 years of male reproductive outcomes in the western world may contribute to the negative population growth in many European countries [92].

At the present time, it is well known that wastewater treatment plants can not completely remove every pharmaceutical compound or hormone [93,94]. Estrogenic substances have been monitored in a wide range of municipal effluents at a concentration level of 1.8 and 17 ng/L for 17β-estradiol and estrone, respectively. These estrogens are naturally produced and excreted by human beings [95], both in urban and agricultural runoff [96]. An extensive mapping of US surface waters was also performed [97] in which 139 streams were sampled across 30 states and reproductive hormones were found in approximately 40% of the streams sampled, indicating the failure of conventional wastewater treatments for removing estrogens [98].

The chemical properties of hormones with their occurrence in wastewater and related removal by CAS reactors are summarized in Table 2.

It is evident from Table 2 that hormones have been successfully removed from wastewaters even without using membrane filtration. In addition, their concentrations are quite low and easily removed by sorption [99], for their removal leads to good efficiencies. Only octylphenol was not well removed by conventional treatments, but there are evidence for each reported compound for achieving good removal rates just by using biological processes.

Table 2. Chemical properties of hormones and their occurrence in wastewaters.

Common Name	Use	Formula	Molecular Weight	CAS	Concentration in WWs, μg/L	Conventional AS efficiency [%]	Ref.
17β-estradiol	Sex hormone	$C_{18}H_{24}O_2$	272,38	50-28-2	0,01 - 0,03	85 - 95	[24,100,101]
17-β-estradiol – 17 acetate	Steroid hormone	$(C_{20}H_{26}O_3)$	314,42	1743-60-8	-	-	[20]
17-α ethinylestradiol	Steroidal hormone	$(C_{20}H_{24}O_2)$	269,40	57-63-6	-	-	[20]
Androstenedione	Steroid hormone	$C_{19}H_{26}O_2$	286,4	63-05-8	0,02 - 0,07	71 - 95	[24,101]
Androsterone	Steroid hormone	$C_{19}H_{30}O_2$	290,44	53-41-8	0,12 - 0,3	90 - 95	[101]
Bisphenol a	Industrial compound w/ estrogenic activity	$C_{15}H_{16}O_2$	228,29	80-05-7	0,2 - 2,5	75 - 90	[100,102,103]
Estriol	Estrogenic hormone	$C_{18}H_{24}O_3$	288,38	50-27-1	0,06 - 0,3	40 - 95	[77,100,102]
Estrone	Estrogenic hormone	$C_{18}H_{22}O_2$	270,366	53-16-7	0,03 - 0,2	80 - 90	[77,100,102]
Ethynylestradiol	Bio-active estrogen	$C_{20}H_{24}O_2$	296,403	57-63-6	0,002 - 0,006	50 - 90	[77,100,104]
Etiocholanolone	Testosterone metabolite	$C_{19}H_{30}O_2$	290,445	53-42-9	0,2 - 0,3	95 - 99	[105]
Nonylphenol	Industrial compound w/ estrogenic activity	$C_{15}H_{24}O$	220,35	25154-52-3	0,09 - 10	60 - 91	[100,102,106]
Octylphenol	Industrial compound w/ estrogenic activity	$C_{14}H_{22}0$	206,36	27193-28-8	0,05 - 4	30 - 65	[100,102]
Testosterone	Steroid hormone	$C_{19}H_{28}O_2$	288,42	58-22-0	0,005 - 0,3	70 - 80	[24,101]

4 ECS removal by MBRs

In the following paragraphs, the performance of MBRs in removing ECs from both real and synthetic wastewater is discussed.

4.1 Removal of pesticides

Pesticides contained in wastewater have been partially removed by MBRs [107]. However, these compounds are quite persistent and somewhere recalcitrant to membrane filtration. Instead of pharmaceuticals, where much research has already been done, pesticides have not deeply analyzed. As stated, pesticides are often unlikely to be removed by MBRs, as shown in Table 3.

Table 3. Pesticides removal efficiencies by MBRs

Compound	MBR system	Membrane porosity	Initial concentration [μg/L]	Sample	Scale	Removal [%]	References
2,4-D	FS	UF	10	R	P	90	[17]
2,4-DP	FS	UF	10	R	P	82	[17]
4-chlorophenol	HF	UF	50000	S	L	99	[108]
Atrazine	HF - FS	MF - UF	0,024 - 200	R - S	L - P	2 - 16	[79,109–112]
	-	MF	5	S	L	> 30	[78]
	HF	UF	5	S	L	< 10	[16]
Ametryn	-	MF	5	S	L	> 90	[78]
	HF	UF	5	S	L	< 80	[16]
	HF	MF	840 - 1351	S	L	36 - 66	[27]
Bayrepel acid	FS	MF	0,403	R	L	93	[79]
Bentazone	FS	MF	0,03	R	L	16	[79]
Clofibric acid	-	MF	5	S	L	> 80	[78]
	HF	UF	5	S	L	< 20	[16]
DEET	HF - FS	MF - UF	0,018 - 2	R - S	L - P	5 - 62	[109,79,87,112]
Fenoprop	HF	MF	5	S	L	< 20	[20]
	-	MF	5	S	L	> 80	[78]
	HF	UF	5	S	L	< 40	[16]
Isoproturon	FS	MF	0,038	R	L	25	[79]
Linuron	HF	UF	1,4 - 2	S	L	19 - 21	[109,111,112]
MCPA	FS	UF	10	R	P	93	[17]
MCPP	FS	MF - UF	0,092 - 10	R	L - P	50 - 97	[17,79]
Pentachlorophenol	HF	-	12000	S	P	99	[113]
	HF	MF	5	S	L	< 60	[20]
	-	MF	5	S	L	< 90	[78]
	HF	UF	5	S	L	< 60	[16]
Propoxur	-	MF	5	S	L	< 60	[78]
	HF	UF	5	S	L	< 20	[16]

HF: Hollow fibre; FS: Flat sheet; UF: Ultrafiltration; MF: Microfiltration; S: Synthetic wastewater; R: Real wastewater; L: Laboratory scale; P: Pilot-scale; F: Full scale.

As evident from data presented in Table 3, about 50% of reported works expressed a substantial inability in removing pesticides from both synthetic and real wastewater (removal efficiency < 20%). A possible explanation of poor performance of CAS treatments in removing pesticides may be the lack of complete biodegradability of pesticides. On the other hand, MBRs may not be able to remove these because of low zeta potential of wastewater, leading to poor aggregation of molecules.

Further research is needed in order to test MBRs with a wider range of pesticides, trying to introduce adequate pretreatments to achieve physical or chemical degradation of recalcitrant compounds, before of the MBR step.

In Fig. 4, information about the performance of different configurations of MBR$_S$ are shown.

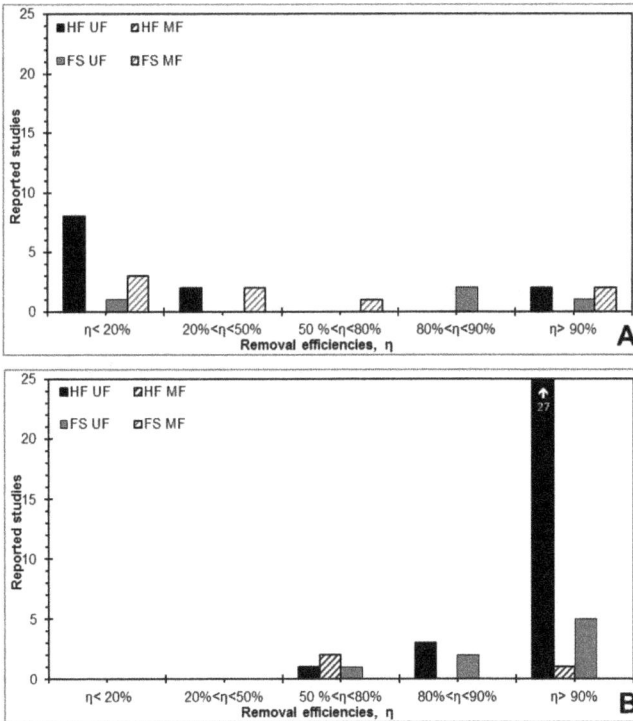

Figure 4. Removal efficiencies of: A-pesticides and B-hormones by MBRs, related to reported studies presented in this paper.

Only 5 reports showed removal efficiencies exceeding 90%, equally distributed between HF and FS configurations, when degrading 4-chlorophenol, aldicarb, Bayrepel acid, MCPA, MCPP (only at high concentrations) and pentachlorophenol. Atrazine, a chloro-triazine herbicide, was removed at very low rates [111, 112]. It has been reported to be poorly removed both in CAS reactors and MBRs [79] and that a major removal mechanism was sorption onto withdrawn sludge [110]. Linuron, a dichloro-phenylurea herbicide, is widely used herbicide but very few reports on its removal of linuron in CAS reactors or MBRs could be found [109,112]. However, its slow natural attenuation rate in various soils and the evolution of more toxic and persistent chloroaniline intermediates in the process have been reported [114]. Again, very low removal rates were achieved by MBRs with regard to linuron. Diethyltoluamide (DEET) is a widely used toluamide compound and is the most common active ingredient in insect repellants: its achieved maximum removal rates of 18% is in agreement with previous studies [3,87].

In conclusion, MBRs are not very effective in removing pesticides from wastewater, even if they perform slightly better than conventional treatments. However, different solutions should be defined when treating pesticide-rich wastewaters.

4.2   Removal of hormones

Hormones and other compounds with estrogen activity are almost always degraded with excellent efficiencies, both with hollow fibre (HF) or flat sheet (FS), and ultrafiltration or microfiltration. It must be said that CAS reactors were already able to remove these compounds at high rates, as previously stated.

In Table 4 data about MBR-degraded hormones are presented which indicate that only 3 compounds (nonylphenol, estrone, and ethinylestradiol) are degraded – the last two only at very low concentrations with efficiencies lower than 80%. The reports show that the removal efficiencies of hormones are higher than 50%.

Therefore, MBRs are very effective against hormones and can be regarded as an elective treatment; it must be noted that only a few works dealt with real wastewater, while the majority of studies applied MBRs to synthetic wastewater, probably because of high complexity of real wastewater matrices and related synergetic effects.

From Fig. 4 it is clear that a significant share of presented works used hollow fiber membranes, with porosity in the ultrafiltration range; high removal rates indicate MBRs as a viable treatment option.

Table 4. Hormones removal efficiencies by MBRs.

Compound	MBR system	Membrane porosity	Initial concentration, [µg/L]	Sample	Scale	Removal, [%]	Ref
17β-estradiol	HF	MF-UF	0,017 - 5	R - S	L - P - F	90-99	[16,20,78,109,112,115,116]
17-β-estradiol – 17 acetate	HF	MF-UF	5	S	L	97-99	[16,20,78]
17-α ethinylestradiol	HF	MF-UF	5	S	L	97-99	[16,20,78]
Androstenedione	HF - FS	MF-UF	0,1 - 2,1	R - S	L - P - F	93 - 99	[87,109,112,116]
Androsterone	HF	MF-UF	1 - 2	S	L - F	99	[109,112,116]
Bisphenol a	HF - FS	MF - UF	0,09 - 1,8	S	L - P - F	90 - 99	[109,115–117]
Estriol	HF - FS	MF - UF	0,318 - 5	R - S	L – P-F	98 - 99	[16,87,109,112,116]
Estrone	HF - FS	MF - UF	0,02 - 2,4 - 5	R - S	L - P - F	64 - 99	[16,20,78,109,112,115–117]
Ethynylestradiol	HF	MF - UF	0,001 - 3,6	S	L - P	71 - 94	[109,112,115]
Etiocholanolone	HF	UF	1,75 - 2	S	L	99	[109,112]
Nonylphenol	HF - FS	MF - UF	1,65 - 3,2	S	L - P	55 - 99	[109,115,117]
Octylphenol	HF - FS	UF	0,024 - 2,7	S	L - P	95 - 99	[109,115,117]
Testosterone	HF - FS	MF-UF	0,06 - 2	R - S	L - P - F	83 - 99	[87,109,112,116]

HF: Hollow Fibre; FS: Flat Sheet; UF: Ultrafiltration; MF: Microfiltration; S: Synthetic Wastewater; R: Real Wastewater; L: Laboratory scale; P: Pilot scale; F: Full scale.

This good performance may probably be due to faster biological transformation, which starts almost immediately, meaning that lag phases and the time required for expressing specific enzymes are negligible [64]. It is noteworthy that all of these compounds possess significant hydrophobicity and bear similar molecular backbone structures [109], which may, in part, explain the similarities of their removal efficiencies.

MBR technology is often considered a promising development in wastewater treatment which combines biological degradation with membrane filtration, and indeed steroids removal rates of greater than 90% were achieved in MBRs. However, enough information suggests that CAS treatments are already able to remove many hormones and other endocrine disruptors from wastewaters, probably because of the biological degradation achieved in the oxidation tank. Biological degradation has been cited as the most important factor in the removal of estrogens and other endocrine disrupters in membrane bioreactor [118]. While microfiltration membranes themselves are not able to enhance the degree of estrogens removal, it has been reported that estrogens adsorption to particulate matter that is retained by the membrane could reduce estrogens concentration in the effluent [119]. Some researchers have found that microfiltration membranes were able to display some retention of smaller particles or colloidal material onto which estrogens may adsorb [118]. Since pore sizing of membrane material is not uniform between manufacturers, it is possible that a difference in membrane material may explain some of the discrepancies in colloid retention [119]. Improved removals exhibited by these systems have also been attributed to low sludge load amongst the previously mentioned factors i.e. high sludge retention time (SRT) and hydraulic retention time (HRT) [120].

In conclusion, conventional treatments are already able to remove hormones and other endocrine disruptors, but the implementation of MBR processes would lead without any doubt to an increase in the removal rates, as it happened with ethinylestradiol, octylphenol and testosterone.

5. Concluding remarks

The adverse effects of ECs to ecosystems, specifically to aquatic life and their correspondence to humans, have been highlighted by a number of studies. Their impacts on health and environment have urged the removal of these compounds, not only from drinking water but also from wastewater treatment processes to avoid the release into surface and ground waters. The aim of this chapter was to evaluate the occurrence of pesticides and hormones as ECs in wastewater and to examine their fate into the environment. Moreover, the applications of conventional treatments and MBRs to wastewater in order to remove ECs, focusing on pesticides and hormones were discussed. Finally, an effort was made to evaluate the performance of different MBR configurations and to summarize environmental effects of ECs. This chapter underlines that MBRs generally outperform CAS treatments in removing ECs from wastewater. MBRs offer advantages compared to conventional systems: membranes allow the detention of particulate matter leading to an effluent free of suspended solids. MBRs achieve high SRTs within compact reactor volumes and as degradation is a function of the operated SRT, this fact representing another advantage of MBRs in comparison to conventional systems in relation to ECs removal rates. Especially in regions with no suitable receiving waters or where a reuse of the treated wastewater is planned, MBRs represent an attractive solution due to many advantages.

There is a lack of regulation on ECs, therefore there is no driving force in removing these from wastewater; threshold values must be defined for many ECs both for discharge and reclamation, hence even for reclaimed wastewater, many ECs are not limited.

The occurrence of ECs in municipal wastewater is a relevant topic among the scientific community in recent years, and in the present chapter concentrations which range between few nanograms per liter and several micrograms per liter were reported.

Additional research is thus required firstly to recognize ECs transformation byproducts and secondly to assess the effect of these micropollutants on the environment. Several hybrid processes need to be investigated, such as MBR-activated carbon, MBR-AOP or MBR-RO, but the economic sustainability of these processes must be analyzed in order to compare removal rates as well as operation and maintenance costs of each treatment solution. Ecotoxicological tests are needed to define effluent toxicity: there is a lack of knowledge regarding environmental effects of byproducts and synergistic effects of complex mixtures.

References

[1]   N. Bolong, A.F. Ismail, M.R. Salim, T. Matsuura, A review of the effects of emerging contaminants in wastewater and options for their removal, Desalination 239 (2009) 229–246. https://doi.org/10.1016/j.desal.2008.03.020

[2]   K.E. Murray, S.M. Thomas, A.A. Bodour, Prioritizing research for trace pollutants and emerging contaminants in the freshwater environment, Environ. Pollut. 158 (2010) 3462–3471. https://doi.org/10.1016/j.envpol.2010.08.009

[3]   Q. Sui, J. Huang, S. Deng, G. Yu, Q. Fan, Occurrence and removal of pharmaceuticals, caffeine and DEET in wastewater treatment plants of Beijing, China, Water Res. 44 (2010) 417–426. https://doi.org/10.1016/j.watres.2009.07.010

[4]   M. Macova, B.I. Escher, J. Reungoat, S. Carswell, K.L. Chue, J. Keller, J.F. Mueller, Monitoring the biological activity of micropollutants during advanced wastewater treatment with ozonation and activated carbon filtration, Water Res. 44 (2010) 477–492. https://doi.org/10.1016/j.watres.2009.09.025

[5]   Z. Liu, Y. Kanjo, S. Mizutani, A review of phytoestrogens: Their occurrence and fate in the environment, Water Res. 44 (2010) 567–577. https://doi.org/10.1016/j.watres.2009.03.025

[6]   R. Rosal, A. Rodríguez, J.A. Perdigón-Melón, A. Petre, E. García-Calvo, M.J. Gómez, A. Agüera, A.R. Fernández-Alba, Occurrence of emerging pollutants in urban wastewater and their removal through biological treatment followed by ozonation, Water Res. 44 (2010) 578–588. https://doi.org/10.1016/j.watres.2009.07.004

[7]   J.L. Oaks, M. Gilbert, M.Z. Virani, R.T. Watson, C.U. Meteyer, B.A. Rideout, H.L. Shivaprasad, S. Ahmed, M.J. Iqbal Chaudhry, M. Arshad, S. Mahmood, A. Ali, A. Ahmed Khan, Diclofenac residues as the cause of vulture population decline in Pakistan, Nature. 427 (2004) 630–633. https://doi.org/10.1038/nature02317

[8]   C.J. Houtman, Emerging contaminants in surface waters and their relevance for the production of drinking water in Europe, J. Integr. Environ. Sci. 7 (2010) 271–295. https://doi.org/10.1080/1943815X.2010.511648

[9]   C. Adams, Y. Wang, K. Loftin, M. Meyer, Removal of antibiotics from surface and distilled water in conventional water treatment processes, J. Environ. Eng. 128 (2002) 8. https://doi.org/10.1061/(ASCE)0733-9372(2002)128:3(253)

[10]   A.A. Khan, Inamuddin, M.M. Alam, Determination and separation of Pb2+ from aqueous solutions using a fibrous type organic–inorganic hybrid cation-exchange material: Polypyrrole thorium(IV) phosphate, React. Funct. Polym. 63 (2005) 119–133. https://doi.org/10.1016/j.reactfunctpolym.2005.02.001

[11]   L.N. Nguyen, F.I. Hai, S. Yang, J. Kang, F.D.L. Leusch, F. Roddick, W.E. Price, L.D. Nghiem, Removal of trace organic contaminants by an MBR comprising a mixed culture of bacteria and white-rot fungi, Bioresour. Technol. 148 (2013) 234–241. https://doi.org/10.1016/j.biortech.2013.08.142

[12]   I. Oller, S. Malato, J.A. Sánchez-Pérez, Combination of Advanced Oxidation Processes and biological treatments for wastewater decontamination—A review, Sci. Total Environ. 409 (2011) 4141–4166. https://doi.org/10.1016/j.scitotenv.2010.08.061

[13]   L.A. Pérez-Estrada, S. Malato, A. Agüera, A.R. Fernández-Alba, Degradation of dipyrone and its main intermediates by solar AOPs: Identification of intermediate products and toxicity assessment, Catal. Today. 129 (2007) 207–214. https://doi.org/10.1016/j.cattod.2007.08.008

[14] W. De Schepper, J. Dries, L. Geuens, J. Robbens, R. Blust, Conventional and (eco) toxicological assessment of batch partial ozone oxidation and subsequent biological treatment of a tank truck cleaning generated concentrate, Water Res. 43 (2009) 4037–4049. https://doi.org/10.1016/j.watres.2009.06.014

[15] A. Santos, W. Ma, S.J. Judd, Membrane bioreactors: Two decades of research and implementation, Desalination. 273 (2011) 148–154. https://doi.org/10.1016/j.desal.2010.07.063

[16] H.V. Phan, F.I. Hai, J. Kang, H.K. Dam, R. Zhang, W.E. Price, A. Broeckmann, L.D. Nghiem, Simultaneous nitrification/denitrification and trace organic contaminant (TrOC) removal by an anoxic–aerobic membrane bioreactor (MBR), Bioresour. Technol. 165 (2014) 96–104. https://doi.org/10.1016/j.biortech.2014.03.094

[17] S. González, M. Petrovic, D. Barceló, Removal of a broad range of surfactants from municipal wastewater – Comparison between membrane bioreactor and conventional activated sludge treatment, Chemosphere. 67 (2007) 335–343. https://doi.org/10.1016/j.chemosphere.2006.09.056

[18] P. Verlicchi, A. Galletti, M. Petrovic, D. Barceló, Hospital effluents as a source of emerging pollutants: An overview of micropollutants and sustainable treatment options, J. Hydrol. 389 (2010) 416–428. https://doi.org/10.1016/j.jhydrol.2010.06.005

[19] M. Petrović, S. Gonzalez, D. Barceló, Analysis and removal of emerging contaminants in wastewater and drinking water, TrAC Trends Anal. Chem. 22 (2003) 685–696. https://doi.org/10.1016/S0165-9936(03)01105-1

[20] L.N. Nguyen, F.I. Hai, J. Kang, W.E. Price, L.D. Nghiem, Removal of emerging trace organic contaminants by MBR-based hybrid treatment processes, Int. Biodeterior. Biodegrad. 85 (2013) 474–482. https://doi.org/10.1016/j.ibiod.2013.03.014

[21] O.T. Komesli, M. Muz, M.S. Ak, S. Bakırdere, C.F. Gökçay, Comparison of EDCs removal in full and pilot scale membrane bioreactor plants: Effect of flux rate on EDCs removal in short SRT, J. Environ. Manage. 2013 (2017) 847-852. https://doi.org/10.1016/j.jenvman.2016.06.004

[22] M. Schriks, M.B. Heringa, M.M.E. van der Kooi, P. de Voogt, A.P. van Wezel, Toxicological relevance of emerging contaminants for drinking water quality, Water Res. 44 (2010) 461–476. https://doi.org/10.1016/j.watres.2009.08.023

[23] H. Yamamoto, Y. Nakamura, S. Moriguchi, Y. Nakamura, Y. Honda, I. Tamura, Y. Hirata, A. Hayashi, J. Sekizawa, Persistence and partitioning of eight selected pharmaceuticals in the aquatic environment: Laboratory photolysis, https://doi.org/10.1016/j.watres.2008.10.039biodegradation, and sorption experiments, Water Res. 43 (2009) 351–362.

[24] S.A. Snyder, S. Adham, A.M. Redding, F.S. Cannon, J. DeCarolis, J. Oppenheimer, E.C. Wert, Y. Yoon, Role of membranes and activated carbon in the removal of endocrine disruptors and pharmaceuticals, Desalination. 202 (2007) 156–181. https://doi.org/10.1016/j.desal.2005.12.052

[25] W.H. Organization, others, The world medicines situation, (2004). http://apps.who.int/iris/handle/10665/68735 (accessed March 27, 2017).

[26] J. Jiang, N. Wan, A model for ecological assessment to pesticide pollution management, Ecol. Model. 220 (2009) 1844–1851. https://doi.org/10.1016/j.ecolmodel.2009.04.028

[27] D. Navaratna, L. Shu, V. Jegatheesan, Evaluation of herbicide (persistent pollutant) removal mechanisms through hybrid membrane bioreactors, Bioresour. Technol. 200 (2016) 795–803. https://doi.org/10.1016/j.biortech.2015.10.041

[28] R. Das, A. Steege, S. Baron, J. Beckman, R. Harrison, Pesticide-related illness among migrant farm workers in the United States, Int. J. Occup. Environ. Health. 7 (2001) 303–312. https://doi.org/10.1179/oeh.2001.7.4.303

[29] N. Yang, M. Matsuda, M. Kawano, T. Wakimoto, PCBs and organochlorine pesticides (OCPs) in edible fish and shellfish from China, Chemosphere. 63 (2006) 1342–1352. https://doi.org/10.1016/j.chemosphere.2005.09.029

[30] World Health Organization, ed., Guidelines for drinking-water quality, 4th ed, World Health Organization, Geneva, 2011.

[31] E.D. Ongley, F. and A.O. of the U. Nations, Control of Water Pollution from Agriculture, Food & Agriculture Org., 1996.

[32] Y.J. Wang, J.K. Lin, Estimation of selected phenols in drinking water with in situ acetylation and study on the DNA damaging properties of polychlorinated phenols, Arch. Environ. Contam. Toxicol. 28 (1995) 537–542. https://doi.org/10.1007/BF00211639

[33] Y.-J. Wang, M.-C. Yang, M.-H. Pan, Dihydrolipoic acid inhibits tetrachlorohydroquinone-induced tumor promotion through prevention of

oxidative damage, Food Chem. Toxicol. Int. J. Publ. Br. Ind. Biol. Res. Assoc. 46 (2008) 3739–3748. https://doi.org/10.1016/j.fct.2008.09.064

[34] M.R. Milnes, D.S. Bermudez, T.A. Bryan, T.M. Edwards, M.P. Gunderson, I.L.V. Larkin, B.C. Moore, L.J. Guillette Jr., Contaminant-induced feminization and demasculinization of nonmammalian vertebrate males in aquatic environments, Environ. Res. 100 (2006) 3–17. https://doi.org/10.1016/j.envres.2005.04.002

[35] J.P. Sumpter, Feminized responses in fish to environmental estrogens, Toxicol. Lett. 82–83 (1995) 737–742. https://doi.org/10.1016/0378-4274(95)03517-6

[36] A.M. Vajda, L.B. Barber, J.L. Gray, E.M. Lopez, A.M. Bolden, H.L. Schoenfuss, D.O. Norris, Demasculinization of male fish by wastewater treatment plant effluent, Aquat. Toxicol. 103 (2011) 213–221. https://doi.org/10.1016/j.aquatox.2011.02.007

[37] J. Gibs, P.E. Stackelberg, E.T. Furlong, M. Meyer, S.D. Zaugg, R.L. Lippincott, Persistence of pharmaceuticals and other organic compounds in chlorinated drinking water as a function of time, Sci. Total Environ. 373 (2007) 240–249. https://doi.org/10.1016/j.scitotenv.2006.11.003

[38] J. CHEN, C. LIU, Z. YANG, J. WANG, Residues and characteristics of organochlorine pesticides in the surface water in the suburb of Beijing, Earth Sci. Front. 15 (2008) 242–247. https://doi.org/10.1016/S1872-5791(09)60005-1

[39] S. Judd, The MBR Book: Principles and Applications of Membrane Bioreactors for Water and Wastewater Treatment, Elsevier, 2011.

[40] L. Innocenti, D. Bolzonella, P. Pavan, F. Cecchi, Effect of sludge age on the performance of a membrane bioreactor: influence on nutrient and metals removal, Desalination 146 (2002) 467–474. https://doi.org/10.1016/S0011-9164(02)00551-9

[41] C. Brepols, E. Dorgeloh, F.-B. Frechen, W. Fuchs, S. Haider, A. Joss, K. de Korte, C. Ruiken, W. Schier, H. van der Roest, M. Wett, T. Wozniak, Upgrading and retrofitting of municipal wastewater treatment plants by means of membrane bioreactor (MBR) technology, Desalination. 231 (2008) 20–26. https://doi.org/10.1016/j.desal.2007.11.035

[42] A. Drews, Membrane fouling in membrane bioreactors—Characterisation, contradictions, cause and cures, J. Membr. Sci. 363 (2010) 1–28. https://doi.org/10.1016/j.memsci.2010.06.046

[43] A.N.L. Ng, A.S. Kim, A mini-review of modeling studies on membrane bioreactor (MBR) treatment for municipal wastewaters, Desalination 212 (2007) 261–281. https://doi.org/10.1016/j.desal.2006.10.013

[44] F. Zanetti, G. De Luca, R. Sacchetti, Performance of a full-scale membrane bioreactor system in treating municipal wastewater for reuse purposes, Bioresour. Technol. 101 (2010) 3768–3771. https://doi.org/10.1016/j.biortech.2009.12.091

[45] J. Arévalo, G. Garralón, F. Plaza, B. Moreno, J. Pérez, M.Á. Gómez, Wastewater reuse after treatment by tertiary ultrafiltration and a membrane bioreactor (MBR): a comparative study, Desalination. 243 (2009) 32–41. https://doi.org/10.1016/j.desal.2008.04.013

[46] J. Ottoson, A. Hansen, B. Björlenius, H. Norder, T.A. Stenström, Removal of viruses, parasitic protozoa and microbial indicators in conventional and membrane processes in a wastewater pilot plant, Water Res. 40 (2006) 1449–1457. https://doi.org/10.1016/j.watres.2006.01.039

[47] M.D. Williams, M. Pirbazari, Membrane bioreactor process for removing biodegradable organic matter from water, Water Res. 41 (2007) 3880–3893. https://doi.org/10.1016/j.watres.2007.06.010

[48] E. Marti, H. Monclús, J. Jofre, I. Rodriguez-Roda, J. Comas, J.L. Balcázar, Removal of microbial indicators from municipal wastewater by a membrane bioreactor (MBR), Bioresour. Technol. 102 (2011) 5004–5009. https://doi.org/10.1016/j.biortech.2011.01.068

[49] F.J. Simmons, D.H.-W. Kuo, I. Xagoraraki, Removal of human enteric viruses by a full-scale membrane bioreactor during municipal wastewater processing, Water Res. 45 (2011) 2739–2750. https://doi.org/10.1016/j.watres.2011.02.001

[50] I.-S. Chang, P. Le Clech, B. Jefferson, S. Judd, Membrane fouling in membrane bioreactors for wastewater treatment, J. Environ. Eng. 128 (2002) 1018–1029. https://doi.org/10.1061/(ASCE)0733-9372(2002)128:11(1018)

[51] F. Meng, S.-R. Chae, A. Drews, M. Kraume, H.-S. Shin, F. Yang, Recent advances in membrane bioreactors (MBRs): Membrane fouling and membrane material, Water Res. 43 (2009) 1489–1512. https://doi.org/10.1016/j.watres.2008.12.044

[52] M. Henze, M.C.M. van Loosdrecht, G.A. Ekama, D. Brdjanovic, Biological Wastewater Treatment, IWA Publishing, 2008.

[53]   P. Schoeberl, M. Brik, M. Bertoni, R. Braun, W. Fuchs, Optimization of operational parameters for a submerged membrane bioreactor treating dyehouse wastewater, Sep. Purif. Technol. 44 (2005) 61–68. https://doi.org/10.1016/j.seppur.2004.12.004

[54]   A.L. Lim, R. Bai, Membrane fouling and cleaning in microfiltration of activated sludge wastewater, J. Membr. Sci. 216 (2003) 279–290. https://doi.org/10.1016/S0376-7388(03)00083-8

[55]   H. Choi, K. Zhang, D.D. Dionysiou, D.B. Oerther, G.A. Sorial, Influence of cross-flow velocity on membrane performance during filtration of biological suspension, J. Membr. Sci. 248 (2005) 189–199. https://doi.org/10.1016/j.memsci.2004.08.027

[56]   H.C. Chua, T.C. Arnot, J.A. Howell, Controlling fouling in membrane bioreactors operated with a variable throughput, Desalination. 149 (2002) 225–229. https://doi.org/10.1016/S0011-9164(02)00764-6

[57]   S.P. Hong, T.H. Bae, T.M. Tak, S. Hong, A. Randall, Fouling control in activated sludge submerged hollow fiber membrane bioreactors, Desalination 143 (2002) 219–228. https://doi.org/10.1016/S0011-9164(02)00260-6

[58]   I.-S. Chang, S.-O. Bag, C.-H. Lee, Effects of membrane fouling on solute rejection during membrane filtration of activated sludge, Process Biochem. 36 (2001) 855–860. https://doi.org/10.1016/S0032-9592(00)00284-3

[59]   T. Ueda, K. Hata, Y. Kikuoka, Treatment of domestic sewage from rural settlements by a membrane bioreactor, Water Sci. Technol. 34 (1996) 189–196.

[60]   P. Côté, H. Buisson, M. Praderie, Immersed membranes activated sludge process applied to the treatment of municipal wastewater, Water Sci. Technol. 38 (1998) 437–442.

[61]   K.-H. Ahn, K.-G. Song, E. Choa, J. Cho, H. Yun, S. Lee, J. Me, Enhanced biological phosphorus and nitrogen removal using a sequencing anoxic/anaerobic membrane bioreactor (SAM) process, Desalination 157 (2003) 345–352. https://doi.org/10.1016/S0011-9164(03)00415-6

[62]   S.L. Khor, D.D. Sun, Y. Liu, J.O. Leckie, Biofouling development and rejection enhancement in long SRT MF membrane bioreactor, Process Biochem. 42 (2007) 1641–1648. https://doi.org/10.1016/j.procbio.2007.09.009

[63]   H. Monclús, J. Sipma, G. Ferrero, I. Rodriguez-Roda, J. Comas, Biological nutrient removal in an MBR treating municipal wastewater with special focus on

biological phosphorus removal, Bioresour. Technol. 101 (2010) 3984–3991. https://doi.org/10.1016/j.biortech.2010.01.038

[64]   C. Abegglen, A. Joss, C.S. McArdell, G. Fink, M.P. Schlüsener, T.A. Ternes, H. Siegrist, The fate of selected micropollutants in a single-house MBR, Water Res. 43 (2009) 2036–2046. https://doi.org/10.1016/j.watres.2009.02.005

[65]   P. Battistoni, F. Fatone, D. Bolzonella, P. Pavan, Full scale application of the coupled alternate cycles-membrane bioreactor (AC-MBR) process for wastewater reclamation and reuse, Water Pract. Technol. 1 (2006) wpt2006077. https://doi.org/10.2166/wpt.2006.077

[66]   G. Li, L. Wu, C. Dong, G. Wu, Y. Fan, Inorganic nitrogen removal of toilet wastewater with an airlift external circulation membrane bioreactor, J. Environ. Sci. China. 19 (2007) 12–17. https://doi.org/10.1016/S1001-0742(07)60002-3

[67]   A.M. Rodrigues, V. Ferreira, V.V. Cardoso, E. Ferreira, M.J. Benoliel, Determination of several pesticides in water by solid-phase extraction, liquid chromatography and electrospray tandem mass spectrometry, J. Chromatogr. A. 1150 (2007) 267–278. https://doi.org/10.1016/j.chroma.2006.09.083

[68]   M. Kuster, M.J. López de Alda, M.D. Hernando, M. Petrovic, J. Martín-Alonso, D. Barceló, Analysis and occurrence of pharmaceuticals, estrogens, progestogens and polar pesticides in sewage treatment plant effluents, river water and drinking water in the Llobregat river basin (Barcelona, Spain), J. Hydrol. 358 (2008) 112–123. https://doi.org/10.1016/j.jhydrol.2008.05.030

[69]   R. Loos, B.M. Gawlik, G. Locoro, E. Rimaviciute, S. Contini, G. Bidoglio, EU-wide survey of polar organic persistent pollutants in European river waters, Environ. Pollut. 157 (2009) 561–568. https://doi.org/10.1016/j.envpol.2008.09.020

[70]   P. Palma, M. Kuster, P. Alvarenga, V.L. Palma, R.M. Fernandes, A.M.V.M. Soares, M.J. López de Alda, D. Barceló, I.R. Barbosa, Risk assessment of representative and priority pesticides, in surface water of the Alqueva reservoir (South of Portugal) using on-line solid phase extraction-liquid chromatography-tandem mass spectrometry, Environ. Int. 35 (2009) 545–551. https://doi.org/10.1016/j.envint.2008.09.015

[71]   Z. Vryzas, G. Vassiliou, C. Alexoudis, E. Papadopoulou-Mourkidou, Spatial and temporal distribution of pesticide residues in surface waters in northeastern Greece, Water Res. 43 (2009) 1–10. https://doi.org/10.1016/j.watres.2008.09.021

[72] M. Babut, B. Corinne, B. Marc, F. Patrick, G. Jeanne, G. Geneviève, Developing environmental quality standards for various pesticides and priority pollutants for French freshwaters, J. Environ. Manage. 69 (2003) 139–147. https://doi.org/10.1016/S0301-4797(03)00133-6

[73] H. Chin, P. Elefsiniotis, N. Singhal, Biodegradation of 2,4-dicholophenoxyacetic acid using an acidogenic anaerobic sequencing batch reactor, J. Environ. Eng. Sci. 4 (2005) 57–63. https://doi.org/10.1139/s04-044

[74] H.R. Rogers, Sources, behaviour and fate of organic contaminants during sewage treatment and in sewage sludges, Sci. Total Environ. 185 (1996) 3–26. https://doi.org/10.1016/0048-9697(96)05039-5

[75] P.B. Kurt-Karakus, T.F. Bidleman, D.C.G. Muir, J. Struger, E. Sverko, S.J. Cagampan, J.M. Small, L.M. Jantunen, Comparison of concentrations and stereoisomer ratios of mecoprop, dichlorprop and metolachlor in Ontario streams, 2006-2007 vs. 2003-2004, Environ. Pollut. Barking Essex 1987. 158 (2010) 1842–1849. https://doi.org/10.1016/j.envpol.2009.11.003

[76] G. Teijon, L. Candela, K. Tamoh, A. Molina-Díaz, A.R. Fernández-Alba, Occurrence of emerging contaminants, priority substances (2008/105/CE) and heavy metals in treated wastewater and groundwater at Depurbaix facility (Barcelona, Spain), Sci. Total Environ. 408 (2010) 3584–3595. https://doi.org/10.1016/j.scitotenv.2010.04.041

[77] S.A. Snyder, P. Westerhoff, Y. Yoon, D.L. Sedlak, Pharmaceuticals, Personal Care Products, and Endocrine Disruptors in Water: Implications for the Water Industry, Environ. Eng. Sci. 20 (2003) 449–469. https://doi.org/10.1089/109287503768335931

[78] K.C. Wijekoon, F.I. Hai, J. Kang, W.E. Price, W. Guo, H.H. Ngo, L.D. Nghiem, The fate of pharmaceuticals, steroid hormones, phytoestrogens, UV-filters and pesticides during MBR treatment, Bioresour. Technol. 144 (2013) 247–254. https://doi.org/10.1016/j.biortech.2013.06.097

[79] M. Bernhard, J. Müller, T.P. Knepper, Biodegradation of persistent polar pollutants in wastewater: Comparison of an optimised lab-scale membrane bioreactor and activated sludge treatment, Water Res. 40 (2006) 3419–3428. https://doi.org/10.1016/j.watres.2006.07.011

[80] M. Bach, M. Letzel, U. Kaul, S. Forstner, G. Metzner, J. Klasmeier, S. Reichenberger, H.G. Frede, Measurement and modeling of bentazone in the river

Main (Germany) originating from point and non-point sources, Water Res. 44 (2010) 3725–3733. https://doi.org/10.1016/j.watres.2010.04.010

[81]   H. Singer, S. Jaus, I. Hanke, A. Lück, J. Hollender, A.C. Alder, Determination of biocides and pesticides by on-line solid phase extraction coupled with mass spectrometry and their behaviour in wastewater and surface water, Environ. Pollut. Barking Essex 1987. 158 (2010) 3054–3064. https://doi.org/10.1016/j.envpol.2010.06.013

[82]   A. Wick, G. Fink, A. Joss, H. Siegrist, T.A. Ternes, Fate of beta blockers and psycho-active drugs in conventional wastewater treatment, Water Res. 43 (2009) 1060–1074. https://doi.org/10.1016/j.watres.2008.11.031

[83]   D.R. Buhler, M.E. Rasmusson, H.S. Nakaue, Occurrence of hexachlorophene and pentachlorophenol in sewage and water, Environ. Sci. Technol. 7 (1973) 929–934. https://doi.org/10.1021/es60082a006

[84]   E. Stumm-Zollinger, G.M. Fair, Biodegradation of Steroid Hormones, J. Water Pollut. Control Fed. 37 (1965) 1506–1510.

[85]   C.E. Purdom, P.A. Hardiman, V.V.J. Bye, N.C. Eno, C.R. Tyler, J.P. Sumpter, Estrogenic Effects of Effluents from Sewage Treatment Works, Chem. Ecol. 8 (1994) 275–285. https://doi.org/10.1080/02757549408038554

[86]   J. Zha, L. Sun, Y. Zhou, P.A. Spear, M. Ma, Z. Wang, Assessment of 17α-ethinylestradiol effects and underlying mechanisms in a continuous, multigeneration exposure of the Chinese rare minnow (Gobiocypris rarus), Toxicol. Appl. Pharmacol. 226 (2008) 298–308. https://doi.org/10.1016/j.taap.2007.10.006

[87]   S.D. Kim, J. Cho, I.S. Kim, B.J. Vanderford, S.A. Snyder, Occurrence and removal of pharmaceuticals and endocrine disruptors in South Korean surface, drinking, and waste waters, Water Res. 41 (2007) 1013–1021. https://doi.org/10.1016/j.watres.2006.06.034

[88]   A.M. Andersson, N.E. Skakkebaek, Exposure to exogenous estrogens in food: possible impact on human development and health, Eur. J. Endocrinol. 140 (1999) 477–485. https://doi.org/10.1530/eje.0.1400477

[89]   M. Kuster, D.A. Azevedo, M.J. López de Alda, F.R. Aquino Neto, D. Barceló, Analysis of phytoestrogens, progestogens and estrogens in environmental waters from Rio de Janeiro (Brazil), Environ. Int. 35 (2009) 997–1003. https://doi.org/10.1016/j.envint.2009.04.006

[90] B. Lei, S. Huang, Y. Zhou, D. Wang, Z. Wang, Levels of six estrogens in water and sediment from three rivers in Tianjin area, China, Chemosphere. 76 (2009) 36–42. https://doi.org/10.1016/j.chemosphere.2009.02.035

[91] G.-G. Ying, R.S. Kookana, A. Kumar, M. Mortimer, Occurrence and implications of estrogens and xenoestrogens in sewage effluents and receiving waters from South East Queensland, Sci. Total Environ. 407 (2009) 5147–5155. https://doi.org/10.1016/j.scitotenv.2009.06.002

[92] A.C. Vidaeff, L.E. Sever, In utero exposure to environmental estrogens and male reproductive health: a systematic review of biological and epidemiologic evidence, Reprod. Toxicol. Elmsford N. 20 (2005) 5–20.

[93] E. Zuccato, S. Castiglioni, R. Fanelli, G. Reitano, R. Bagnati, C. Chiabrando, F. Pomati, C. Rossetti, D. Calamari, Pharmaceuticals in the environment in Italy: causes, occurrence, effects and control, Environ. Sci. Pollut. Res. Int. 13 (2006) 15–21. https://doi.org/10.1065/espr2006.01.004

[94] T.A. Ternes, Occurrence of drugs in German sewage treatment plants and rivers1, Water Res. 32 (1998) 3245–3260. https://doi.org/10.1016/S0043-1354(98)00099-2

[95] M.R. Servos, D.T. Bennie, B.K. Burnison, A. Jurkovic, R. McInnis, T. Neheli, A. Schnell, P. Seto, S.A. Smyth, T.A. Ternes, Distribution of estrogens, 17β-estradiol and estrone, in Canadian municipal wastewater treatment plants, Sci. Total Environ. 336 (2005) 155–170. https://doi.org/10.1016/j.scitotenv.2004.05.025

[96] M. Hewitt, M. Servos, An overview of substances present in Canadian aquatic environments associated with endocrine disruption, Water Qual. Res. J. Can. 36 (2001) 191–213.

[97] D.W. Kolpin, E.T. Furlong, M.T. Meyer, E.M. Thurman, S.D. Zaugg, L.B. Barber, H.T. Buxton, Pharmaceuticals, Hormones, and Other Organic Wastewater Contaminants in U.S. Streams, 1999−2000: A National Reconnaissance, Environ. Sci. Technol. 36 (2002) 1202–1211. https://doi.org/10.1021/es011055j

[98] S. Zorita, P. Hallgren, L. Mathiasson, Steroid hormone determination in water using an environmentally friendly membrane based extraction technique, J. Chromatogr. A. 1192 (2008) 1–8. https://doi.org/10.1016/j.chroma.2008.03.030

[99] Y. Yoon, P. Westerhoff, S.A. Snyder, E.C. Wert, J. Yoon, Removal of endocrine disrupting compounds and pharmaceuticals by nanofiltration and ultrafiltration membranes, Desalination. 202 (2007) 16–23. https://doi.org/10.1016/j.desal.2005.12.033

[100] Y. Zhang, J.L. Zhou, Occurrence and removal of endocrine disrupting chemicals in wastewater, Chemosphere. 73 (2008) 848–853. https://doi.org/10.1016/j.chemosphere.2008.06.001

[101] S. Liu, G.-G. Ying, J.-L. Zhao, F. Chen, B. Yang, L.-J. Zhou, H. Lai, Trace analysis of 28 steroids in surface water, wastewater and sludge samples by rapid resolution liquid chromatography–electrospray ionization tandem mass spectrometry, J. Chromatogr. A. 1218 (2011) 1367–1378. https://doi.org/10.1016/j.chroma.2011.01.014

[102] N. Nakada, T. Tanishima, H. Shinohara, K. Kiri, H. Takada, Pharmaceutical chemicals and endocrine disrupters in municipal wastewater in Tokyo and their removal during activated sludge treatment, Water Res. 40 (2006) 3297–3303. https://doi.org/10.1016/j.watres.2006.06.039

[103] J. Sánchez-Avila, J. Bonet, G. Velasco, S. Lacorte, Determination and occurrence of phthalates, alkylphenols, bisphenol A, PBDEs, PCBs and PAHs in an industrial sewage grid discharging to a Municipal Wastewater Treatment Plant, Sci. Total Environ. 407 (2009) 4157–4167. https://doi.org/10.1016/j.scitotenv.2009.03.016

[104] L. Clouzot, B. Marrot, P. Doumenq, N. Roche, 17α-Ethinylestradiol: An endocrine disrupter of great concern. Analytical methods and removal processes applied to water purification. A review, Environ. Prog. 27 (2008) 383–396. https://doi.org/10.1002/ep.10291

[105] B.L.L. Tan, D.W. Hawker, J.F. Müller, F.D.L. Leusch, L.A. Tremblay, H.F. Chapman, Comprehensive study of endocrine disrupting compounds using grab and passive sampling at selected wastewater treatment plants in South East Queensland, Australia, Environ. Int. 33 (2007) 654–669. doi:10.1016/j.envint.2007.01.008.

[106] J. Lian, J.X. Liu, Y.S. Wei, Fate of nonylphenol polyethoxylates and their metabolites in four Beijing wastewater treatment plants, Sci. Total Environ. 407 (2009) 4261–4268. https://doi.org/10.1016/j.scitotenv.2009.03.022

[107] N. Tadkaew, F.I. Hai, J.A. McDonald, S.J. Khan, L.D. Nghiem, Removal of trace organics by MBR treatment: The role of molecular properties, Water Res. 45 (2011) 2439–2451. https://doi.org/10.1016/j.watres.2011.01.023

[108] A. Carucci, S. Milia, G. Cappai, A. Muntoni, A direct comparison amongst different technologies (aerobic granular sludge, SBR and MBR) for the treatment of wastewater contaminated by 4-chlorophenol, J. Hazard. Mater. 177 (2010) 1119–1125. https://doi.org/10.1016/j.jhazmat.2010.01.037

[109] N. Tadkaew, F.I. Hai, J.A. McDonald, S.J. Khan, L.D. Nghiem, Removal of trace organics by MBR treatment: The role of molecular properties, Water Res. 45 (2011) 2439–2451. https://doi.org/10.1016/j.watres.2011.01.023

[110] H. Bouju, G. Buttiglieri, F. Malpei, Perspectives of persistent organic pollutants (POPS) removal in an MBR pilot plant, Desalination. 224 (2008) 1–6. https://doi.org/10.1016/j.desal.2007.04.071

[111] F.I. Hai, N. Tadkaew, J.A. McDonald, S.J. Khan, L.D. Nghiem, Is halogen content the most important factor in the removal of halogenated trace organics by MBR treatment?, Bioresour. Technol. 102 (2011) 6299–6303. https://doi.org/10.1016/j.biortech.2011.02.019

[112] A.A. Alturki, N. Tadkaew, J.A. McDonald, S.J. Khan, W.E. Price, L.D. Nghiem, Combining MBR and NF/RO membrane filtration for the removal of trace organics in indirect potable water reuse applications, J. Membr. Sci. 365 (2010) 206–215. https://doi.org/10.1016/j.memsci.2010.09.008

[113] C. Visvanathan, L.N. Thu, V. Jegatheesan, J. Anotai, Biodegradation of pentachlorophenol in a membrane bioreactor, Desalination. 183 (2005) 455–464. https://doi.org/10.1016/j.desal.2005.03.046

[114] W. Dejonghe, E. Berteloot, J. Goris, N. Boon, K. Crul, S. Maertens, M. Höfte, P. De Vos, W. Verstraete, E.M. Top, Synergistic degradation of linuron by a bacterial consortium and isolation of a single linuron-degrading variovorax strain, Appl. Environ. Microbiol. 69 (2003) 1532–1541. https://doi.org/10.1128/AEM.69.3.1532-1541.2003

[115] L. Gunnarsson, M. Adolfsson-Erici, B. Björlenius, C. Rutgersson, L. Förlin, D.G.J. Larsson, Comparison of six different sewage treatment processes— Reduction of estrogenic substances and effects on gene expression in exposed male fish, Sci. Total Environ. 407 (2009) 5235–5242. https://doi.org/10.1016/j.scitotenv.2009.06.018

[116] T. Trinh, B. van den Akker, H.M. Coleman, R.M. Stuetz, J.E. Drewes, P. Le-Clech, S.J. Khan, Seasonal variations in fate and removal of trace organic chemical contaminants while operating a full-scale membrane bioreactor, Sci. Total Environ. 550 (2016) 176–183. https://doi.org/10.1016/j.scitotenv.2015.12.083

[117] V. Cases, V. Alonso, V. Argandoña, M. Rodriguez, D. Prats, Endocrine disrupting compounds: A comparison of removal between conventional activated sludge and

membrane bioreactors, Desalination. 272 (2011) 240–245.
https://doi.org/10.1016/j.desal.2011.01.026

[118] Y.K.K. Koh, T.Y. Chiu, A. Boobis, E. Cartmell, M.D. Scrimshaw, J.N. Lester, Treatment and removal strategies for estrogens from wastewater, Environ. Technol. 29 (2008) 245–267. https://doi.org/10.1080/09593330802099122

[119] C. Scruggs, G. Hunter, E. Snyder, B. Long, S. Snyder, EDCS IN WASTEWATER: WHAT'S THE NEXT STEP?, Proc. Water Environ. Fed. 2004 (2004) 642–664. https://doi.org/10.2175/193864704784138403

[120] J. Sipma, B. Osuna, N. Collado, H. Monclús, G. Ferrero, J. Comas, I. Rodriguez-Roda, Comparison of removal of pharmaceuticals in MBR and activated sludge systems, Desalination. 250 (2010) 653–659.
https://doi.org/10.1016/j.desal.2009.06.073

Chapter 7

Pesticides: Positive and Negative Aspects and their Treatment in Wastewater

Qamruzzaman, Abu Nasar*

Department of Applied Chemistry, Faculty of Engineering and Technology, Aligarh Muslim University, Aligarh – 202 002, India

*abunasaramu@gmail.com

Abstract

The use of a wide variety of pesticides to prevent, destroy, repel or control pests has become an essential practice in the agricultural field as well as in our daily domestic life. Nowadays, a variety of pesticides is available to suit specific requirement for the agricultural and domestic purposes. In order to fulfill the increasing demand of agricultural foods, pesticides are being used all over the world. Unquestionably, the use of pesticides has increased crop yield and provided an economic benefit to the farmers. However, the extensive use of these pesticides has resulted in their presence in foods, water resources, soils and atmosphere in many forms. Consequently, this has led to concern over the harmful effects of pesticides on human and other living organisms. In the present chapter, the classification of pesticides on the basis of target organisms as well on the basis of sources has been discussed. Various advantages and disadvantages of the pesticides have been highlighted. As the pesticides are generally formulated before being sent to market, the formulations of pesticides have also been briefly discussed. The available treatment methods employed for the detoxification of pesticides from the contaminated wastewater are reviewed in light of current available literature.

Keywords

Pesticides, Pesticide Formulation, Treatment, Wastewater

Contents

1. Introduction

Pesticides are natural or synthetic materials used to prevent, destroy, repel or control pests. The pests may be any destructive insects, undesirable plants (weeds), fungi, mosquitos, flies, microorganisms such as viruses and bacteria which are capable to damage or disturb the growth of living organisms such as crops, food and livestock etc. Pesticides are the only toxic materials discharged deliberately into our environment because they are an unavoidable part of agricultural, domestic and social activities. Keeping in mind to fulfill the growing requirement for crop production pesticides are being used all over the world. These pesticides have played a strong role in the commercial production of wide varieties of foods such as cereals, vegetables, oil crops and fruits etc. Additionally, pesticides are used everywhere such as agricultural fields, homes, parks, schools, swimming pools, buildings, forests, and roads.

The widespread use of the various pesticides directed to increasing public attention due to their existence in food products and environmental matrices such as air, soil, and water. The contamination of crop products, soil and water resources due to increasing use of pesticides in agricultural, domestic and social activities is one of the major challenges of the 21st century. The presence of pesticides in water, food or other matrices, even in traces, has negative impacts on human health. Since the consumption of pesticides is associated with serious human health risks, the removal of these compounds is necessary to reduce their negative impacts. In fact, the fate of a pesticide is controlled by the transformation associated with the decomposing of molecules by chemical, photochemical or biological degradation. In view of the above facts, there has been increasing interest in developing economical and efficient methods for the treatment of pesticides after their applications. In the present chapter, the general classification, advantages, disadvantages, and treatment methods of pesticides have been discussed in the light of current developments in the area.

2. Classification of pesticides

Although pesticides may be classified in a number of ways the easiest classification is on the basis of target organisms and their sources:

2.1 Classification of pesticides on the basis of target organism

In fact, the pesticides comprise a large variety of compounds or mixture of compounds and may easily be classified as insecticides, herbicides, fungicides, algaecides, rodenticides, and antimicrobials etc. according to the target organism. The classification based on the target organism is summarized in Table 1.

Table 1: Classification of pesticides on the basis of the target organism.

Pesticide category	Target Organism	Examples
Insecticide	Insects	Acephate, aldrin, acetamiprid, bendiocarb, carbaryl, chlordane, DDT, diazinon, deltamethrin, fenitrothion, fenthion, fosthiazate, malathion, methamidophos, methidathion, methomyl, mevinphos monocrotophos, naled, omethoate, oxydemeton-methyl, parathion, and phostebupirim etc.
Herbicides	Unwanted plants and weeds	Acetochlor, alachlor, asulam, benfluralin, butachlor, 2,4-D, hexazinone, prometon, metribuzin, diuron and bispyribac sodium etc.
Fungicides	Fungi or fungal spores	Tea tree oil, cinnamaldehyde, thiabendazole, iprodione, triflumizole and tricyclazole.
Algaecides	Algae	Benzalkonium chloride, bethoxazin, copper sulfate, cybutryne, dichlone, dichlorophen, diuron, endothal, fentin, hydrated lime, isoproturon, methabenzthiazuron, nabam, oxyfluorfen, pentachlorophenyl laurate, quinoclamine and quinonamid.
Rodenticides	Rodents (rats, mice, squirrels, woodchucks, chipmunks, porcupines, etc.)	Warfarin, coumatetralyl, chlorophacinone, diphacinone, bromadiolone, difethialone, brodifacoum, bromethalin, zinc phosphide and strychnine
Antimicrobials	Microorganisms	Sulphonamides, zithromax, peridox,

	such as bacteria and viruses	chloramphenicol
Fumigants	Domestic and non-domestic pests	Methyl bromide, dazomet, chloropicrin, formaldehyde, hydrogen cyanide, iodoform, methyl isocyanate, and phosphine, etc.
Miticides	Mites	Bifenazate, chlorfenapyr, clofentezine, cyflumetofen, dicofol, propargite, pyridaben, spiromesifen, spirotetramat, etc.
Nematicides	Nematodes	Aldicarb and nematophagous fungi
Molluscicides	slugs, snails	Iron(III) phosphate and aluminum sulfate, metaldehyde and methiocarb,

2.2 Classification of pesticides on the basis of sources

On the basis of sources, the pesticides may be classified as chemical pesticides and biopesticides. Chemical pesticides are further divided into three categories, namely, inorganic, organic and organometallic pesticides. In the context of chemical structure, the organic pesticides are more complex than inorganic ones.

2.2.1 Inorganic pesticides

Inorganic pesticides are simpler compounds and usually water soluble. The earliest known inorganic pesticides were derived from sulfur and lime. The commonly used inorganic pesticides contain arsenic, antimony, boron, copper, fluorine, lead, manganese, mercury, selenium, sulfur, thallium, tin and zinc as their active ingredients. The inorganic pesticides are less popular because of their poor pesticidal activity, non-biodegradability and long persistence in environments. In some cases, they have shown extreme harmful effects on soil and water resources resulting in adverse impacts on human health, the fertility of soil and crop yields. Some common representative examples of inorganic pesticides are borax (insecticide), calcium polysulphide also known as lime sulfur (fungicide and insecticide), mercurous chloride (fungicide), potassium thiocyanate (fungicide), calcium arsenate (herbicide and insecticide) and potassium arsenite (insecticide, fungicide and herbicide) etc.

2.2.2 Organic pesticides

Most of the pesticides used for agricultural and domestic activities come under this category. On the basis of chemical composition, important types of organic pesticides are compiled in Table 2 and briefly discussed below:

Table 2: Classification of organic pesticides on the basis of chemical composition.

Pesticide type	Example	Application type
Organochlorines	Aldrin, chlordane, DDT, dieldrin, endosulfan, endrin and heptachlor	Insecticide
	Dicofol	Miticide
Organophosphate	Acephate, chlorpyrifos, diazinon, dichlorvos, ethion, fenthion and malathion	Insecticide
	Merphos and tribufos,	Herbicide
Carbamates	Bendiocarb, carbaryl, methiocarb, methomyl and thiodicarb	Insecticide
	Chlorprocarb, chlorprocarb, dichlormate and terbucarb	Herbicide
	Benthiavalicarb, iodocarb, propamocarb and thiophanate-methyl	Fungicide
Triazines	Atrazine, hexazinone and metribuzin	Herbicide
Pyrethroids	Permethrin, cypermethrin and deltamethrin	Insecticide
Substituted urea	Diuron, isoproturon and linuron	Herbicide
Carboxylic acid derivatives	2,4-D (2,4-dichloro-phenoxy acetic acid) and bispyribac sodium	Herbicide
Neem based formulations	*Azadirachta indica*	Insecticide

2.2.3 Organometallic pesticides

The organometallic group of pesticides covers a wide range of compounds which can be used as insecticides, herbicides, fungicides and acaricides etc. The activity of these pesticides depends on the chelating action of the metal ion as well as the activity of the organic matrix. Details of some important organic pesticides containing metal ions are presented in Table 3.

Table 3: Organometallic pesticides.

Pesticide name	Metal ion present	Application type
Aluminium phosethyl	Al	Fungicide
Azocyclotin	Sn	Acaricide
Alloxydim-sodium	Na	Herbicide
Ferbam	Fe	Fungicide
Mancozeb	Mn, Zn	Fungicide
Maneb	Mn	Fungicide
Metham-sodium	Na	Herbicide and fungicide
Metiram	Zn	Fungicide
Nabam	Na	Fungicide
Propineb	Zn	Fungicide
Zineb	Zn	Fungicide
Ziram	Zn	Fungicide

2.2.4 Biopesticides

Biopesticides are a special category of pesticides found from natural resources such as animals, plants, bacteria and minerals. For example, baking soda and canola oil have pesticidal uses and therefore they may be categorized to be biopesticides. Biopesticides are generally less poisonous than the pesticides of other categories and usually attack the specified target pests or organisms. These compounds being biodegradable are environment-friendly. Because of high specificity and low speed of action, biopesticides are less popular than the chemical pesticides.

3.   Advantages of pesticides

Prominent attention regarding the negative impacts of pesticides towards the humankind has been widely emphasized. However, the actual fact is too far from the perception which assumes that pesticides are the enemy of humanity. In fact, the benefits of pesticides are much more than the risk associated with them. During recent years, pesticides have gained prominence for their increasing uses due to two stages of benefits: primary and secondary. The primary benefits are direct achievements so obtained and secondary benefits are based on their long-term effects. As pointed out by Cooper and Dobson [1], that there are 26 primary benefits (immediate and incontrovertible) and 31 secondary benefits (as resulted from the primary benefits) associated with the use of pesticides. However, the most prominent benefits of pesticides are based on the direct crop returns [2]. The economic impact on agricultural activity is very fruitful because for any amount spent on pesticides results in at least four times gains on crop yields [3]. Some common advantages of the pesticides are briefly listed below:

3.1   Improved crop production

In order to accomplish the increasing demand of food production due to the continuous increase in world population, the use of pesticides has become essential. The world crop yield has already been lost due to increasing effects of pests along with decreasing areas of agricultural land. The crop losses would be much more if pesticides were not used.

3.2   Pest control

The major role of pesticides is to destroy or repel target pests faster than other pest control methods. This is because pesticides are formulated to act as target specific.

3.3   Decreased cost of food

Because of the improved crop yield by the use of pesticides, the cost of food has become decreased. If these chemicals were not used, the production of many agricultural foods, fruits and vegetables were in short supply and prices were drastically up.

3.4   Vector disease control

Vector-borne illnesses have been successfully undertaken by destroying the vectors. Insecticides are the useful material to govern the insects that spread serious diseases such as dengue, yellow fever, malaria and chikungunya, etc.

3.5   Preventive measures

Pesticides are generally applied to control the spread of pests during import and export businesses, prevent harmful weeds and protect household goods and appliances.

3.6   Protection of pets and humans

A lot of pesticide formulations of different trade names are available in the form of a spray, strips, lotion, liquid vaporizer and baits etc. to control a variety of domestic pests viz., pet flea, fly, mosquitoes, wasps, lice, ant and termite etc. Such products are used to make life more comfortable and pleasant.

4.   Disadvantages of pesticides

Since pesticides are deliberately formulated to destroy living organisms (target species) and therefore they can hurt humans, animals and plants and environment as well. The persons who are involved in the manufacturing or applications of pesticides are more prone to adverse health effects. Some common negative effects of pesticides are discussed below:

4.1   Acute poisoning

Acute poisoning happens when the pesticide planned to control a pest, attacks non-target bodies. There are three types of pesticide poisoning ranging from – a short-term and high-level dose (e.g. persons committing suicide), long-term high-level regular exposure (e.g. persons working in manufacturing unit and agricultural field), long-term low-level exposure (e.g. individuals consuming pesticide from residue in food, water, or inhaling from the contaminated air). In the year 2002, it has been projected that the worldwide usage of pesticides resulted in about 220,000 deaths and 26 million cases of acute poisonings annually [4,5].

4.2   Reduction of beneficial species

Pesticides not only harm the target species such as insects and pests but also affect the non-target species like predators, parasites which are a scavenger of natural importance [6]. Thus, useful species and biodiversity are adversely affected by pesticides. The reduction of these beneficial species can disturb the natural biological balance.

4.3 Drift of sprays and vapors

Pesticides can disturb other areas during their applications and can originate severe problems to non-target bodies and the general environments. The drift is the airborne movement of pesticides from the application area to other unwanted location. Drift usually observed during pesticide application time because droplets or dust are traveled by the wind from the target area to the non-target sites. However, drift is also possible after the use of pesticides when some chemicals vaporize and move to other sites. It is very difficult to manage pesticide drift because the full range of drift cannot be detected. The drift can provide the foundation of accidental exposure to people, animals, plants and property. Furthermore, wildlife and aquatic organisms are affected as well.

4.4 Residues in food

Pesticide residue in food products can cause substantial harm to the persons who consume such products. The residues adversely affect the human health particularly when such contaminated products are used frequently for a longer time. The possibility always exists for the presence of pesticide residues in food.

4.5 Contamination of water resources

The major worry for groundwater contamination with pesticides is that as soon as the groundwater is polluted, the pesticide residues persist for a long time. Most of the pesticides have high water solubility and therefore have potential to contaminate the water resources nearby the application areas.

4.6 Resistance

The widespread usage of pesticides frequently results in the development resistance in target pests. The development pesticide resistance in pest causes more and more additional use of pesticides to maintain crop yields.

5. Formulations of pesticides

Pesticides are very useful for human but because of their negative effects, a lot of precautions are required during their manufacturing, handling and applications. Pesticides are therefore suitably formulated before sending to market and accordingly, they are available in different types of formulations. Formulating a pesticide allows a small quantity to be mixed with a larger quantity of carrier so that they can be effectively applied more uniformly to a larger area. The pesticides are formulated by mixing with surfactants as well as some inert ingredients which may or may not be toxic. The ingredients are added to improve the effectiveness of the pesticides. These ingredients as mentioned below generally decide the method of application.

5.1 Dust

Dust is generally obtained by the sorption of an active ingredient on a finely-powdered inert material like talc, clay or chalk. They are easy to employ because they are lightweight and can be handled by simple equipment such as hand bellows, bulb dusters etc. Dust provides outstanding coverage but due to small particle size they also create an inhalation and drift dangers. These are usually used as a spot treatment for insects.

5.2 Wettable powders

Wettable powders are fine powdered inert solid (e.g. mineral clay), to which a pesticide constituent is sorbed. Wettable powder is generally diluted by water to make suspension and sprayed to the target sites. The wettable powders deliver the best method to spray the active ingredient pesticide which is not easily soluble in water.

5.3 Dry flowable/Water-dispersible granules

Dry flowable or water-dispersible granules are just similar to wettable powder and are obtained by impregnating the active ingredient on a diluent or carrier like talc or clay. The major advantages of dry flowable/water dispersible granules over wettable powders include easier monitoring and the negligible possibility of inhalation problem during measuring and intimate mixing. Flowable formulations are possible with pesticides which are available in solid or semi-solid form.

5.4 Microencapsulates

These formulations are made of an active ingredient surrounded by a plastic or starch coating. The capsules are generally made available as dispersible granules (dry flowable) or as liquid formulations. Encapsulation provides the safety operator by giving timely release of the active ingredient. Liquid microencapsulates are generally diluted with water and spread as sprays.

5.5 Attractants

Attractants include pheromones, sugar, hydrolysate syrup, yeast and meat etc. Pest workers utilise the attractants in gummy traps and arrest bags. Attractants can also be mixed with pesticides and sprayed on the target area.

5.6　Fumigants

Fumigants are the type of pesticides which form a poisonous gas when applied to target areas. They may be used as gas, liquid or solid formulations but reach to target species in the gaseous phase by vaporisation. Fumigants act either as respiratory poisons or as suffocants.

5.7　Aerosols

Aerosols hold one or more active ingredients and a suitable solvent. The aerosols generally contain low concentration of the active ingredients. The aerosol formulations are generally available as ready-to-use and sold as pressurised sealed containers. Alternatively, products are used in electrical or gasoline powered aerosol generators which discharge the formulation as a smoke or fog.

5.8　Wetting agents or surfactants

Surfactants (surface-active agents) are commonly used in pesticide formulations. These agents lower the surface tension of water and make the pesticides solution to become more dispersible and spreadable. Thus, the inclusion of a surfactant in pesticides formulation results in a very economical consumption of pesticides, particularly in waxy and hairy leaves. The use of surfactants lowers down the consumption of total pesticides by permitting efficient application and therefore also minimise their negative impacts. Surfactants are very commonly used in the formulation of pesticides.

6.　Treatment of pesticides

The soil and water (surface and ground) contamination caused by pesticides have been documented as a worldwide big problem due to their persistence in water bodies and the subsequent bad effects on human health [7,8]. Pesticides have been commonly found to micro levels in ground and/or surface water nearby the application area. They come from intensive agricultural, domestic and industrial activities. In most cases, after the application of pesticides in the agricultural field, their residues remain in vessels or containers. The application containers are washed by rinsing water and the washing is generally thrown in the nearby area which results in the formation of highly contaminated wastewater. Since pesticides severely contaminate the water, soil, environment and agricultural products, they create severe public health hazards due to their high toxicity and long persistence. The contaminated wastewater can adversely affect people, domestic animals, pets, and other beneficial organisms [9,10]. Thus, the treatment of pesticides containing wastewater is essential to eliminate or minimize its negative effects of pesticides.

In fact, in most cases, immediately after its application, a pesticide tends to break down or degrade into simpler components which are usually less toxic. Each pesticide has its own rate of degradation which depends on its chemical/photochemical stability, nature of active gradients, formulations and environmental conditions. The process may take anywhere from hours or days to years. Although the pesticides decomposed readily usually don't persist in the environment for a longer time and have little harmful effects but they provide only short-term pest control. In fact, pesticides degradation is associated with transformation based on the breaking of bonds which comprises biological, photochemical and chemical decomposition. Pesticides are generally degraded by a number of ways. The microbial breakdown is the decomposition of molecules by microorganisms such as fungi and bacteria [11,12]. However, the extent of biodegradation depends on many factors such as climatic temperature, pH of the soil, soil moisture, availability of oxygen and soil fertility etc. Photodegradation is the decomposition of pesticides by sunlight [13,14]. Photodegradation can destroy pesticide on foliage, on the soil surface or even in the air. Most pesticides are susceptible to photodegradation and the rate of degradation is controlled by many factors such as intensity of light, exposure time, nature of pesticides, etc.

Chemical degradation is based on the chemical reactions that occur in soil [15,16]. The rate of degradation is controlled by factors such as binding of pesticides with soil, climatic temperature, pH of the soil and moisture etc. At low temperatures, the chemical reaction may be relatively slow, but at high temperatures degradation is accelerated.

The different treatment methods have been employed to decontaminate the water and wastewater containing a variety of impurities and thereby lower down the health dangers linked to consumption of contaminated waters and agricultural food as well. Physical methods like nanofiltration [17-20], slow sand filtration [21] and adsorption [22-25] etc. have established to be beneficial for the handling of toxic residues during water treatment. However, chemical oxidation is an important alternative for the degradative treatment of pesticides. A considerable effort has been devoted to advancement in this direction and variety of oxidants such as ClO_2, Cl_2, O_3, $KMnO_4$, MnO_2, H_2O_2 etc. have been employed for this purpose [26-28].

It is established that humic acid and organic materials undergo degradation by manganese compounds. Manganese exists in variable oxidation states and involves in redox functions of systems. Permanganates are known to be broad range oxidants and also act as a disinfectant. Permanganate ions are effective oxidizing agents and have high affinity to oxidize organic compounds containing olefin and carbonyl groups. In most cases, the extent of oxidation of organic compounds by permanganate ions is very high in acidic medium. The permanganate redox reactions involve in different oxidation states such as

Mn(II), Mn(III), Mn(IV) and Mn(V). It has also been suggested that the species containing Mn(IV) might be a soluble form of colloidal manganese dioxide [31-33]. Manganese oxides are extremely reactive mineral phases and influence biogeochemical cycles. Furthermore, it has been observed that Mn(IV) oxides i.e. MnO_2 present in earth's crust and natural water frequently involved in the oxidation of humic acid and organic materials. In fact, oxidizing capacity of MnO_2 is restricted due to its water insolubility. However, Perez-Benito and coworkers [34-37] have successfully prepared a transparent solution of colloidal MnO_2 by employing the redox reaction between sodium thiosulphate and potassium permanganate. After the preparation of the water-soluble form of colloidal MnO_2, it has been successfully exploited for the oxidative degradation of a number of pesticides such as methomyl [38], metribuzin [39], tricyclazole [40], acetamiprid [41] and acephate [42] etc.

The effect of surfactants on the redox reactions between a formic acid and colloidal MnO_2 has been reported for the first time by Tuncay et al. [43]. They observed that surfactants, particularly non-ionic, have a catalytic effect on the rate of redox reactions. In the light of pioneer work of Tuncay et al. [43], the authors' group studied the catalytic role of surfactants such as sodium dodecyl sulfate, cetyl trimethyl ammonium bromide and Triton X-100 on the oxidative degradation of a number of pesticides [38-42] and observed that the reaction is catalyzed by catalyst Triton X-100 in all the pesticides undertaken.

7. Conclusions

In order to fulfill the increasing demand of agricultural food, pesticides are being commonly used throughout the world. However, the extensive usage of pesticidal products has encouraged public interest due to their presence in foods as well as in environmental matrices such as soils, ground and surface water. Most of the pesticides are water soluble and have the potential to leach and contaminate surface and ground waters. The associated risks can be reduced by optimum and minimal use of pesticides which can be achieved by formulating a pesticide by taking a small quantity to be mixed with a larger quantity of inert carrier so that they can be effectively applied more uniformly to a larger area. Further, their toxic effect can be eliminated or minimized by the degradative treatment of pesticides. Different available methods of treatment of pesticides from wastewater have been reviewed in light of current development in this area.

Acknowledgements

The authors are grateful to the Chairman, Department of Applied Chemistry, Faculty of Engineering and Technology, Aligarh Muslim University for providing necessary facilities.

References

[1]   J Cooper, H. Dobson, The benefits of pesticides to mankind and the environment, Crop Prot. 26 (2007) 1337–1348. https://doi.org/10.1016/j.cropro.2007.03.022

[2]   D. Pimental, Environmental and economic costs of the application of pesticides primarily in the United States, Environ. Dev. Sustain. 7 (2005) 229–252. https://doi.org/10.1007/s10668-005-7314-2

[3]   D. Pimental, H. Acquay, M. Biltonen, P. Rice, M. Silva, J. Nelson, V. Lipner, S. Giordano, A. Horowitz, A., M.D. Amore, Environmental and economic costs of pesticide use, BioScience 42 (1992)750–760. https://doi.org/10.2307/1311994

[4]   E.D. Richter, 2002, 'Acute human pesticide poisonings', in D. Pimentel (ed.), Encyclopaedia of Pest Management, New York, Marcel Dekker, pp. 3–6. https://doi.org/10.1201/NOE0824706326.ch2

[5]   K. Hart, D. Pimentel, 2002, 'Public health and costs of pesticides', in D. Pimentel (ed.), Encyclopaedia of Pest Management, New York, Marcel Dekker, pp. 677–679. https://doi.org/10.1201/NOE0824706326.ch313

[6]   D. Pimentel, H., Acquay, M. Biltonen, P. Rice, M. Silva, J. Nelson, V. Lipner, S. Giordana, A. Horowitz, M. D'Amore, 1993, Assessment of environmental and economic impacts of pesticide use', in D. Pimentel and H. Lehman (eds.), The Pesticide Question: Environment, Economics and Ethics. New York, Chapman & Hall, pp. 47–84. https://doi.org/10.1007/978-0-585-36973-0_3

[7]   K. Ikehata, M.G. El-Din, Aqueous pesticide degradation by hydrogen peroxide/ultraviolet irradiation and Fenton-type advanced oxidation processes: a review, J. Environ. Eng. Sci. 5 (2006) 81–135. https://doi.org/10.1139/s05-046

[8]   M.K.K. Pillai, Pesticide pollution of soil, water and air in Delhi area, India, Sci. Total Environ. 55 (1986) 321–327. https://doi.org/10.1016/0048-9697(86)90189-0

[9]   T.J. Centner, Unwanted agricultural pesticides: state disposal programs, J. Environ. Quality 27 (1998) 736-742. https://doi.org/10.2134/jeq1998.00472425002700040002x

[10]   R.L. Ridgway, J.C. Tinney, J.T. MacGregor, N.J. Starler, Pesticide use in agriculture, Environ. Health Perspect. 27 (1978) 103-112. https://doi.org/10.1289/ehp.7827103

[11]   K.A. Fenlon, K.C. Jonesa, K.T. Semple, Development of microbial degradation of cypermethrin and diazinon in organically and conventionally managed soils, J. Environ. Monit. 9 (2007) 510-515. https://doi.org/10.1039/b700668c

[12]   A. Ghafoor, J. Moeys, J. Stenström, G. Tranter, N.J. Jarvis, Modeling spatial variation in microbial degradation of pesticides in soil, Environ. Sci. Technol. 45 (2011) 6411-6419. https://doi.org/10.1021/es2012353

[13]   I.K. Konstantinou, T.M. Sakellarides, V.A. Sakkas, T.A. Albanis, Photocatalytic degradation of selected s-triazine herbicides and organophosphorus insecticides over aqueous TiO_2 suspensions, Environ. Sci. Technol. 35 (2001) 398–405. https://doi.org/10.1021/es001271c

[14]   M. Tamimi, S. Qourzal, A. Assabbane, J.M. Chovelon, C. Ferronato, Y. Ait-Ichou, Photocatalytic degradation of pesticide methomyl: determination of the reaction pathway and identification of intermediate products, Photochem. Photobiol. Sci. 5 (2006) 477-482. https://doi.org/10.1039/b517105a

[15]   K.R. Raju, S.R.K.M. Akella, J.V.S. Murthy, U.T. Bhalerao, Spectrophotometric determination of Isoproturon and Metoxuron using ethylacetoacetate and application to technical and formulation grade samples, Talanta 43 (1996) 577–581. https://doi.org/10.1016/0039-9140(95)01773-9

[16]   F. Norberto, S. Santos, D. Silva, P. Hervés, A.S. Miguel, F. Vilela, Reactivity of N-pyridylcarbamates in basic media, J. Chem. Soc. Perkin Trans. 2 (2002) 1162-1165. https://doi.org/10.1039/b200445n

[17]   K.M. Agbekodo, B. Legube S. Dard, Atrazine and simazine removal mechanisms by nanofiltration: Influence of natural organic matter concentration, Water Res. 30 (1996) 2535-2542. https://doi.org/10.1016/S0043-1354(96)00128-5

[18]   P. Berg, G. Hagmeyer, R. Gimbel, Removal of pesticides and other micropollutants by nanofiltration, Desalination 113 (1997) 205-208. https://doi.org/10.1016/S0011-9164(97)00130-6

[19]   B. V. Bruggen, J. Schaep, W. Maes, D. Wilms, C. Vandecasteele, Nanofiltration as a treatment method for the removal of pesticides from ground waters, Desalination 117 (1998) 139-147. https://doi.org/10.1016/S0011-9164(98)00081-2

[20] R. Boussahel, S. Bouland, K.M. Moussaoui, A. Montiel, Removal of pesticide residues in water using the nanofiltration process, Desalination 132 (2000) 205-209. https://doi.org/10.1016/S0011-9164(00)00151-X

[21] Sukru Aslan, Hatice Cakici, Biological denitrification of drinking water in a slow sand filter, J. Hazard. Mater. 148 (2007) 253–258. https://doi.org/10.1016/j.jhazmat.2007.02.012

[22] S. Shakoor, A. Nasar, Adsorptive treatment of hazardous methylene blue dye from artificially contaminated water using cucumis sativus peel waste as a low-cost adsorbent, Groundw. Sustain. Dev. 5 (2017) 152–159. https://doi.org/10.1016/j.gsd.2017.06.005

[23] N.P. Thacker, M.V. Vaidya, M. Sipani, A. Kalra, Removal technology for pesticide contaminants in potable water, J. Environ. Sci. Health Part B 32 (1997) 483-496. https://doi.org/10.1080/03601239709373099

[24] S. Shakoor, A. Nasar, Removal of methylene blue dye from artificially contaminated water using citrus limetta peel waste as a very low cost adsorbent, J. Taiwan Inst. Chem. Eng. 66 (2016) 154–163. https://doi.org/10.1016/j.jtice.2016.06.009

[25] A. Nasar, S. Shakoor, Remediation of dyes from industrial wastewater using low-cost adsorbents, Chapter in Book "Appl. Adsorpt. Ion Exch. Chromatogr. Wastewater Treat". Materials Research Foundations 15(2017) 1–33. https://doi.org/10.21741/9781945291333-1

[26] S. Chiron, A. Fernandez-Alba, A. Rodriguez, E. Garcia-Calvo, Pesticide chemical oxidation: state-of-the-art, Water Res. 34 (2000) 366–377. https://doi.org/10.1016/S0043-1354(99)00173-6

[27] V Camel, A Bermond, The use of ozone and associated oxidation processes in drinking water treatment, Water Res. 32 (1998) 3208–3222. https://doi.org/10.1016/S0043-1354(98)00130-4

[28] H.M. Coleman, C.P. Marquis, J.A. Scott, S.S. Chin, R. Amal, Bactericidal effects of titanium dioxide-based photocatalysts, Chem. Eng. J. 113 (2005) 55–63. https://doi.org/10.1016/j.cej.2005.07.015

[29] E. John, E. (2001). "Manganese" Nature's Building Blocks: An A-Z Guide to the Elements. Oxford, UK: Oxford University Press. p. 249.

[30] V.M. Goldschmidt, (1958). Geochemistry. Oxford University Press, London, p. 621.

[31] F. Freeman, J.C. Kappos, Permanganate ion oxidations, Additional evidence for formation of soluble (colloidal) manganese dioxide during the permanganate ion oxidation of carbon-carbon double bonds in phosphate-buffered solutions, J. Am. Chem. Soc. 107 (1985) 6628–6633. https://doi.org/10.1021/ja00309a034

[32] F. Mata-Perez, J.F. Perez-Benito, Identification of the product from the reduction of permanganate ion by trimethylamine in aqueous phosphate buffers, Canad. J. Chem. 63(1985) 988-992. https://doi.org/10.1139/v85-165

[33] J.F. Perez-Benito, D.G. Lee, Oxidation of hydrocarbons, A study of the oxidation of alkenes by methyltributylammonium permanganate Canad. J. Chem. 63 (1985) 3545-3550. https://doi.org/10.1139/v85-582

[34] J.F. Perez-Benito, E. Brillas, R. Pouplana, Identification of a soluble form of colloidal manganese(IV), Inorg. Chem. 28 (1989) 390–392. https://doi.org/10.1021/ic00302a002

[35] J.F. Perez-Benito, C. Arias, A kinetic study of the reaction between soluble (colloidal) manganese dioxide and formic, J. Colloid Interface Sci. 149 (1992) 92-97. https://doi.org/10.1016/0021-9797(92)90394-2

[36] J.F. Perez-Benito, C. Arias, E. Amat, A Kinetic study of the reduction of colloidal manganese dioxide by oxalic acid, J. Colloid Interface Sci. 177 (1996) 288-297. https://doi.org/10.1006/jcis.1996.0034

[37] J.F. Perez-Benito, Reduction of colloidal manganese dioxide by manganese (II), J. Colloid Interf. Sci. 248 (2002) 130-135. https://doi.org/10.1006/jcis.2001.8145

[38] Qamruzzaman, A. Nasar, Degradation of methomyl by colloidal manganese dioxide in acidic medium, Chem. Sci. Rev. Lett. 1 (2012) 113-119.

[39] Qamruzzaman, A. Nasar, Kinetics of metribuzin degradation by colloidal manganese dioxide in absence and presence of surfactants, Chem. Pap. 68 (2014) 65–73. https://doi.org/10.2478/s11696-013-0424-7

[40] Qamruzzaman, A. Nasar, Degradation of tricyclazole by colloidal manganese dioxide in the absence and presence of surfactants J. Indust. Engg. Chem. 20 (2014) 897–902. https://doi.org/10.1016/j.jiec.2013.06.020

[41] Qamruzzaman, A. Nasar, Treatment of acetamiprid insecticide from artificially contaminated water by colloidal manganese dioxide in the absence and presence of surfactants, RSC Adv. 4 (2014) 62844–62850.

[42]   Qamruzzaman, A. Nasar, Degradation of acephate by colloidal manganese dioxide in the absence and presence of surfactants, Desal. Water Treat. 55 (2015) 2155–2164. https://doi.org/10.1080/19443994.2014.937752

[43]   M. Tuncay, N. Yuce, B. Arlkan, S. Gokturk, A kinetic study of the reaction between colloidal manganese dioxide and formic acid in aqueous perchloric acid solution in the presence of surface active agents, Colloid. Surface. A 149 (1999) 279-284. https://doi.org/10.1016/S0927-7757(98)00520-2

Chapter 8

Wastewater Biological Treatment Technologies: An Eminent Hybrid Approach to Treat Textile Wastewater

Rabia Shoukat[1], Sher Jamal Khan[1,*], Yousuf Jamal[1]

[1]Institute of Environmental Sciences and Engineering, School of Civil and Environmental Engineering, National University of Sciences and Technology (NUST), Islamabad, Pakistan

*s.jamal@iese.nust.edu.pk; sherjamal77@gmail.com

Abstract

Wastewater treatment has been practiced for a long time. However, in recent years the focus has shifted to treating textile wastewater because of the environmental degradation effects of the toxic industrial discharge. Many studies have been conducted on types of numerous technologies in treating textile wastewater with the aim of improving system efficiency and environmental sustainability. Among biological treatment technologies, hybrid anaerobic-aerobic membrane bioreactor (MBR) would be the best approach due to its cost effectiveness and efficient treatment processing. This chapter critically evaluates the potential of the hybrid wastewater treatment systems for textile wastewater and discusses various biological treatment systems with their strengths and limitations and possibility of hybrid technology adoption in developing countries.

Keywords

Textile Wastewater, Membrane Bioreactor, Biological Treatment, Hybrid Systems, Environmental Sustainability

Contents

1.   Introduction

With the inevitable increase in population, rapid urbanization and industrial developments have created an alarming situation of environmental degradation in developing countries. Wastewater discharge from industries containing toxic chemicals like acids, dyes, surfactants, metals and dispersing agents has become an environmental challenge for the industries and the population [1]. Textile industry encompasses innumerable processing steps that require use of huge amount of water for various processes. Therefore, the textile industry discharges vast amount of wastewater that have detrimental impacts on the environment [2]. The textile processing technology involves steps like sizing, desizing, scouring, bleaching, mercerizing, dyeing and finishing, that are associated with the discharging of various pollutants. The types of pollutants allied to the operational steps are illustrated schematically in Fig. 1. Notably, cotton dyeing involves chromophore groups like azo, carbonyl and nitro which chiefly enhance the unacceptable toxicity to textile wastewater [3].

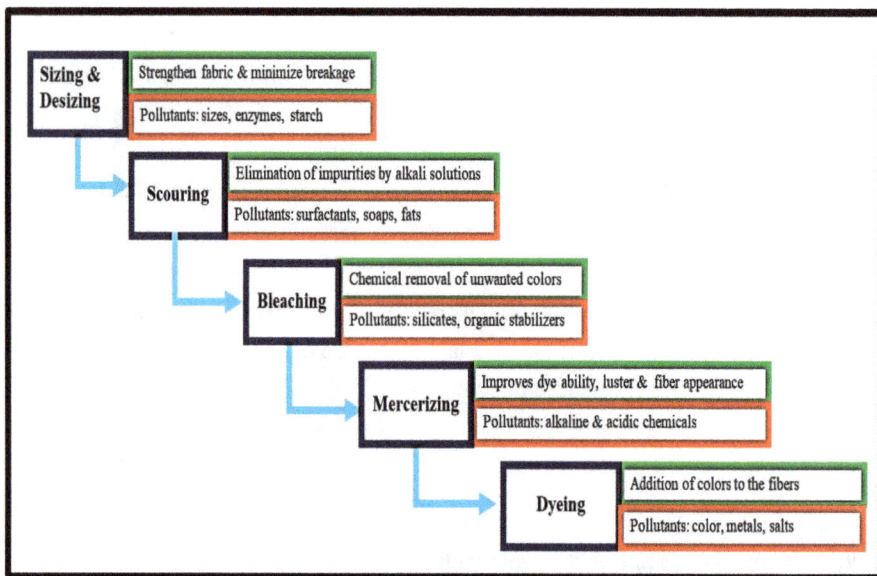

Figure 1. Schematic illustration of textile industry unit operations and pollutants generated at each processing level (Adopted from [3]).

Textile wastewater classically composed of organic matter, suspended solids, dyes and numerous other chemicals like polycyclic aromatic hydrocarbons, solvents, detergents, recalcitrant compounds and heavy metal ions [4]. Variation in type of processes results in various nature of wastewaters streams generated from the textile industry [5] . In this chapter wastewater discharge from textile industries is not focused for potable usage but for reuse of water for agriculture, horticulture, landscaping and floor cleaning [6].

Developing countries' economy principally depend upon agriculture and almost 70% of the total water is consumed for agricultural purposes [7]. In addition to high organic loads from textile effluents, the primary target pollutant is color which if not removed, can be chaotic for the receiving water bodies and agriculture use [8]. Colorants releasing in textile outflow is mainly from the dyeing processing step which comprises the use of water as a standard medium for dye application [9]. The dyes used in dyeing operations are usually complex organic compounds which are resilient to degrade and released as target pollutant in the water bodies [10].

United States Environmental Protection Agency (USEPA) has reported that a high volume of water consumption is mainly due to the use of numerous technologies in processing steps as shown in Fig. 2. For a textile fabric production unit of 20,000 lb/day, a quantity of 36,000 liters of water is consumed for cotton processing [11].

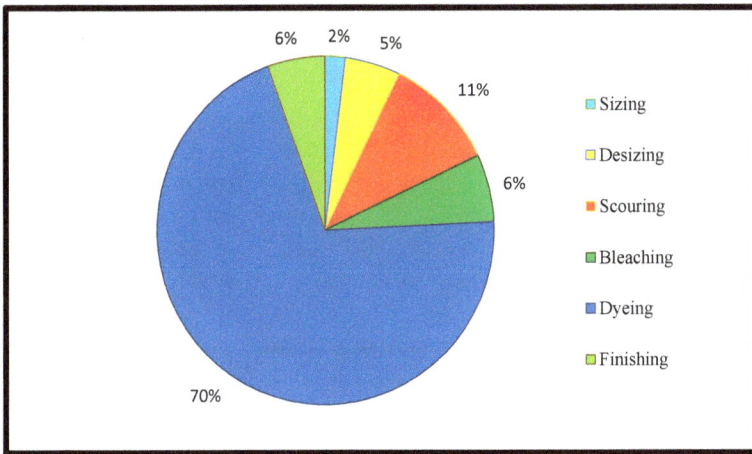

Figure 2. Volume of water required per unit operation of cotton industry (Adopted from [11]).

Many advanced technologies have been practiced in technically advanced countries to treat textile wastewater irrespective of the wastewater source, a conventional chemical treatment is adopted until now [12]. Wastewater treatment technologies especially chemical treatment technologies, encompasses the coagulation processes and sedimentation followed by extended aeration for the treatment [13]. The use of process water in the textile industries should satisfy the stringent quality standards before the effluent can reuse in other applications. Therefore, the vision of textile wastewater treatment is to implement sustainably viable environmental practices to promote industrial sector.

2. Textile industry effluent discharge in developing countries

Industrialization in developing countries though lead to many technical advancements but it has also brought many environmental concerns regarding environmental degradation [14]. Wastewater discharge from industries can be cut down to certain limits but in case of the textile industries, major practices are exceedingly dependent on processes that generate a huge amount of wastewater [13]. Practicing direct discharge of highly polluted industrial wastewater is worst in developing countries where treatment of industrial wastewater is not in practice relative to domestic wastewater treatment [15]. For wastewater reclamation and management, an approach to develop feasible treatment technologies by reducing environmental footprints has become an urge in textile industrial sector [16].

An existing affordable trend of treatment technology is the wet land or waste stabilization ponds [17]. Adoption of anaerobic technology in developing countries including Pakistan is in practice especially in arid and semi-arid regions where the warm climates provide an efficient level of treatment of wastewater [18]. Fig. 3 demonstrates the evolution of global water usage for industrial sectors as reported by United nations environment program (UNEP) [19].

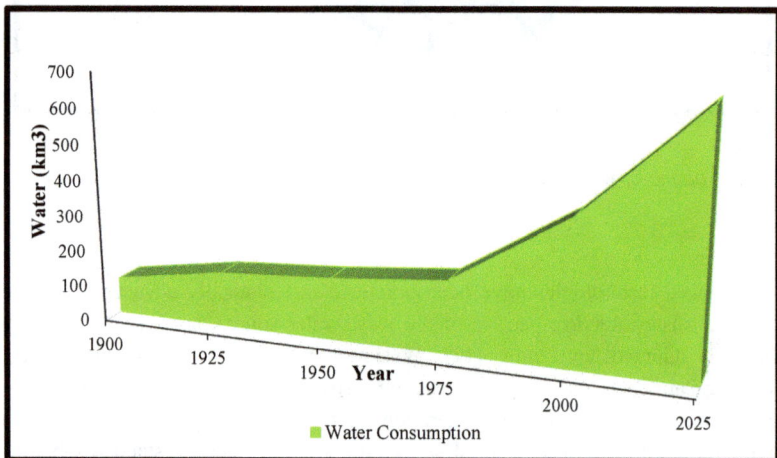

Figure 3. Evolution of global water usage for industrial sector (Adopted from [19]).

The unplanned industrial growth in developing countries has put a burden on the ecology in an alarming way. Bangladesh is known among developing countries facing serious crisis of environmental damage due to untreated textile discharges [4,20].

Textile wastewater standards are usually overlooked in under developed countries and the reason of this incomprehension is that either the standards are too stringent or too easy-going that don't even promise the environmental protection. The stringent compliance requirements lead to adopt alternative technologies [21]. In China, textile industry has set its standards to save the local environment which sets a rousing example for other highly populated countries to take steps in adapting the effluent treatment technologies [22,23]. For example, adaption of:

- Wastewater equalization and pH adjustment approach for complying with conserving conditions for aquatic life in receiving water bodies
- Chemical treatment methods to achieve high degree of treatment efficiency

3. Biological treatment technologies

Since the 19th century, the use of microorganisms, to degrade organic and inorganic pollutants present in water, is in practice with advancement in technologies like trickling filters, activated sludge and septic tanks [24]. A typical wastewater flow in textile industries is shown in Fig. 4. In developing countries, for complete degradation of toxic pollutants, an approach towards biological treatment is under adaptation [15].

Biological treatment methods treating wastewater involve microbial community responsible for pollutant degradation [25]. Developing countries practice biological treatment technologies to treat industrial wastewater more proficiently than the physico-chemical methods because of their high chemical costs and ineffectiveness over target pollutant removal [26,27]. When wastewater is subjected to the conventional biological treatment plant, it involves a secondary treatment in which organic removal takes place [28].

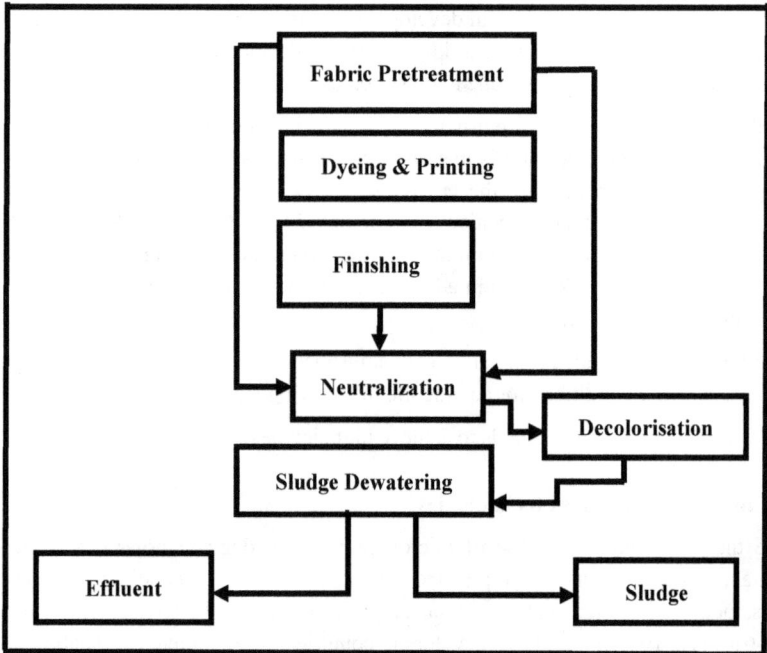

Figure 4. Wastewater flowchart from textile processing entities (Adopted from [15]).

3.1 Biological trends towards removal of textile dyes from industrial effluents

Biological treatment can be aerobic and/or anaerobic type for the treatment of textile wastewater. Both technologies coupled with one another can be a mainstream of biological treatment in comparison to the separate treatments exclusively. In textile wastewater, dye removal is of major concern. Typically, the amount of dyes present in textile effluent ranges between 10 to 50 mg/L [29]. Color removal from dyebath discharge can be effectively achieved by the use of acclimatized microbial consortium. The reported trend of biological activity on color reduction is shown in Fig. 5 [30,31].

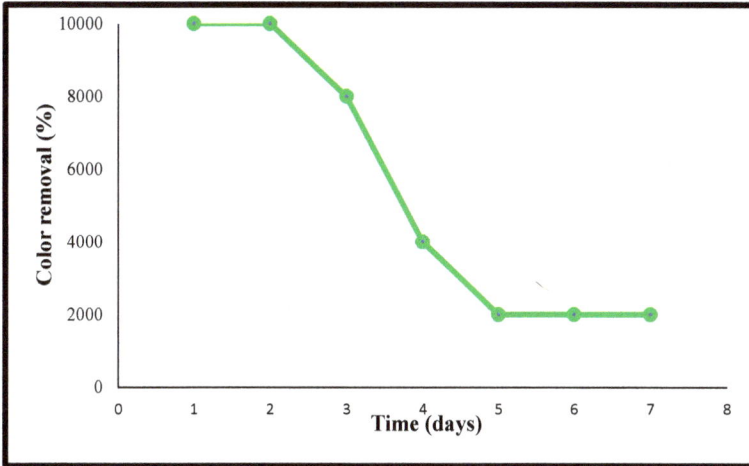

Figure 5. Color reduction profile due to biological activity in anaerobic treatment system (Adopted from [30,31]).

Dyes, resilient to light and oxidative degradation are most likely eliminated during biological processes involving bacterial groups having xenobiotic nature e.g. genera of *Basidiomycetes* [32]. In addition, an anaerobic microbial consortium consisting of methanogenic and acetogenic species is responsible for wastewater decolorization [33].

3.2 Classification of biological treatment processes

Recent biological treatment processes are generally classified in two categories, conventional activated sludge process (CASP) and membrane bioreactor (MBR). Both treatment processes are compared in Table 1 [34, 35].

Table 1: Comparison of conventional activated sludge process vs. MBR (Adopted from [34,35]).

	Conventional Activated Sludge	Membrane Bioreactors
COD Removal Efficiency	94.5%	99.0 %
Dissolved Organic Carbon Removal	92.7%	96.9 %
Floc Size	Large	Small
Filamentous organisms	Higher amount	Smaller amount

Biomass Fraction variability	High	Low
Settelability	Good	Inferior

The activated sludge process consists of aerobic microbes which are responsible to provide necessary treatment [35]. The provision of oxygen to maintain microbial growth in a biological tank coupled with membrane filtration, in an advanced technology of membrane bioreactor is the key consideration in achieving acceptable level of treatment [36].

3.2.1 Conventional activated sludge process (CASP)

Conventional activated sludge processes (CASPs) demonstrate limited potential to remove pathogens from the effluents. This technology was principally adopted to treat domestic or municipal wastewater i.e. low strength and colorless. Studies have reported that amongst conventional biological treatment technologies, activated sludge technology is not effective for degradation of azo compounds [25,37]. Fig. 6 shows the schematic diagram of CASP which consists of an aerobic process occurring in an biological aeration tank which is followed by a clarifier for the removal of flocs resulting from bacterial growth [38].

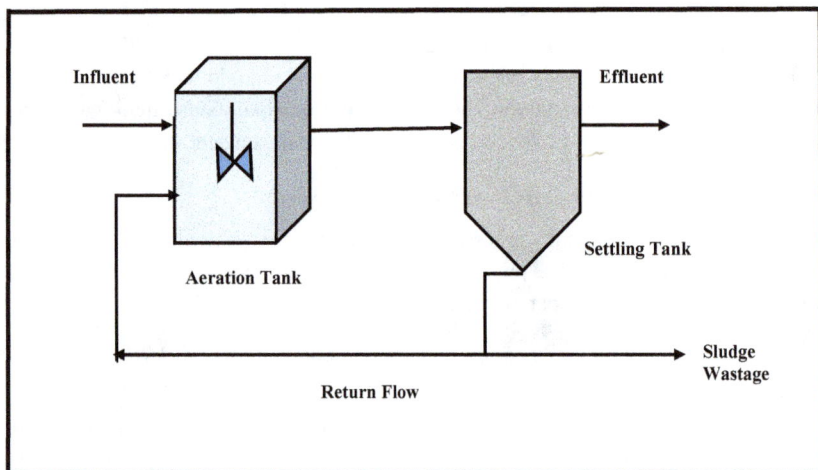

Figure 6. Schematic of conventional activated sludge process (Adopted from [38]).

3.2.1.1   Limitations of CASP

CASP is not acclaimed for treating textile wastewater as it involves high organic loading. Since aerobic bacteria are accountable for treating in CASP, they are inefficient and incapable for degrading complex dye compounds [39,40]. In addition to sensitivity to industrial wastewater, CASP requires high operational cost and large scale sludge disposal unit, and can therefore not be considered as a flexible technology.

3.2.2 Membrane bioreactors (MBRs)

Among all the biological treatment technologies, the most advanced and efficient treatment technology for treating wastewater is a membrane bioreactor (MBR). MBRs composed of two primary processes constitute a combination of biological treatment and filtration as illustrated in Fig. 7 [41,42].

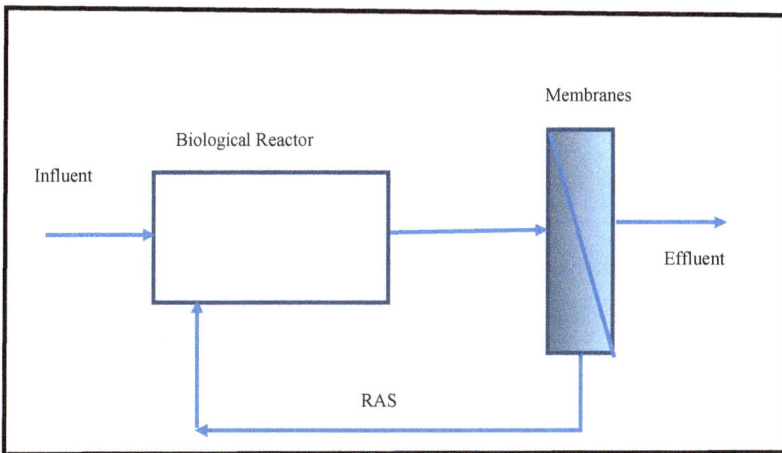

Figure 7. Schematic of membrane bioreactor process (Adopted from [41,42]).

The effluent quality of MBR in terms of chemical oxygen demand (COD) is better than any conventional technologies due to the absorbance and enhanced degradation of colloidal compounds [41]. The potential of treatment and operational flexibility of MBR technology are perhaps the chief advantages that drag its importance of installation especially for the treatment of high strength wastewaters [43]. MBR technology professed novelty, supports the investment decisions for the treatment of wastewater as it provides the enhanced benefits of energy efficiency, improved operational procedures and longer warranty [12]. The MBR adoption in developed countries has been popular because of their small footprint treatment processes and 80-95% nutrient reduction [44]. Worldwide, large capacity installations of MBR are in operation, manufacturing or design phase. Fig. 8 shows that MBRs installation of varying capacity are increasing in Europe since 1998 [45].

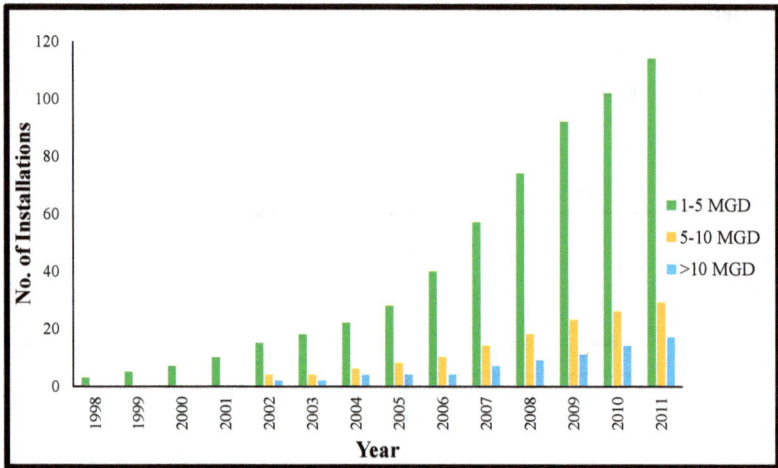

Figure 8. Cumulative capacity installations of MBRs (Adopted from [45]).

4.   Categories of membrane bioreactors (MBRs)

MBRs are designed based on type of target pollutant elimination from wastewater discharges. Despite of the energy limitations, MBRs in Europe have been used for treatment of industrial wastewater [46]. The aerobic MBR (AeMBR) is energy intensive as compared to anaerobic MBR (AnMBR) configuration because of the recovery of biogas at mesophilic temperatures for wastewaters of strength 4-5 g-COD/L [47]. In comparison to CASP, COD removal in MBR can be achieved up to 96-99% [48] .

4.1   Aerobic MBR (AeMBR)

AeMBR is typically applied for the treatment of organics containing municipal wastewater. By-products of aerobic treatment process are generally carbon dioxide and bio-solids. The operating cost and greenhouse gases increase proportional with the increase of organics in wastewater [49]. Aerobic MBRs require high demand of oxygen for the bacterial activity responsible for organic matter degradation and membrane scouring [50].

4.1.1 Configurations of aerobic MBR (AeMBR)

For the treatment of diverse types of wastewaters, variety of membranes like hollow fiber, flat sheet and tubular membrane can be used. Membrane process configurations can also be divided into categories of internal submerged (immersed) and external submerged (side stream) MBR [51,52]. Each membrane configuration has inherent advantages and disadvantages like operational flux, clogging propensity, energy usage and capital cost. In immersed type of membrane configuration, the membranes are operated under suction and the biologically active sludge is retained at the outside of the membrane [53]. This conformation is likely opposite to the side stream type in which sludge is allowed to pump into the membrane modules and the membrane unit is operated entirely separated from the biological tank [54].

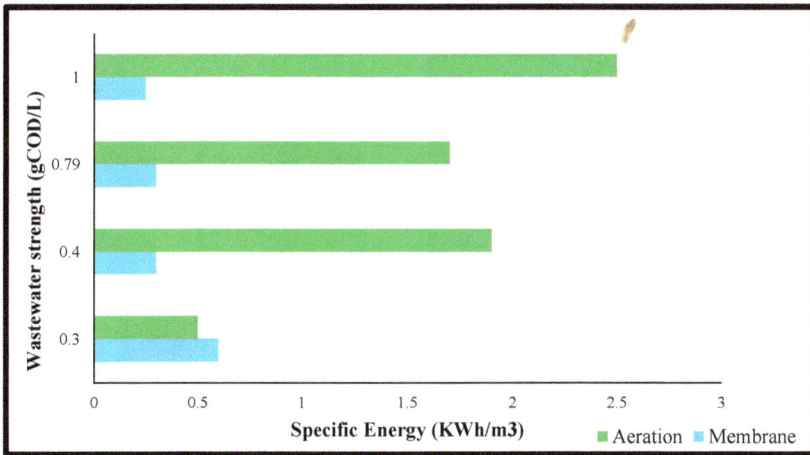

Figure 9. AeMBRs with sludge retention: Membrane and aeration energy demands (Adopted from [47,55]).

Fig. 9 shows an energy demand estimation for membrane (blue bars) and aeration (green bars), which has been reported through operational data and modeling, in which it is described that in AeMBRs, irrespective of the applied organic load, the energy demand approaches 2.5 KWh/m^3 or more for variable strength wastewaters [47]. The energy demand assessment of AeMBRs and AnMBRs revealed that even though the AnMBR required lower energy in terms of aeration but the membrane capital cost would be greater due to lower flux and wide variation of gas demands utilized in anaerobic systems [55].

4.2 Anaerobic MBR (AnMBRs)

AnMBRs are the best option for exploitation of biomass retention on membrane [18]. AnMBRs are suitably recommended for textile wastewater with strength of 4-5 g-COD/L, as heat is required to achieve the mesophillic temperature (35°C) in the reactor this is only possible with high strength textile industrial wastewater [47]. AnMBRs though facilitates reducing energy demand and sludge production but several studies reported that AnMBRs operational parameters affected the biological treatment performance. In AnMBRs, the end products are usually in the form of biogas. Table 2 shows the flexible degrees of methanisation in AnMBR for wastewater treatment. The extreme methane yield perceived from research experimentations for AnMBRs from domestic wastewater was in the range 0.29-0.33 L-CH$_4$/g [56].

Table 2: Variable degrees of methanisation in AnMBRs for actual wastewater influent (Adopted from [56]).

Wastewater types	Degree of Methanisation (CH$_4$-L/g)
Screened Wastewater	0.20-.23
Raw Wastewater	0.27
Black Water	0.09-0.12

4.2.1 Configurations of anaerobic MBR (AnMBR)

Anaerobic digestion process is considered as a unique type of biological treatment which is best suited for the treatment of high strength industrial wastewater specifically textile wastewaters. Different types of wastewaters have been subjected to the anaerobic treatment with MBRs [57]. AnMBRs are actually categorized on the basis of two main configurations:

1. Anaerobic side stream/external membrane bioreactor (AnsMBR)

2. Anaerobic immersed/submerged membrane bioreactor (AniMBR)

For the former configuration, the membrane module is operated individually by placing it outside the bioreactor and this type of design is more common in comparison to later configuration. AnsMBRs demand high cross flow velocity which results in sludge floc breakage that leads to the declination of sludge activity thus affecting the economic feasibility [56].

In the AniMBR configuration, the membrane is operated under a vacuum and the system runs under very mild operating conditions. Energy demand for this arrangement is two times lower than the AnsMBRs [58]. One dominant characteristic of AniMBR is that this system allows self-cleaning of the membrane surface by providing biogas recirculation [57]. Both AnsMBR and AniMBR can be best implemented for treating synthetic, municipal and industrial wastewaters depicting the importance of anaerobic biological treatment of a wide range of wastewaters. Studies revealed that significant methane production occurred at mesophilic temperature that depicted temperature as an important parameter for submerged AnMBR design. But AnMBR operated at ambient temperature resulted in lowered biogas production [59]. Fig. 10 shows membrane energy demand (green bars) and energy from biogas converted to heat energy (blue bars) for wastewater of variable strength in submerged AnMBRs [47]. The energy required to heat the influent from 15 to 35°C in a reactor refers that 50% energy can be recovered due to the heating of the influent from the permeate [55].

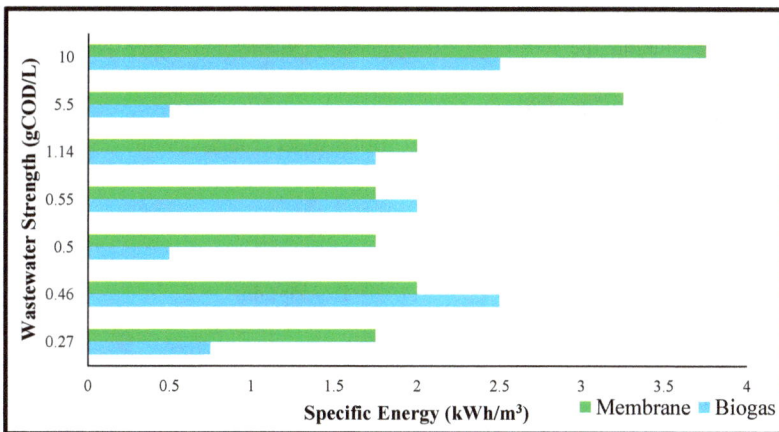

Figure 10. Submerged AnMBRs with sludge retention: Membrane energy demand and biogas converted to heat energy (Adopted from [47 55]).

4.2.2 Limitations of membrane bioreactors (MBRs)

Membrane fouling involves the attachment, entrapment, adsorption and accumulation of certain materials on the surface of the membrane. This results in the increase of the rate of hydraulic resistance [60]. Pore blocking and cake formation are the two categories of membrane fouling that usually occur during filtration process. Table 3 lists different membrane fouling types along with their characterization [61].

Table 3: Various types of fouling (Adopted from [61]).

Fouling Types	Characterization
Nondominant fouling	Membrane scaling due to mineral deposition
Bio-fouling	Microbial biofilm formation on membrane surface
Organic fouling	Connotation of macromolecules with membrane i.e. Vander Waals forces
Pore blocking	Pores blockage by means of colloids

5. Anaerobic biological treatment technologies

For the treatment of low strength domestic wastewater, the anaerobic technology has been extensively used [18]. The perception of introducing anaerobic treatment systems in treating textile wastewaters is mainly because of the resource recovery, reduction of aeration cost making the technology economically viable and above all less sludge production relative to aerobic treatment which decreases the disposal cost and requirement of landfills [62]. Anaerobic treatment systems are better for use as effective alternatives to treat high strength industrial wastewaters at mespophillic temperatures in addition to the other advantages that are enlisted in Table 4 [63].

Table 4: Advantageous characteristics of anaerobic systems for wastewater treatment (Adopted from [63]).

Individualities	Description
Budget	Cost effective relative to physico-chemical treatment
Working Operation	Simple in operational procedures and processing
Energy Demand	Zero energy consumption or maximum possibility of energy minimization
Sludge	Less sludge production volume which makes ease of disposal

For high strength textile wastewater, the adoption of anaerobic treatment allows the energy resource recovery i.e. methane, that not only ensure the heat balance but also subsidize in an electric power generation for export [64].

5.1 Upflow anaerobic sludge blanket (UASB)

Anaerobic digestion involves the consortium of microorganisms which are responsible for better color removal in case of high strength wastewater. Upflow anaerobic sludge blanket (UASB) technology is the treatment technique for treating high strength industrial wastewater. This technology (Fig. 11) consists of expanded granular sludge bed in which reactions of hydrolysis, acidification, and methanogenesis take place under different hydraulic retention times (HRTs) [65]. An approach to degradation and decolorization of complex dyes along with the COD removal during anaerobic conditions makes this technology widely acceptable [66].

Decolorization efficiency of UASB for textile wastewater containing azo compounds is reported as approximately 95% [67]. The removal of microbes i.e. helminthes in UASB reactors has been reported on an average of 60-90% which makes the treated wastewater inadequate for direct agricultural usage [68]. UASB provides the benefit of less sludge production, 30-50% lower than the trickling filter technology [18]. Conventionally, the UASB technology encompasses the use of granular sludge or films to retain biomass [62].

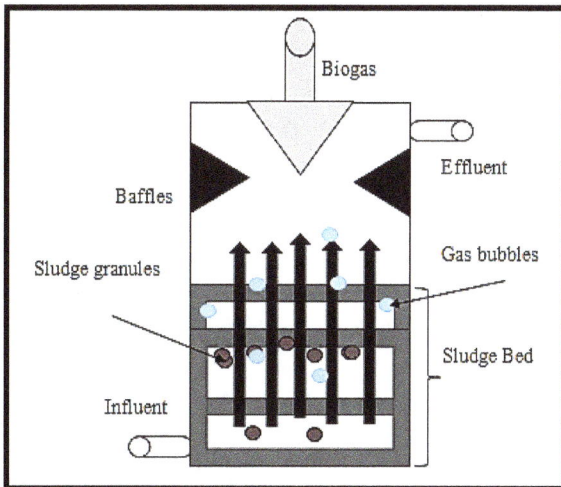

Figure 11. Schematic of upflow anaerobic sludge blanket (UASB) reactor (Adopted from [65]).

5.2   Continuously stirred tank reactor (CSTR)

Anaerobic CSTR has benefits of modest designing, easy operation, low capital costs and importantly self-regulating biomass for treating high strength wastewater [69]. The CSTR configuration as shown in Fig. 12, for activated sludge can be the best approach for several industrial applications by permitting direct control over pH and temperature [70]. For the treatment of textile wastewater subjected to CSTR, the influence of bacterial activity fluctuates the performance i.e. increase in wastewater loading rates provides an average of 90% COD removal in addition to the reduction in color intensity [71]. By providing modifications in existing wastewater treatment system designs, high stability, greater loading capacities and desired process efficiencies can be achieved.

The CSTRs have been highly preferred wastewater reactors for effective wastewater treatment, specifically for the elimination of recalcitrant compounds from textile wastewater [72]. CSTR ordinarily has certain limitations like solids settling especially in treating municipal wastewater where COD is of low strength and solids can go down excessively; the wastewater instantly dilutes in the reactor and all the microorganisms are exposed to food scarce conditions [73].

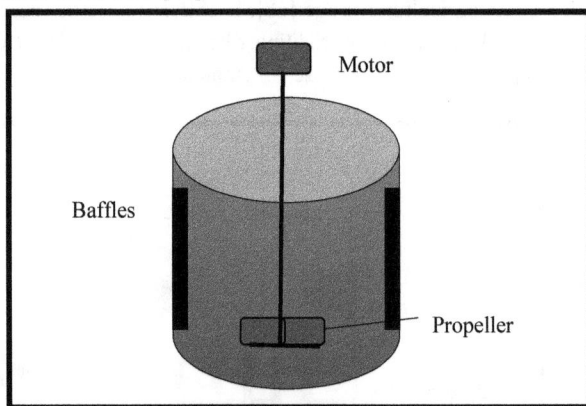

Figure 12. Schematic representation of continuously stirred tank reactor (CSTR) (Adopted from [74]).

5.3   Limitation of anaerobic biological treatment

There are certain limitations that have been observed throughout the research studies while examining anaerobic treatment systems individually. These limitations can be due to:

- slow degradation rate in anaerobic treatment; most of the constituents removed are slightly at lower loading conditions which lead to the requirement of secondary treatment [75].

- though, the anaerobic degradation of pollutants is a cost-effective natural process but its effluent discharge is typically not in compliance with the environmental standards which means this treatment cannot be recommended exclusively [18].

- temperature is an important aspect as it plays an important role in degradation of pollutant kinetics. The limitations of anaerobic system is at low temperatures and longer SRT, where the degradation rate becomes slow suggesting the better potential of this treatment in warmer climates [76].

6. Hybrid treatment advancements

6.1 Coupled anaerobic-aerobic MBR

For the stringent regulations of textile wastewater discharge, MBR technology is the most viable option to adopt [77]. The textile wastewater characteristics and national environmental quality standards (NEQS) of Pakistan environmental protection agency (Pak-EPA) are shown in Table 5 [78].

Table 5: Characteristics of raw textile wastewater and NEQS for effluent discharge (Adopted from [78]).

No.	Parameters	Influent	NEQS
1	Temperature	$33.4 - 42.4^{\circ}C$	$40^{\circ}C$
2	pH	11.2 – 12.0	6-9
3	Total Dissolved Solids (TDS)	2600 – 8900 mg/L	3500 mg/L
4	Total Suspended Solids (TSS)	70 – 300 mg/L	200 mg/L
5	Total Nitrogen (TN)	14 – 110 mg/L	-
6	Chemical Oxygen Demand (COD)	1500 – 4500 mg/L	150 mg/L
7	Biochemical Oxygen Demand (BOD)	300 – 1750	80 mg/L

In view of the strict regulations and standards of industrial effluent wastewater, both aerobic and anaerobic treatments can be used in combination to achieve maximum degradation. The two-phase anaerobic-aerobic treatment eliminates the limitations of each treatment independently. For the color removal from textile industrial discharge, a hybrid system can be composed of two stage treatment, anaerobic CSTR coupled with aerobic MBR. Aromatic amines that usually show complexity in degradation, when subjected to aerobic treatment after the anaerobic stage, their removal efficiency surge up to 87% demonstrating the potential effectiveness of a coupled treatment technology [68]. Nevertheless, conventionally, the typical arrangement adopted for textile wastewater treatment is UASB i.e. anaerobic treatment followed by activated sludge process i.e. aerobic treatment [40].

The combined anaerobic-aerobic treatment has the following advantages which makes this hybrid technology attractive to treat textile wastewater proficiently [18].

- For pretreatment with UASB, the primary sedimentation tank requirement can be eliminated.
- Lower volumes of sludge from anaerobic system provides an ease of disposal of inclusive collective sludge.
- Cost effectiveness and environmental feasibility also adds up in compensations.

9.   Conclusions and future research directions

The reuse of treated industrial effluent can lower the pressure on freshwater consumption for agriculture activities. A number of treatment technologies have been investigated to evaluate the performance efficiency for the treatment of textile wastewater. The treatment through hybrid systems would be better in the framework of sustainable development goals (SDG). The justifiable approach to implement hybrid anaerobic – aerobic biological system to treat textile wastewater would be very environmental friendly. Textile wastewater is considered to be of very high strength relative to municipal wastewater, so the treatment implementation will result in sustainable outcomes and recovery of the resources i.e. biogas production. Maximum removal of high load organics from wastewater can be effectually accomplished by anaerobic treatment in order to recover methane gas (biogas production). When this anaerobic treatment is coupled with aerobic membrane based treatment, the treatment level would reach its maximum efficiency as membrane thoroughly eliminates viruses and unsolicited bacterial organisms from wastewater.

When aerobic treatment is coupled with anaerobic treatment, the aeration cost is cut down to half or less due to the reduced organic load in comparison to self-regulating aerobic biological system with or without membrane filtration. Aerobic membrane bioreactor (AeMBR) subjected to high organic load requires enhanced aeration operational cost. Textile industrial wastewater containing increased organics, if treated directly through an activated sludge process (ASP) or AeMBR would involve increase in the operational cost in addition to the maintenance cost. The hybrid system would be the best option in which anaerobic treatment is followed by aerobic treatment for efficiency enhancement. The aerobic post treatment of anaerobic permeate provides the reduction of residual organics and ammonium nitrification.

It is recommended that laboratory scale studies should be up-scaled to the level of pilot and commercial scale. The current condition of depletion of water resources dictates the importance of adopting hybrid anaerobic-aerobic treatment technologies to treat textile wastewater in order to reclaim water and protect the environment from further degradation. The pioneering approaches of recommended wastewater treatment technologies would contribute a lot towards industrial sector of a country along with the suggestion of conservation of energy resources like biogas that could bring new innovations in the industrial setup.

References

[1]   C.R. Holkar, A.J. Jadhav, D.V. Pinjari, N.M. Mahamuni, A.B. Pandit, A critical review on textile wastewater treatments: possible approaches, J. Environ. Manage. 182 (2016) 351-366. https://doi.org/10.1016/j.jenvman.2016.07.090

[2]   R. Shamey, T. Hussein, Critical solutions in the dyeing of cotton textile materials, Text. Prog. 37 (2005) 1-84. https://doi.org/10.1533/tepr.2005.0001

[3]   C.S. Rodrigues, L.M. Madeira, R.A. Boaventura, Synthetic textile dyeing wastewater treatment by integration of advanced oxidation and biological processes–Performance analysis with costs reduction, J. Environ. Chem. Eng. 2 (2014) 1027-1039. https://doi.org/10.1016/j.jece.2014.03.019

[4]   M.S. Miah, Cost-effective treatment technology on textile industrial wastewater in Bangladesh, J. Chem. Eng. 27 (2013) 32-36. https://doi.org/10.3329/jce.v27i1.15855

[5]   I. Bisschops, H. Spanjers, Literature review on textile wastewater characterisation. Environ. Technol. 24 (2003) 1399-411. https://doi.org/10.1080/09593330309385684

[6]   M. Meneses, J.C. Pasqualino, F. Castells, Environmental assessment of urban wastewater reuse: treatment alternatives and applications, Chemosphere. 81 (2010) 266-272. https://doi.org/10.1016/j.chemosphere.2010.05.053

[7]   M.W. Rosegrant, X. Cai, S.A. Cline, World water and food to 2025: dealing with scarcity. Int. Food Policy Res. Inst. (2002)

[8]   G. Ciardelli, L. Corsi, M. Marcucci, Membrane separation for wastewater reuse in the textile industry, Resour. Conserv. Recycl. 31 (2001) 189-197. https://doi.org/10.1016/S0921-3449(00)00079-3

[9]   A.K. Verma, R.R. Dash, P. Bhunia, A review on chemical coagulation/flocculation technologies for removal of colour from textile wastewaters, J. Environ. Manage. 93 (2012) 154-168. https://doi.org/10.1016/j.jenvman.2011.09.012

[10]   S. Ding, Z. Li, R. Wang, Overview of dyeing wastewater treatment technology, Water Resour. Prot. 26 (2010) 73-78.

[11]   M.A. Shaikh, Water conservation in textile industry. Pakistan Textile Journal (2009) 48-51.

[12]   B. Lesjean, E.H. Huisjes, Survey of the European MBR market: trends and perspectives, Desalination 231 (2008) 71-81. https://doi.org/10.1016/j.desal.2007.10.022

[13]   H. Seif, M. Malak, Textile wastewater treatment. InSixth International Water Technology Conference, IWTC (2001) 608-614.

[14]   V. Jegatheesan, B.K. Pramanik, J. Chen, D. Navaratna, C.Y. Chang, L. Shu, Treatment of textile wastewater with membrane bioreactor: a critical review, Bioresour. Technol. 204 (2016) 202-212. https://doi.org/10.1016/j.biortech.2016.01.006

[15]   A.I. Ohioma, N.O. Luke, O. Amraibure, Studies on the pollution potential of wastewater from textile processing factories in Kaduna, Nigeria, J. Toxicol. Environ. Health Sci. 1 (2009) 034-037.

[16]   M.K. Kazi, F. Eljack, N.A. Elsayed, M.M. El-Halwagi, Integration of energy and wastewater treatment alternatives with process facilities to manage industrial flares during normal and abnormal operations: Multiobjective extendible optimization framework, Ind. Eng. Chem. Res. 55 (2016) 2020-2034. https://doi.org/10.1021/acs.iecr.5b03938

[17]   J.M. Dalu, J. Ndamba, Duckweed based wastewater stabilization ponds for wastewater treatment a low cost technology for small urban areas in Zimbabwe, Phys. Chem. Earth. 28 (2003) 1147-1160. https://doi.org/10.1016/j.pce.2003.08.036

[18]   C.D. Chernicharo, Post-treatment options for the anaerobic treatment of domestic wastewater, Rev. Environ. Sci. Bio/Technol. 5 (2006) 73-92. https://doi.org/10.1007/s11157-005-5683-5

[19]   E. Corcoran, Sick water?: the central role of wastewater management in sustainable development: a rapid response assessment, UNEP/Earthprint (2010)

[20]   S. Dey, A. Islam, A review on textile wastewater characterization in Bangladesh, Resour. Environ. 5 (2015) 15-44.

[21]   C. Visvanathan, R.B. Aim, K. Parameshwaran, Membrane separation bioreactors for wastewater treatment, Crit. Rev. Environ. Sci. Technol. 30 (2000) 1-48. https://doi.org/10.1080/10643380091184165

[22]   P. Li, X. Wang, G. Allinson, X. Li, X. Xiong, Risk assessment of heavy metals in soil previously irrigated with industrial wastewater in Shenyang, China. J. Hazard. Mater. 161 (2009)516-521. https://doi.org/10.1016/j.jhazmat.2008.03.130

[23]   Z. Wang, M. Xue, , K. Huang, Z. Liu, Textile dyeing wastewater treatment. In Advances in treating textile effluent. In Tech. (2011). https://doi.org/10.5772/22670

[24]   Y. Luo, W. Guo, , H.H. Ngo, L.D. Nghiem, F.I. Hai, J. Zhang, S. Liang, X.C. Wang, A review on the occurrence of micropollutants in the aquatic environment and their fate and removal during wastewater treatment, Sci. Total Environ. 473 (2014) 619-641. https://doi.org/10.1016/j.scitotenv.2013.12.065

[25]   K. Sarayu, S. Sandhya, Current technologies for biological treatment of textile wastewater–a review, Appl. Biochem. Biotechnol. 167 (2012) 645-661. https://doi.org/10.1007/s12010-012-9716-6

[26]   T.A. Rangreez, Inamuddin, A.M. Asiri, B.G. Alhogbi, M. Naushad, Synthesis and ion-exchange properties of graphene th(IV) phosphate composite cation exchanger: Its applications in the selective separation of lead metal ions, Int. J. Environ. Res. Public Health. 14 (2017) 828. https://doi.org/10.3390/ijerph14070828

[27]   G.G. Ying, R.S. Kookana, Triclosan in wastewaters and biosolids from Australian wastewater treatment plants, Environ. Int. 33 (2007) 199-205. https://doi.org/10.1016/j.envint.2006.09.008

[28]   C. Silva, S. Quadros, P. Ramalho, H. Alegre, M.J. Rosa, , Translating removal efficiencies into operational performance indices of wastewater treatment plants, Water Res. 57 (2014) 202-214. https://doi.org/10.1016/j.watres.2014.03.025

[29]   M.T. Yagub, T.K. Sen, S. Afroze, H.M. Ang, Dye and its removal from aqueous solution by adsorption: a review, Adv. Colloid Interface Sci. 209 (2014) 172-184. https://doi.org/10.1016/j.cis.2014.04.002

[30] T.L. Petzoldt, The effect of azo textile dyes on gross primary production and community respiration in an artificial environment (2014).

[31] M.T. Dao, H.A. Le, T.K.T. Nguyen, V.C.N. Nguyen, Effectiveness on color and COD of textile wastewater removing by biological material obtained from Cassia fistula seed, J. Viet. Env. 8 (2016) 121-128.

[32] A. Pandey, P. Singh, L. Iyengar, Bacterial decolorization and degradation of azo dyes, Int. Biodeterior. Biodegrad. 59 (2007) 73-84. https://doi.org/10.1016/j.ibiod.2006.08.006

[33] S. Şen, G.N. Demirer, Anaerobic treatment of real textile wastewater with a fluidized bed reactor, Water Res. 37 (2003) 1868-1878. https://doi.org/10.1016/S0043-1354(02)00577-8

[34] N. Çiçek, , J.P. Franco, M.T. Suidan, V. Urbain, J. Manem, Characterization and comparison of a membrane bioreactor and a conventional activated-sludge system in the treatment of wastewater containing high-molecular-weight compounds, Water Environ. Res. 71 (1999) 64-70. https://doi.org/10.2175/106143099X121481

[35] M.B. Ahmed, J.L. Zhou, H.H. Ngo, W. Guo, N.S. Thomaidis, J. Xu, Progress in the biological and chemical treatment technologies for emerging contaminant removal from wastewater: A critical review, J. Hazard. Mater. 323 (2017) 274-298. https://doi.org/10.1016/j.jhazmat.2016.04.045

[36] J. Radjenović, M. Petrović, D. Barceló, Fate and distribution of pharmaceuticals in wastewater and sewage sludge of the conventional activated sludge (CAS) and advanced membrane bioreactor (MBR) treatment, Water Res. 43 (2009) 831-841. https://doi.org/10.1016/j.watres.2008.11.043

[37] M. Inyang, R. Flowers, D. McAvoy, E. Dickenson, Biotransformation of trace organic compounds by activated sludge from a biological nutrient removal treatment system, Bioresour.Technol. 216 (2016) 778-784. https://doi.org/10.1016/j.biortech.2016.05.124

[38] A.D. Kotzapetros, P.A. Paraskevas, A.S. Stasinakis, Design of a modern automatic control system for the activated sludge process in wastewater treatment, Chin. J. Chem. Eng. 23 (2015) 1340-1349. https://doi.org/10.1016/j.cjche.2014.09.053

[39] T. El Moussaoui, Y. Jaouad, L. Mandi, B. Marrot, N. Ouazzani, Biomass behaviour in a conventional activated sludge system treating olive mill wastewate, Environ. Technol. (2017) 1-13.

[40] Y.K. Oh, Y.J. Kim, Y. Ahn, S.K. Song, S. Park, Color removal of real textile wastewater by sequential anaerobic and aerobic reactors, Biotechnol. Bioprocess Eng. 9 (2004) 419-422. https://doi.org/10.1007/BF02933068

[41] S. Judd, The MBR book: principles and applications of membrane bioreactors for water and wastewater treatment. Elsevier (2010)

[42] W. Yang, N. Cicek, J. Ilg, State-of-the-art of membrane bioreactors: Worldwide research and commercial applications in North America, J. Membr. Sci. 270 (2006) 201-211. https://doi.org/10.1016/j.memsci.2005.07.010

[43] Z. Hirani, J. Oppenheimer, J. DeCarolis, A. Kiser, B. Rittmann, Membrane bioreactor effluent water quality and technology–organics, nutrients and microconstituents removal, Proc. Water Environ. Fed. 10 (2010) 5880-5890. https://doi.org/10.2175/193864710798193987

[44] M. Kraume, U. Bracklow, M. Vocks, A. Drews, Nutrients removal in MBRs for municipal wastewater treatment, Water Sci. Technol. 51 (2005) 391-402.

[45] J. Oppenheimer, B. Rittmann, J. DeCarolis, Z. Hirani, A. Kiser, Investigation of membrane bioreactor effluent water quality and technology, Wate Reuse Research Foundation (2012)

[46] Y. Xiao, D.J. Roberts, A review of anaerobic treatment of saline wastewater, Environ. Technol. 31 (2010) 1025-1043. https://doi.org/10.1080/09593331003734202

[47] I. Martin, M. Pidou, A. Soares, S. Judd, Jefferson, B., Modelling the energy demands of aerobic and anaerobic membrane bioreactors for wastewater treatment, Environ. Technol. 32 (2011) 921-932. https://doi.org/10.1080/09593330.2011.565806

[48] P. Bérubé, Membrane bioreactors: Theory and applications to wastewater reuse, Sustainability Sci. Engn. 2 (2010) 255-292. https://doi.org/10.1016/S1871-2711(09)00209-8

[49] S.H. Baek, K.R. Pagilla, Aerobic and anaerobic membrane bioreactors for municipal wastewater treatment. Water Environ. Res. 78 (2006) 133-140. https://doi.org/10.2175/106143005X89599

[50] P. Le-Clech, Membrane bioreactors and their uses in wastewater treatments, Appl. Microbiol. Biotechnol. 88 (2010) 1253-1260. https://doi.org/10.1007/s00253-010-2885-8

[51]   B. Verrecht, S. Judd, G. Guglielmi, C. Brepols, J.W. Mulder, An aeration energy model for an immersed membrane bioreactor, Water Res. 42 (2008) 4761-4770. https://doi.org/10.1016/j.watres.2008.09.013

[52]   H. Ozgun, R.K. Dereli, M.E. Ersahin, C. Kinaci, H. Spanjers, J.B. van Lier, A review of anaerobic membrane bioreactors for municipal wastewater treatment: integration options, limitations and expectations, Sep. Purif. Technol. 118 (2013) 89-104. https://doi.org/10.1016/j.seppur.2013.06.036

[53]   A. Sofia, W.J. Ng, S.L. Ong, Engineering design approaches for minimum fouling in submerged MBR, Desalination 160 (2004) 67-74. https://doi.org/10.1016/S0011-9164(04)90018-5

[54]   M.W.D. Brannock, H. De Wever, Y. Wang, G. Leslie, Computational fluid dynamics simulations of MBRs: Inside submerged versus outside submerged membranes, Desalination 236 (2009) 244-251. https://doi.org/10.1016/j.desal.2007.10.073

[55]   M.D. Seib, K.J. Berg, D.H. Zitomer, Low energy anaerobic membrane bioreactor for municipal wastewater treatment, J. Membr. Sci. 514 (2016) 450-457. https://doi.org/10.1016/j.memsci.2016.05.007

[56]   A.Y. Hu, D.C. Stuckey, Treatment of dilute wastewaters using a novel submerged anaerobic membrane bioreactor, J. Environ. Eng. 132 (2006) 190-198. https://doi.org/10.1061/(ASCE)0733-9372(2006)132:2(190)

[57]   B.Q. Liao, J.T. Kraemer, D.M. Bagley, Anaerobic membrane bioreactors: applications and research directions, Crit. Rev. Environ. Sci. Technol. 36 (2006) 489-530. https://doi.org/10.1080/10643380600678146

[58]   I.S. Chang, P.L. Clech, B. Jefferson, S. Judd, Membrane fouling in membrane bioreactors for wastewater treatment, J. Environ. Eng. 128 (2002) 1018-1029. https://doi.org/10.1061/(ASCE)0733-9372(2002)128:11(1018)

[59]   A. Saddoud, M. Ellouze, A. Dhouib, S. Sayadi, A comparative study on the anaerobic membrane bioreactor performance during the treatment of domestic wastewaters of various origins, Environ. Technol. 27 (2006) 991-999. https://doi.org/10.1080/09593332708618712

[60]   M. Simonič, Perspectives of Textile Wastewater Treatment Using MBR: A review, Text. Light. Ind Sci. Tech. (2013).

[61]   T. Jiang, S. Myngheer, D.J. De Pauw, H. Spanjers, I. Nopens, M.D. Kennedy, G. Amy, P.A. Vanrolleghem, Modelling the production and degradation of soluble

microbial products (SMP) in membrane bioreactors (MBR), Water Res. 42 (2008) 4955-4964. https://doi.org/10.1016/j.watres.2008.09.037

[62] H.J. Lin, K. Xie, B. Mahendran, D.M. Bagley, K.T. Leung, S.N. Liss, B.Q. Liao, Sludge properties and their effects on membrane fouling in submerged anaerobic membrane bioreactors (SAnMBRs), Water Res. 43 (2009) 3827-3837. https://doi.org/10.1016/j.watres.2009.05.025

[63] S. McHugh, M. Carton, G. Collins, V. O'flaherty, S. McHugh, Reactor performance and microbial community dynamics during anaerobic biological treatment of wastewaters at 16–37 C, FEMS Microbiol. Ecol. 48 (2004) 369-378. https://doi.org/10.1016/S0168-6496(04)00080-7

[64] A.E. Maragkaki, M. Fountoulakis, A. Gypakis, A. Kyriakou, K. Lasaridi, T. Manios, Pilot-scale anaerobic co-digestion of sewage sludge with agro-industrial by-products for increased biogas production of existing digesters at wastewater treatment plants, Waste Manage. 59 (2017) 362-370. https://doi.org/10.1016/j.wasman.2016.10.043

[65] X. Lu, G. Zhen, A.L. Estrada, M. Chen, J. Ni, T. Hojo, K. Kubota, Y.Y. Li, Operation performance and granule characterization of upflow anaerobic sludge blanket (UASB) reactor treating wastewater with starch as the sole carbon source, Bioresour.Technol. 180 (2015) 264-273. https://doi.org/10.1016/j.biortech.2015.01.010

[66] A.B. Dos Santos, F.J. Cervantes, J.B. van Lier, Review paper on current technologies for decolourisation of textile wastewaters: perspectives for anaerobic biotechnology, Bioresour. Technol. 98 (2007) 2369-85. https://doi.org/10.1016/j.biortech.2006.11.013

[67] T. Robinson, G. McMullan, R. Marchant, P. Nigam, Remediation of dyes in textile effluent: a critical review on current treatment technologies with a proposed alternative, Bioresour. Technol. 77 (2001) 247-255. https://doi.org/10.1016/S0960-8524(00)00080-8

[68] M. Işık, D.T. Sponza, Anaerobic/aerobic treatment of a simulated textile wastewater, Sep. Purif. Technol. 60 (2008) 64-72. https://doi.org/10.1016/j.seppur.2007.07.043

[69] Y.J. Chan, M.F. Chong, C.L. Law, D.G. Hassell, A review on anaerobic–aerobic treatment of industrial and municipal wastewater, Chem. Eng. J. 155 (2009) 1-18. https://doi.org/10.1016/j.cej.2009.06.041

[70]   Z. Yuan, H. Yang, , X. Zhi, J. Shen, Increased performance of continuous stirred tank reactor with calcium supplementation, Int. J. Hyd. Energ. 35 (2010) 2622-2626. https://doi.org/10.1016/j.ijhydene.2009.04.018

[71]   E. Khelifi, H. Bouallagui, Y. Touhami, J.J. Godon, M. Hamdi, Bacterial monitoring by molecular tools of a continuous stirred tank reactor treating textile wastewater, Bioresour. Technol. 100 (2009) 629-633. https://doi.org/10.1016/j.biortech.2008.07.017

[72]   T.T. Firozjaee, G.D. Najafpour, A. Asgari, M. Khavarpour, Biological treatment of phenolic wastewater in an anaerobic continuous stirred tank reactor, Chem. Ind. Chem. Eng. Q. 19 (2013) 173-179. https://doi.org/10.2298/CICEQ120216052F

[73]   B. Gargouri, F. Karray, N. Mhiri, F. Aloui, S. Sayadi, Application of a continuously stirred tank bioreactor (CSTR) for bioremediation of hydrocarbon-rich industrial wastewater effluents, J. Hazard. Mater. 189 (2011) 427-434. https://doi.org/10.1016/j.jhazmat.2011.02.057

[74]   S. Liu, Bioprocess Engineering: Kinetics, Sustainability and Reactor Design. Elsevier (2016)

[75]   U. Marchaim, Biogas processes for sustainable development. Food & Agriculture Org. (1992).

[76]   M. Kamali, T. Gameiro, M.E.V. Costa, I. Capela, Anaerobic digestion of pulp and paper mill wastes–An overview of the developments and improvement opportunities, Chem. Eng. J. 298 (2016) 162-182. https://doi.org/10.1016/j.cej.2016.03.119

[77]   N.O. Yigit, N. Uzal, H. Koseoglu, I. Harman, H. Yukseler, U. Yetis, G. Civelekoglu, M. Kitis, Treatment of a denim producing textile industry wastewater using pilot-scale membrane bioreactor, Desalination 240 (2009) 143-150. https://doi.org/10.1016/j.desal.2007.11.071

[78]   S. Hyder, A. Bari, Characterization and study of correlations among major pollution parameters in textile wastewater, Mehran Univ. Res. J. Eng. Technol. 30 (2011) 577-582.

Chapter 9

Multi-Walled Carbon Nanotubes (MWNTs) and their Composites for the Treatment of Organic Compounds in Industrial and Municipal Wastewater

Qammer Zaib[1] and Farrukh Ahmad[1]*

[1]Department of Chemical and Environmental Engineering, Institute Center for Water and Environment (iWATER), Masdar Institute of Science and Technology, PO Box 54224, Abu Dhabi, United Arab Emirates

*fahmad@masdar.ac.ae

Abstract

The presence of common industrial organic compounds and emerging micropollutants (e.g., pharmaceuticals and personal care products) in industrial and municipal wastewaters, respectively, has been recognized as a serious environmental problem owing to the recalcitrance and persistence of these compounds in the aquatic environment. In the past two decades, carbon nanotubes and their composites have been gradually applied for the removal of undesirable organic contaminants from water via adsorption. This chapter describes the removal of established toxic chemicals from oil and gas produced water and emerging micropollutants from treated municipal wastewater through the application of multiwalled carbon nanotubes (MWNTs) and their composites with titanium dioxide.

Keywords

MWNTs, TiO_2, Adsorption, Photocatalysis, Organic Compounds

Contents

1.   Introduction

Organic compounds constitute a major portion of chemicals which are being discharged into water bodies through anthropogenic activities. The main sources of these aquatic organic contaminants are: (i) discharge of untreated industrial effluent, (ii) reinjection of wastewater from mining and oilfields, (iii) sewage from municipal communities, and (iv) runoff from agricultural farms [1]. The conventional wastewater treatment processes are not capable to completely remove these organic compounds from contaminated waters [2]. Once discharged into the environment, these find their way into water bodies and ultimately end up in drinking water systems [3] where these organic pollutants are held responsible for most of the waterborne diseases in the developing counties [4]. Therefore, it is important to develop and implement innovative water treatment technologies to effectively remove these organic compounds from water and wastewater. Several water treatment technologies have been developed over the years to remove organic compounds including biological treatment, membrane separation, chemical oxidation/reduction, coagulation/flocculation, ion exchange, and adsorption [5]. Adsorption is one of the most effective and useful water treatment technologies to remove organic contaminants due to its high efficiency, ease of operation, selectivity, regeneration potential, and simplicity [6, 7]. Adsorption, in order to remove toxic chemicals from water, is usually carried out on activated carbons, clays, polymers, and zeolites [7]. In particular, activated carbons are known for their high adsorption capacity for organic contaminants removal from aqueous systems [8]. However, their slow adsorption kinetics (towards certain organic compounds due to narrow pore size distribution), limited availability, and significant loss on regeneration have limited their effective use in removing organic pollutants from water [7, 9, 10]. Nanoscale materials are an emerging class of materials that are highly desired for a broad range of applications, including but not limited to catalysis, optical, biological, microelectronic, and environmental applications, due to their unique superior properties [11]. Carbon nanotubes, in particular, have been successfully used to remove organic pollutants from water and wastewater [12-15]. This chapter focuses on the use of multi-walled carbon nanotubes alone and in combination with TiO_2 (a photocatalyst) (i.e. MWNTs-TiO_2) for the removal of organic contaminates from (i) oil produced wastewater containing BTEX (Benzene, Toluene, Ethylbenzene, and Xylene) and other organic contaminants and (ii) treated water containing pharmaceuticals.

2. Removal of hydrocarbons from produced water using MWNTs

Produced water is a byproduct of oil & gas exploration and production activities. It contains a wide range of chemical constituents, including petroleum hydrocarbons, electrolytes, heavy metals, and radionuclides [16]. Its chemical composition (dissolved organic content and salinity) varies considerably over different oil fields and geological formations [17, 18]. The potential environmental impacts of produced water discharge are a matter of concern due to prevalence of significant concentrations of toxic organic compounds it contains. Upon the release of produced water into aquatic ecosystems, these compounds are accumulated in marine fauna [19]. High concentrations of polyaromatic hydrocarbons (PAHs), alkylphenols, and BTEX compounds (i.e. benzene, toluene, ethylbenzene, and xylenes) have been reported to occur in untreated produced water, which can potentially effect water bodies upon release into the environment [19, 20].

To date, a number of physical, chemical, and biological methods have been proposed for the treatment of the organic constituents of produced water. Some of these methods include chemical oxidation, de-oiling, sand filtration, membrane-based separation, adsorption, cyclone separation, chemical precipitation, electrodialysis, photocatalysis, and trickling filters [18, 21]. The application of nanomaterials to water treatment offers various advantages over conventional technologies owing to their unique material characteristics. Carbon based nanomaterials are among the most widely used nanomaterials for environmental applications [10, 11, 14, 22, 23]. Amongst others, some of the desired properties of carbon based nanomaterials for water treatment include their large surface area, high intraparticle diffusion, chemical recalcitrance and high electrical conductivity [10, 11, 13, 24].

The objective of this work was to assess the effectiveness of multiwalled carbon nanotubes (MWNTs) as sorbents for the treatment of dissolved hydrocarbons in produced water. A special emphasis was put on the human-health-risk-driving BTEX compounds, which often determine treatment endpoints for petroleum-contaminated waters. MWNTs were selected for the experiments, instead of single walled carbon nanotubes, owing to their bigger market share [25], easier bulk production [26], lower cost [26], and comparatively low cytotoxicity [27].

2.1 Methodology

2.1.1 Materials

MWNTs, having specific surface area and electrical conductivity of at least 233 m²/g and 100 S/cm respectively, were purchased from Cheaptubes, Inc., (Brattleboro, VT, USA). Abu Dhabi Company for Onshore Oil Operations (ADCO) provided the crude oil from the onshore Asab oilfield located in the Emirate of Abu Dhabi [28] in a sealed container with no headspace. Pure Asab crude oil sample comprised of the BTEX, naphthalene, other aromatic hydrocarbon derivatives, and aliphatic hydrocarbons. The identities of these compounds were confirmed during gas chromatography – mass spectrometry (GC-MS) analysis by matching the mass spectral library to the Wiley-NIST library. Normal alkane analytical standard (C_8–C_{20}), NaCl, $KHCO_3$, $CaCO_3$, and $MgCl_2.6H_2O$ were purchased from Sigma-Aldrich® (Steinheim, Germany). HC BTEX mix standard, comprising of benzene, toluene, ethyl benzene, o-, m-, p-xylene, was also purchased from Supelco, Sigma-Aldrich® (Steinheim, Germany). Deionized (Milli-Q) water and salt-enriched Milli-Q water were used for generating produced waters in the lab by equilibrating with Asab crude oil. In general, Milli-Q water (~18.2 MΩ.cm) was used throughout the study.

2.1.2 Batch adsorption experiments

The batch adsorption experiments were performed to estimate the equilibrium concentrations of organic compounds in MWNTs-produced water systems. All the adsorption experiments were performed in triplicate and at room temperature (23 ± 1 °C). In a typical experiment, small volumes of MWNTs stock solution were added to a 40 mL amber glass vial containing produced water (with minimal headspace) to acquire desired concentrations of MWNTs in produced water. All the vials were sealed with Teflon lining septa and agitated for up to five days before filtering the equilibrated suspension through GF/C filter papers. The filtrate was collected in 10 mL air tight vials which were sealed and stored at 4°C until analysis.

2.2   Results and discussion

2.2.1   Characterization of MWNTs

SEM images were used to study the morphological structure of MWNTs. The outer diameter of an individual MWNT was estimated between 21-25 nm, however, the MWNTs mostly exist in the form of large aggregates [24]. The EDS analysis showed the presence of some metallic impurities (such as magnesium and aluminum) as reported by the manufacturer. The BET isotherm of N_2 adsorption followed type IV as observed earlier for the similar material [29]. Nitrogen adsorption/desorption were carried out at 77 K. The BET surface area of the MWNTs was also in reasonable agreement with the previously reported experimental [29, 30] and theoretical [31] values. The average diameter of MWNTs aggregates was observed to be in the range of 600-1500 nm. SEM observations along with Dynamic Light Scattering (DLS) data suggested that 50 ± 20 MWNTs aggregated together to form a cluster through a pH range of 5 to10 [15]. The zeta potential of the MWNTs was frequently negative and mostly observed between -40 to -22 mV. The detailed discussion on the surface charge of the MWNTs can be found elsewhere [10].

2.2.2   Adsorption removal of BTEX and other organic compounds by MWNTs

MWNTs were effective to adsorb significantly high quantity of organic compounds from produced water. Fig. 1 exhibits the chromatographs for the total ion current of major compounds in produced water after and before treatment with MWNTs (50 mg/L). The complex mixture of organic compounds in produced water is noticeable from the chromatographs in Fig. 1. The BTEX peak height counts were significantly higher (~2.8 e 7 to 4 e 7) than the other dissolved organics in the produced water. These peak heights significantly decreased after treatment of produced water with MWNTs along with the total number of peaks representing the wide range of compounds. The peak heights of octanol, nonane, decane, undecane, and dodecane were also significantly reduced after treatment. Upon comparing the chromatographs, after and before treatment with MWNTs, it can be inferred that MWNTs possess the ability to remove BTEX and other organics from produced water. Therefore, adsorption kinetics and equilibrium studies were carried out to understand the mechanism of adsorption responsible for the uptake of these organic compounds by MWNTs.

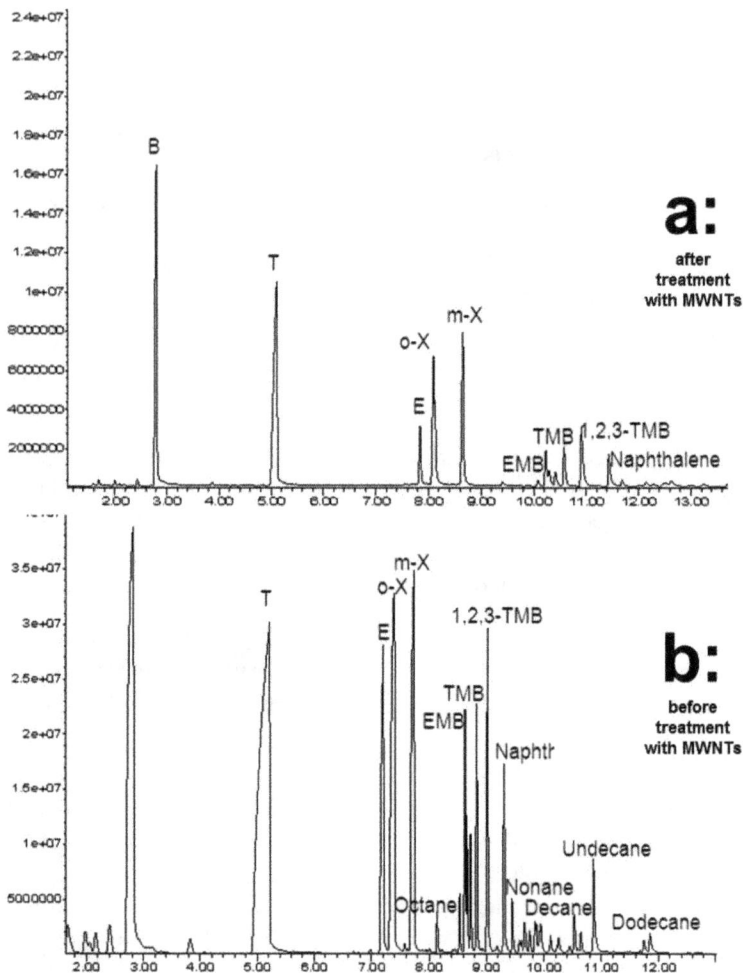

Figure. 1. Comparison of representative chromatograms of produced water samples (a) after and (b) before treatment with MWNTs (50 mg/L) at equilibrium. Aliphatic fractions present in the produced water sample were almost completely removed, while a reduction was observed in the concentration of the BTEX compounds after treatment with MWNTs. Key: o-X (ortho-xylene), m-X (meta- xylene), EMB (ethylmethylbenzene), TMB (trimethylbenzene), 1, 2, 3-TMB (1, 2, 3- trimethylbenzene).

Table 1: Mathematical expressions and nomenclature for the equilibrium adsorption models fitted for the BTEX adsorption on MWNTs.

Adsorption model	Nomenclature			Mathematical expression
Freundlich Isotherm	K_f n	(mg/g) / (mg/L)$^{(1/n)}$	Freundlich affinity coefficient Dimensionless number	$q_e = K_f Ce^{1/n}$
Langmuir Isotherm	Q_0 K_L Ce	mg/g L/mg mg/L	Maximum adsorption capacity Langmuir fitting parameter Equilibrium solution phase concentration	$q_e = \dfrac{Q_0 K_L Ce}{1 + K_L Ce}$

The adsorption kinetics tests were performed for five days where MWNTs concentration was kept constant at 50 mg/L. The adsorption equilibria for benzene, toluene, ethylbenzene, and o-xylene reached earlier (after one day) than m-xylene (after three days). In order to establish the equilibration time, the adsorption kinetics was observed for another two days (total five days). Initially (at 0.5 h contact time), adsorption kinetics proceeded in the following order: toluene (34.6 mg/g-MWNTs) > o-xylene (31.2 mg/g-MWNTs) > benzene (20.6 mg/g-MWNTs) > m-xylene (10.8 mg/g-MWNTs) > ethylbenzene (3.6 mg/g-MWNTs). This order, however, changed to ethylbenzene > o-xylene > toluene > m-xylene > benzene after 4h contact time. Ultimately, after 5 days of contact, the cumulative adsorption of BTEX was: ethylbenzene (183.4 mg/g-MWNTs) > o-xylene (156 mg/g-MWNTs) > m-xylene (155.6 mg/g-MWNTs) > toluene (135 mg/g-MWNTs) > benzene (70 mg/g-MWNTs).

The equilibrium experiments were performed by suspending MWNTs in deionized produced water (DI-PW) and salt enriched produced water (SE-PW). The composition and characteristics of DI-PW and SE-PW can be found elsewhere [15]. After three days of contact time, adsorption equilibria (of BTEX on MWNTs in the presence of DI-PW and SE-PW) were observed. Linearized Freundlich and Langmuir isotherms were used to fit the adsorption data. Table 1 enlists the details of equilibrium models used to evaluate sorption data. The obtained fitting parameters of the two isotherms are presented in Table 2. Langmuir and Freundlich adsorption models have been frequently used to describe adsorption of organic compounds on MWNTs [32-34]. Ideally, the Freundlich isotherm describes multilayer adsorption whereas the Langmuir isotherm assumes a single-layer adsorption. During multilayer adsorption process, the adsorbate molecules initially interact with adsorbent surface before attaching to those of their own kind. Whereas, during mono-layer adsorption, the adsorbate molecules only interacts with the surface of adsorbent [13]. It is clear from the adsorption constants' values listed in Table 2 that the BTEX adsorption on MWNTs could not be explained by Langmuir isotherm model (negative values of Q_0 and K_L). Freundlich isotherm model, on the other hand, reasonably explained the adsorption process. Log K_f values for SE-PW were always lower than that of DI-PW which indicated the suppressed adsorption of BTEX on MWNTs in the presence of background ions. Additionally, the values of correlation coefficient of Freundlich isotherms for SE-PW system were usually lower than that of DI-PW except m-xylene. Hence it can be hypothesized that, in a produced water, BTEX molecules are adsorbed on MWNTs before interacting with each other to reach the adsorption equilibrium state.

Table 2: Equilibrium isotherm model constants and correlation coefficients for the adsorption of BTEX on MWNTs.

	Freundlich			Langmuir		
	SE-PW					
	K_f	n	R^2	Q_0	K_L	R^2
Benzene	1.024	0.436	0.95	-63.29	-0.08	0.87
Toluene	1.011	0.455	0.91	-84.03	-0.07	0.68
Ethylbenzene	1.142	0.535	0.90	-79.37	-0.05	0.67
o-Xylene	1.263	0.425	0.92	-102.04	-0.08	0.76
m-Xylene	1.366	0.436	0.99	-92.59	-0.08	0.98
	DI-PW					
Benzene	1.05	0.339	0.98	-68.87	-0.12	0.83
Toluene	1.10	0.294	0.97	-88.89	-0.14	0.79
Ethylbenzene	22.76	0.469	0.96	-222.22	-0.16	0.87
o-Xylene	2.33	0.318	0.94	-113.77	-0.15	0.71
m-Xylene	1.63	0.274	0.99	-85.69	-0.18	0.85

3. Photo-regenerable MWNT/TiO$_2$ membranes for removing micropollutants from treated wastewater

Several pharmaceutically active compounds are being considered as emerging contaminants because of their frequent detection in the natural environment [2]. They usually enter the aquatic environment either through direct disposal of unused and/or expired medicines or via excretion from humans and animals [35, 36]. Their occurrence in water and wastewater is of concern owing to their ability to adversely affect reproduction, growth, and behavior of non-targeted aquatic organisms [37]. Several researchers have pointed to the inability of wastewater treatment plants to effectively mineralize these pharmaceuticals into innocuous products [38]. The failure of conventional wastewater treatment plants to remove these compounds and their biotransformed daughter products has attracted the attention of researchers around the world over the past two decades [2, 35, 39]. A number of treatments have been proposed, since then, to address this problem such as hydrolysis, chemical oxidation, biodegradation, volatilization, and adsorption [40, 41]. Adsorption is of special importance amongst these processes owing to its ability to remove not only pharmaceutical parent molecules, but also their daughter metabolites, which might cause more harm than the parent compounds [42, 43]. CNTs have been reported to be successfully employed for the removal of heavy metals [44], metalloids [45], polycyclic aromatic hydrocarbons [46], endocrine disrupting compounds [13, 47], and pharmaceuticals [48], chiefly through adsorption [12]. CNTs can be applied for adsorption of target compounds either by suspending freely in solution (as demonstrated in the case of dissolved aqueous contamination [47]) or by modifying them into gas permeable flat sheets or membranes [49].

In this work membranes of various thicknesses were developed from multi-walled carbon nanotubes (MWNTs), which were then grafted with in-house hydrothermally prepared titanium dioxide (TiO$_2$) spheres. Detailed characterization of these membranes was carried out by a number of techniques, including scanning electron microscopy (SEM), electron dispersive x-ray spectroscopy (EDS), zeta potential measurements, thermogravimetric analysis (TGA), and hydrophobicity estimation. Then the efficacy of these membranes for removing common pharmaceuticals from water was evaluated after measuring their permeate flux. Finally, the photocatalytic regeneration potential of these membranes was assessed.

3.1   Methodology

3.1.1   Synthesis and characterization of multi-walled carbon nanotube-titanium dioxide (MWNT-TiO$_2$) membranes

Each MWNTs-TiO$_2$ membrane was prepared in two steps: (1) MWNTs membrane was prepared by following previously developed protocol [50] with some modifications, (2) Grafting of TiO$_2$ spheres was performed by filtration of aqueous dispersion of in-house prepared TiO$_2$ spheres through the previously prepared MWNT membrane. The TiO$_2$ retained on the surface of MWNTs membrane resulting in MWNT-TiO$_2$ membrane. The resulting membrane was washed with copious amount of deionized water to remove surfactant and other soluble impurities. The relationship between permeate flux and thickness of MWNTs-TiO$_2$ membranes is presented in Fig. 2. The membrane thickness was observed to be inversely related to permeate flux. The membrane with MWNTs and

TiO_2 masses of 10 mg each having a thickness of 28 μm and an aqueous permeate flux ~2 L/m².h was selected for further study.

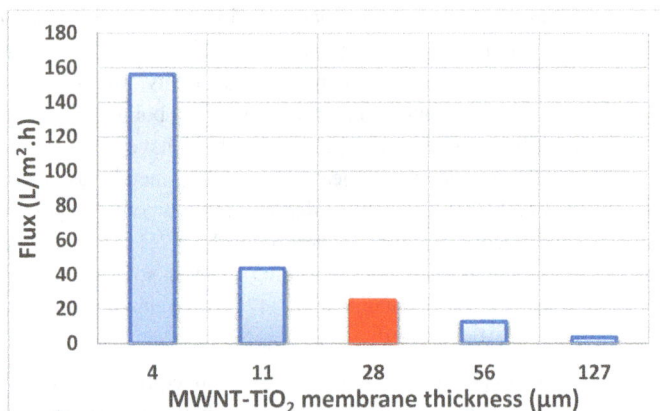

Figure 2. The permeate flux of MWNTs-TiO₂ membranes for water were inversely proportional to their thicknesses. The plot displays the flux of deionized water through MWNTs-TiO₂ membranes of various thicknesses under the effect of gravity. Red bar indicates the selected membrane having 10 mg each of MWNTs and TiO₂ resulting in 28 μm thickness and providing 25 L/m².h permeate. The pressure was kept constant at 3.43 N/cm² by controlling water head.

Thorough characterization of these membranes was carried out by a number of techniques, including scanning electron microscopy (SEM), electron dispersive x-ray spectroscopy (EDS), zeta potential measurements, thermogravemetric analysis (TGA), and hydrophobicity estimation.

Continuous adsorption experiments and photocatalytic regeneration of membranes

Three pharmaceuticals (acetaminophen, ibuprofen, and carbamazepine) were selected on the basis of their high frequency of occurrence in water systems [39, 43, 51-53] and their physicochemical properties (Table 3). A broad range of parameters like log Kow (0.46 to 3.97) and water solubilities (17.7 to 14000 mg/L) were covered by using these compounds as represented in Table 3. The pharmaceuticals were allowed to adsorb on to the MWNTs membrane (control) and MWNTs-TiO₂ membranes. After saturation, the MWNTs-TiO₂ membranes were regenerated (cleaned) using UV 254 nm light.

Table 3. Physical properties of Compounds [59]

Compound	Formulae	Mol. Wt [g/mol]	Structure	Log Kow	Water Solubility (25 °C) [mg/L]	pKa
Acetaminophen	$C_8H_9NO_2$	151.17		0.46	1.40E+04	9.5
Carbamazepine	$C_{15}H_{12}N_2O$	236.28		2.45	17.7	13.9 [60]
Ibuprofen	$C_{13}H_{18}O_2$	206.29		3.97	21	4.9

3.2   Results and discussion

3.2.1   Characterization of MWNTs, TiO₂, and MWNTs-TiO₂ membrane

Electron microscopy, zeta potential technique, and TGA were applied to characterize MWNTs, in-house prepared TiO_2 spheres, and MWNTs-TiO_2 membrane. The morphology of the three materials was analyzed using SEM. The electron micrograph in Fig. 3 (a) shows the overall structure of a loosely packed dense mesh of MWNTs that acts as a water-permeable membrane. Each square millimeter of the membrane contains a large number of MWNTs in which the interstices generated between the randomly ordered MWNTs may allow water and other small molecules to pass through. Fig. 3 (b) represents the successful grafting or embedding of functionalized TiO_2 spheres on the surface of MWNTs membrane. The porous morphology and the ultra-thin, superimposed TiO_2 spheres significantly increased the overall specific surface area of the membrane. The selected area in Fig. 3 (b) shows nanotubes entangling the TiO_2 spheres and effectively binding the two together. This arrangement is expected to offer minimum resistance to fluid flow. The MWNT membranes grafted with TiO_2 spheres were selected for further investigation. The EDS analysis was performed on the MWNTs-TiO_2 membrane to look into the elemental composition of the membrane. Fig. 3(c), the EDS elemental analysis result, indicated the presence of carbon, titanium, and oxygen, thereby confirming the presence of these elemental components of the MWNTs-TiO_2 membrane.

Zeta potential measurements revealed that MWNTs were negatively charged. This negative charge on MWNTs decreased with an increase in concentration of sodium chloride in the background medium (i.e., water). The negative surface charge of

MWNTs decreased from -40.6 ± 1.5 to -22 ± 3.9 mV upon increasing the concentration of sodium chloride from ~0 to 100 mM. The TGA analysis was carried out for both MWNTs and TiO_2 in nitrogen atmosphere in order to monitor their weight loss and thermal stability. In MWNTs, a 10% weight loss was observed between 53-100°C. This can be attributed to moisture, volatiles, and other impurities that have a low flash point [54]. Contact angle measurements were applied to measure extent of hydrophilicity [55]. The contact angles of both MWNTs and MWNTs-TiO_2 membranes were found to be less than the 90° showing that the membranes were hydrophilic. However, the MWNTs-TiO_2 membranes appeared to be more hydrophilic than MWNTs membranes [10].

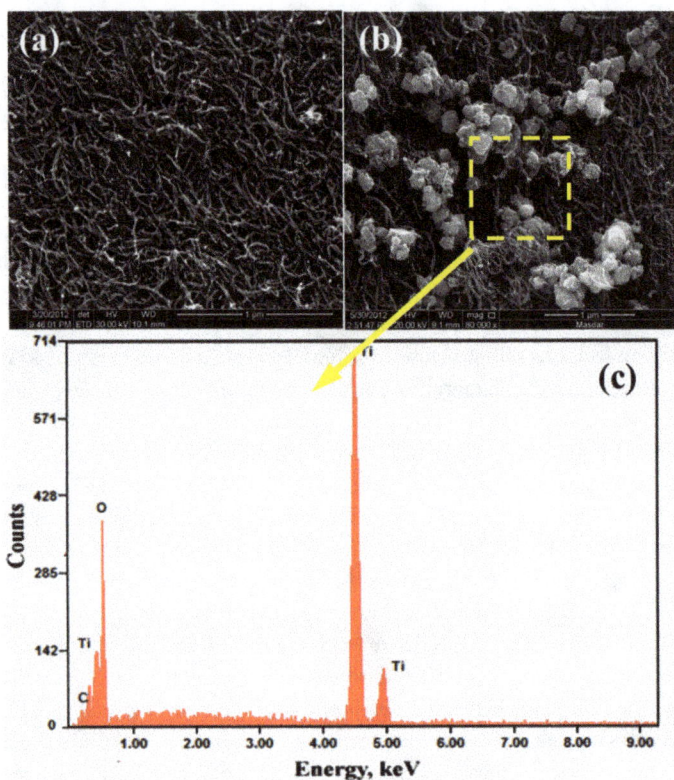

Figure 3. SEM photomicrographs of (a) MWNTs membrane, (b) TiO_2 spheres distributed in MWNTs membrane, and (c) EDS of MWNTs- TiO_2 membrane confirming the presence of carbon, titanium, and oxygen.

3.2.2 Removal of pharmaceuticals from water by MWNTs and MWNTs-TiO$_2$ membranes

The ability of MWNTs-TiO$_2$ membranes to remove pharmaceuticals from water was investigated. Experiments were also conducted with MWNTs-only membranes (i.e., without any grafted TiO$_2$ spheres) to evaluate the superiority of MWNTs-TiO$_2$ membranes over MWNTs membranes in terms of pharmaceuticals removal efficiency. Fig. 4 illustrates the removal of pharmaceuticals from water by MWNTs and MWNTs-TiO$_2$ membranes. The influent pharmaceutical concentration (10 mg/L) and solution flux (40 L/m^2·h) were kept constant, and effluent fractions were analyzed to assess performance of membranes and their saturation over time.

The pharmaceutical removal efficiencies of MWNTs-TiO$_2$ membranes were consistently higher than that of MWNTs-only membranes probably due to simultaneous adsorption and degradation of pharmaceuticals in the presence of TiO$_2$ [56-58]. Acetaminophen exhibited the least affinity towards MWNTs compared to the other two compounds. The removal of acetaminophen was ~10% when 10 mg/L acetaminophen solution was allowed to flow through a MWNTs membrane. However, in the case of the MWNTs-TiO$_2$ membrane, this removal was initially increased to ~24 % (2.4 mg/L from 10 mg/L influent solution). Ultimately, this removal gradually decreased to <1% when 20 mL of acetaminophen solution was passed through. The MWNTs-TiO$_2$ membrane was regenerated by UV light. After regeneration, the initial removal efficiency of MWNTs-TiO$_2$ membrane decreased to ~18% (compared to 24% in the 1st run) for the removal of acetaminophen. It should be noted that, even after regeneration, the adsorption removal efficiency of MWNTs-TiO$_2$ membranes was eighty percent higher than those of MWNTs which supports the hypothesis of continued adsorption and photolysis of acetaminophen on MWNTs-TiO$_2$ membranes. The MWNTs-only and MWNTs-TiO$_2$ membranes were tested for the removal of more hydrophobic pharmaceutical, ibuprofen. The adsorption removal efficiency and saturation time of both (MWNTs-only and MWNTs-TiO$_2$) membranes for ibuprofen were higher than that of acetaminophen. A maximum of 30% ibuprofen removal by the MWNTs membrane for the first 5 mL of 10 mg/L ibuprofen solution was observed. This removal efficiency was gradually decreased below/down to 10% upon the passage of an additional 10 mL of the same solution. Finally, the removal decreased to ~5% and after the continuous flow of (a total) 40 mL ibuprofen solution. In the case of MWNTs-TiO$_2$ membrane, initially ~45% (4.5 mL) ibuprofen was removed. After 20 mL of the solution had passed through the membrane, the removal of ibuprofen was 37%, significantly higher than in the case of the MWNTs-only membrane. The adsorption efficiency of MWNTs-TiO$_2$ membrane decreased to 6% after passing 100 mL ibuprofen (10 mg/L) solution. After regeneration, the membrane was effective in removing 32% of influent ibuprofen, initially. The trends of ibuprofen removal were similar in case of MWNTs and MWNTs-TiO$_2$ (before and after regeneration) membranes. Carbamazepine exhibited high affinity towards both MWNTs-only and MWNTs-TiO$_2$ membranes. Carbamazepine removal by MWNTs-TiO$_2$ started from 80% of the influent concentration (8 mg/L was removed from 10 mg/L solution) which was approximately double that of the MWNTs-only membrane (39% removal). After 20 mL, MWNTs-TiO$_2$ membrane was still able to remove ~45% of the carbamazepine from the solution compared to only 5% removal by MWNTs-only membrane. However, in the case of carbamazepine, the membranes saturated at faster rate compared to ibuprofen. The regenerated MWNTs-TiO$_2$ membrane initially removed 53% of carbamazepine from solution which was 13% higher than the removal by non-photoregenerable MWNTs-only membrane.

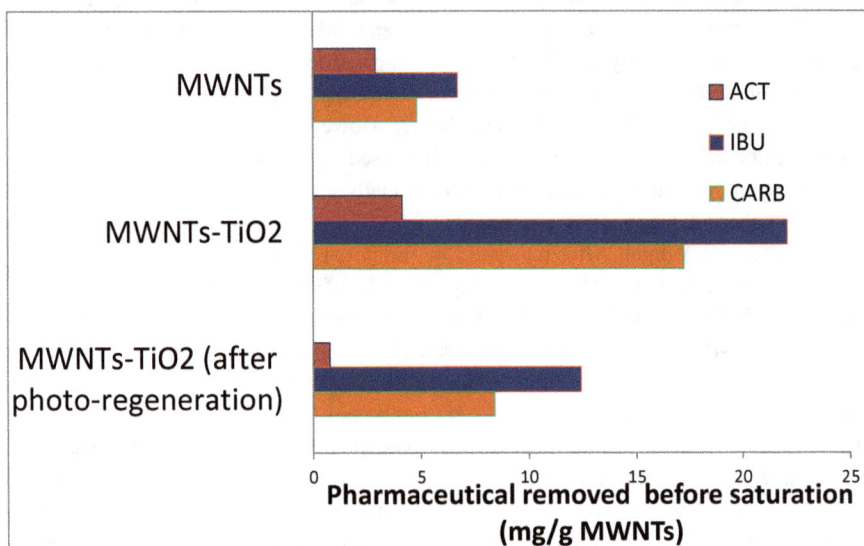

Figure 4 The mass loadings of pharmaceuticals on (i) MWNTs membranes, (ii) MWNTs-TiO$_2$ membranes during 1st run, and (iii) MWNTs-TiO$_2$ membranes after photo-regeneration. The pharmaceutical mass removal was calculated upon complete saturation of membranes. The influent concentration of acetaminophen, ibuprofen, and carbamazepine was 10 mg/L each in deionized water (18 µS/cm).

The membranes exhibited differential trends towards the removal of different pharmaceuticals tested. Fig. 4 shows mass loading of membranes upon complete saturation with pharmaceuticals. The results clearly exhibit superiority of MWNTs-TiO$_2$ membrane over the MWNTs-only membrane for the removal of selected pharmaceuticals. For each gram of MWNTs, a total of 2.9, 6.7, and 4.8 mg of acetaminophen, ibuprofen, and carbamazepine were removed by MWNTs membranes, respectively. This removal increased to 4.13, 22.1, and 17.2 mg in the case of MWNTs-TiO$_2$ membranes for the first run and ultimately decreased to 0.78, 12.4, and 8.4 mg for the MWNTs-TiO$_2$ membranes after photo-regeneration. This adsorption removal of pharmaceuticals is comparable to the reported data of acetaminophen [51], ibuprofen [53], and carbamazepine [52] adsorption by carbonaceous materials. However, these researchers chiefly performed experiments using activated carbon suspensions which were hard to recover from the system and almost impossible to regenerate for further use. In contrast, the MWNTs-only and MWNTs-TiO$_2$ membranes, developed and applied in this work, could be easily recovered and regenerated. This improved adsorption of ibuprofen on MWNTs-only and MWNTs-TiO$_2$ membranes paved a way to further investigate the effect of pH on its adsorption. In addition, the difference in removal between first-time sorption (or 1st run) and post-regeneration sorption for all three sorbates warrants further study, specifically to optimize the efficacy of the photo-regeneration parameters such as regeneration wavelength and exposure time.

4.   Conclusions and future directions

MWNTs have been potential contributer to improved removal of organic contaminants from water. Our experimental investigations showed their effectiveness for removing BTEX compounds and aliphatic hydrocarbons from industrial wastewater (oilfield produced water) with and without high concentrations of salts in the background solution. Along with the removal of known toxins (BTEX) from the industrial wastewater, MWNTs were effective in removing emerging contaminants (pharmaceuticals) from municipal wastewater. MWNTs and their composites with TiO$_2$ effectively adsorbed and photocatalyzed ibuprofen (an ionizable pharmaceutical) and carbamazepine (an antipsychotic drug). The use of MWNTs and MWNTs-metal oxide composites to remove undesirable aqueous organic compounds might play a major role in future water treatment facilities. In developing countries, where the infrastructure to construct advanced water treatment facility is unaffordable, MWNTs and their composites can be used at point of use. Conversely, in developed countries MWNTs can be used as an additional protective measure to ensure water quality and enhance the safety of treated water. To improve the understanding of adsorption and catalysis properties of MWNTs and its composites with various metal oxides further research is required.

References

[1] A. Bhatnagar, M. Sillanpää, Utilization of agro-industrial and municipal waste materials as potential adsorbents for water treatment—a review, Chem. Eng. J. 157 (2010) 277-296. https://doi.org/10.1016/j.cej.2010.01.007

[2] D.W. Kolpin, E.T. Furlong, M.T. Meyer, E.M. Thurman, S.D. Zaugg, L.B. Barber, H.T. Buxton, Pharmaceuticals, Hormones, and Other Organic Wastewater Contaminants in U.S. Streams, 1999–2000: A National Reconnaissance, Environ. Sci. Technol. 36 (2002) 1202-1211. https://doi.org/10.1021/es011055j

[3] B. Halling-Sørensen, S.N. Nielsen, P. Lanzky, F. Ingerslev, H.H. Lützhøft, S. Jørgensen, Occurrence, fate and effects of pharmaceutical substances in the environment-A review, Chemosphere 36 (1998) 357-393. https://doi.org/10.1016/S0045-6535(97)00354-8

[4] L. Haller, G. Hutton, J. Bartram, Estimating the costs and health benefits of water and sanitation improvements at global level, J. Water Health 5 (2007) 467-480. https://doi.org/10.2166/wh.2007.008

[5] H. Ma, C. Burger, B.S. Hsiao, B. Chu, Highly permeable polymer membranes containing directed channels for water purification, ACS Macro Lett. 1 (6) (2012) 723-726. https://doi.org/10.1016/j.matchemphys.2006.02.012

[6] Y.F. Sun, A.M. Zhang, Y.Yin, Y.M. Dong, Y.C. Cui, X. Zhang, J.M. Hong, The investigation of adsorptive performance on modified multi-walled carbon nanotubes by mechanical ball milling, Mater. Chem. Phys. 101 (2007) 30-34.

[7] J.-G. Yu, X.-H. Zhao, H. Yang, X.-H. Chen, Q. Yang, L.-Y. Yu, J.-H. Jiang, X.-Q. Chen, Aqueous adsorption and removal of organic contaminants by carbon nanotubes, Sci. Total Environ. 482 (2014) 241-251.

[8] A. Demirbas, Agricultural based activated carbons for the removal of dyes from aqueous solutions: a review, J. Hazard. Mater. 167 (2009) 1-9. https://doi.org/10.1016/j.jhazmat.2008.12.114

[9] Q. Zaib, Characterization, functionalization, and adsorption behavior of single-walled carbon nanotubes towards bisphenol A and 17 beta-estradiol, Thesis (2011) University of South Carolina.

[10] Q. Zaib, B. Mansoor, F. Ahmad, Photo-regenerable multi-walled carbon nanotube membranes for the removal of pharmaceutical micropollutants from water, Environ. Sci.: Processes Impacts 15 (2013) 1582-1589. https://doi.org/10.1039/c3em00150d

[11] R.H. Baughman, A.A. Zakhidov, W.A. de Heer, Carbon nanotubes--the route toward applications, Science 297 (2002) 787-792. https://doi.org/10.1126/science.1060928

[12] V.K.K. Upadhyayula, S. Deng, M.C. Mitchell, G.B. Smith, Application of carbon nanotube technology for removal of contaminants in drinking water: A review, Sci. Total Environ. 408 (2009) 1-13. https://doi.org/10.1016/j.scitotenv.2009.09.027

[13] L. Joseph, Q. Zaib, I.A. Khan, N.D. Berge, Y.-G. Park, N.B. Saleh, Y. Yoon, Removal of bisphenol A and 17α-ethinyl estradiol from landfill leachate using single-walled carbon nanotubes, Water Res. 45 (2011) 4056-4068. https://doi.org/10.1016/j.watres.2011.05.015

[14] Q. Zaib, I.A. Khan, N.B. Saleh, J.R. Flora, Y.-G. Park, Y. Yoon, Removal of Bisphenol A and 17β-Estradiol by Single-Walled Carbon Nanotubes in Aqueous Solution: Adsorption and Molecular Modeling, Water Air Soil Pollut. (2012) 1-13.

[15] Q. Zaib, O.D. Aina, F. Ahmad, Using multi-walled carbon nanotubes (MWNTs) for oilfield produced water treatment with environmentally acceptable endpoints, Environ. Sci.: Processes Impacts 16 (2014) 2039-2047. https://doi.org/10.1039/C4EM00201F

[16] D.F. Boesch, N.N. Rabalais, Long-term Environmental Effects of Offshore Oil and Gas Development, Taylor & Francis, Oxford, 2005.

[17] E.T. Igunnu, G.Z. Chen, Produced water treatment technologies, Int. J. Low-Carbon Technol. 9 (2012) 155-157.

[18] A. Fakhru'l-Razi, A. Pendashteh, L.C. Abdullah, D.R. Biak, S.S. Madaeni, Z.Z. Abidin, Review of technologies for oil and gas produced water treatment, J. Hazard. Mater. 170 (2009) 530-551. https://doi.org/10.1016/j.jhazmat.2009.05.044

[19] J. Neff, K. Lee, E.M. DeBlois, Produced Water: Overview of Composition, Fates, and Effects, Springer, New York, 2011.

[20] J. Ranck, R. Bowman, J. Weeber, L. Katz, E. Sullivan, BTEX Removal from Produced Water Using Surfactant-Modified Zeolite, J. Environ. Eng. 131 (2005) 434-442. https://doi.org/10.1061/(ASCE)0733-9372(2005)131:3(434)

[21] A. Kumar, M. Naushad, A. Rana, Inamuddin, Preeti, G. Sharma, A.A. Ghfar, F.J. Stadler, M.R. Khan, ZnSe-WO$_3$ nano-hetero-assembly stacked on Gum ghatti for photo-degradative removal of Bisphenol A: Symbiose of adsorption and photocatalysis, Int. J. Biol. Macromol. 104 (2017) 1172–1184. https://doi.org/10.1016/j.ijbiomac.2017.06.116

[22] M.S. Mauter, M. Elimelech, Environmental Applications of Carbon-Based Nanomaterials, Environ. Sci. Technol. 42 (2008) 5843-5859. https://doi.org/10.1021/es8006904

[23] Q. Zaib H. Fath, Application of carbon nano-materials in desalination processes, Desalin. Water Treat. 51 (2013) 627-636. https://doi.org/10.1080/19443994.2012.722772

[24] Q. Zaib, I.A. Khan, Y. Yoon, J.R. Flora, Y.G. Park, N.B. Saleh, Ultrasonication study for suspending single-walled carbon nanotubes in water, J. Nanosci. Nanotechnol. 12 (2012) 3909-3917. https://doi.org/10.1166/jnn.2012.6212

[25] Global Carbon Nanotubes Market – SWCNT, MWCNT, Technology, Applications, Trends & Outlook (2011 – 2016)," www.marketsandmarkets.com, , 2011.

[26] M.F. De Volder, S.H. Tawfick, R.H. Baughman, A.J. Hart, Carbon nanotubes: present and future commercial applications, Science 339 (2013) 535-539.

[27] G. Jia, H. Wang, L. Yan, X. Wang, R. Pei, T. Yan, Y. Zhao, X. Guo, Cytotoxicity of Carbon Nanomaterials: Single-Wall Nanotube, Multi-Wall Nanotube, and Fullerene, Environ. Sci. Technol. 39 (2005) 1378-1383.

[28] A. Alsharhan, Asab Field--United Arab Emirates, Rub Al Khali Basin, Abu Dhabi, 1993

[29] H. Al-Johani, M.A. Salam, Kinetics and thermodynamic study of aniline adsorption by multi-walled carbon nanotubes from aqueous solution, J. Colloid Interface Sci. 360 (2011) 760-767. https://doi.org/10.1016/j.jcis.2011.04.097

[30] J. H. Lehman, M. Terrones, E. Mansfield, K.E. Hurst, V. Meunier, Evaluating the characteristics of multiwall carbon nanotubes, Carbon 49 (2011) 2581-2602. https://doi.org/10.1016/j.carbon.2011.03.028

[31] A. Peigney, C. Laurent, E. Flahaut, R.R. Bacsa, A. Rousset, Specific surface area of carbon nanotubes and bundles of carbon nanotubes, Carbon 39 (2001) 507-514. https://doi.org/10.1016/S0008-6223(00)00155-X

[32]   G.D. Sheng, D.D. Shao, X.M. Ren, X.Q. Wang, J. X. Li, Y.X. Chen, X.K. Wang, Kinetics and thermodynamics of adsorption of ionizable aromatic compounds from aqueous solutions by as-prepared and oxidized multiwalled carbon nanotubes, J. Hazard. Mater. 178 (2010) 505-516. https://doi.org/10.1016/j.jhazmat.2010.01.110

[33]   M. Kragulj, J. Tričković, B. Dalmacija, Á. Kukovecz, Z. Kónya, J. Molnar, S. Rončević, Molecular interactions between organic compounds and functionally modified multiwalled carbon nanotubes, Chem. Eng. J. 225 (2013) 144-152. https://doi.org/10.1016/j.cej.2013.03.086

[34]   W. Chen, L. Duan, D. Zhu, Adsorption of polar and nonpolar organic chemicals to carbon nanotubes, Environ. Sci. Technol. 41 (2007) 8295-8300. https://doi.org/10.1021/es071230h

[35]   A. Nikolaou, S. Meric, D. Fatta, Occurrence patterns of pharmaceuticals in water and wastewater environments, Anal. Bioanal.Chem. 387 (2007) 1225-1234. https://doi.org/10.1007/s00216-006-1035-8

[36]   T. Heberer, Occurrence, fate, and removal of pharmaceutical residues in the aquatic environment: a review of recent research data, Toxicol. Lett. 131 (2002) 5-17.

[37]   S. Monteiro, A.A. Boxall, Occurrence and Fate of Human Pharmaceuticals in the Environment, in D.M. Whitacre, M. Fernanda, F.A. Gunther (Eds.), Reviews of Environmental Contamination and Toxicology, Springer, New York, 2010, pp. 53-154.

[38]   J. Radjenovic, M. Petrovic, D. Barceló, Analysis of pharmaceuticals in wastewater and removal using a membrane bioreactor, Anal. Bioanal. Chem. 387 (2007) 1365-1377. https://doi.org/10.1007/s00216-006-0883-6

[39]   T.A. Ternes, Occurrence of drugs in German sewage treatment plants and rivers, Water Res. 32 (1998) 3245-3260. https://doi.org/10.1016/S0043-1354(98)00099-2

[40]   B. Li, T. Zhang, Biodegradation and Adsorption of Antibiotics in the Activated Sludge Process, Environ. Sci. Technol. 44 (2010) 3468-3473. https://doi.org/10.1021/es903490h

[41]   F.J. Benitez, J.L. Acero, F.J. Real, G. Roldan, F. Casas, Comparison of different chemical oxidation treatments for the removal of selected pharmaceuticals in water matrices, Chem. Eng. J. 168, (2011) 1149-1156.

[42]   T. Urase, T. Kikuta, Separate estimation of adsorption and degradation of pharmaceutical substances and estrogens in the activated sludge process, Water Res. 39 (2005) 1289-1300. https://doi.org/10.1016/j.watres.2005.01.015

[43]   X.-S. Miao, J.-J. Yang, C.D. Metcalfe, Carbamazepine and its metabolites in wastewater and in biosolids in a municipal wastewater treatment plant, Environ. Sci. Technol. 39 (2005) 7469-7475. https://doi.org/10.1021/es050261e

[44]   Z.-C. Di, J. Ding, X.-J. Peng, Y.-H. Li, Z.-K. Luan, J. Liang, Chromium adsorption by aligned carbon nanotubes supported ceria nanoparticles, Chemosphere 62 (2006) 861-865. https://doi.org/10.1016/j.chemosphere.2004.06.044

[45]   X. Peng, Z. Luan, J. Ding, Z. Di, Y. Li, B. Tian, Ceria nanoparticles supported on carbon nanotubes for the removal of arsenate from water, Mater. Lett. 59 (2005) 399-403. https://doi.org/10.1016/j.matlet.2004.05.090

[46]   K. Yang, L. Zhu, B. Xing, Adsorption of Polycyclic Aromatic Hydrocarbons by Carbon Nanomaterials, Environ. Sci. Technol. 40 (2006) 1855-1861. https://doi.org/10.1021/es052208w

[47]   Q. Zaib, I.A. Khan, N. Saleh, J. Flora, Y.-G. Park, Y. Yoon, Removal of Bisphenol A and 17β-Estradiol by Single-Walled Carbon Nanotubes in Aqueous Solution: Adsorption and Molecular Modeling, Water Air Soil Pollut. 223 (2012) 1-13. https://doi.org/10.1007/s11270-011-0833-6

[48]   X.M. Yan, B.Y. Shi, J.J. Lu, C.H. Feng, D.S. Wang, H.X. Tang, Adsorption and desorption of atrazine on carbon nanotubes, J. Colloid Interface Sci. 321 (2008) 30-38. https://doi.org/10.1016/j.jcis.2008.01.047

[49]   R. Smajda, Á. Kukovecz, Z. Kónya, I. Kiricsi, Structure and gas permeability of multi-wall carbon nanotube buckypapers, Carbon 45 (2007) 1176-1184. https://doi.org/10.1016/j.carbon.2007.02.022

[50]   Z. Wu, Z. Chen, X. Du, J.M. Logan, J. Sippel, M. Nikolou, K. Kamaras, J.R. Reynolds, D.B. Tanner, A.F. Hebard, A.G. Rinzler, Transparent, conductive carbon nanotube films, Science 305 (2004) 1273-1276. https://doi.org/10.1126/science.1101243

[51]   A.P. Terzyk, G. Rychlicki, The influence of activated carbon surface chemical composition on the adsorption of acetaminophen (paracetamol) in vitro: The temperature dependence of adsorption at the neutral pH, Colloids Surf. A: Physicochem. Eng. Aspects 163 (2000) 135-150.

[52] Z. Yu, S. Peldszus, P.M. Huck, Adsorption characteristics of selected pharmaceuticals and an endocrine disrupting compound—Naproxen, carbamazepine and nonylphenol—on activated carbon, Water Res. 42 (2008) 2873-2882. https://doi.org/10.1016/j.watres.2008.02.020

[53] A.S. Mestre, J. Pires, J.M.F. Nogueira, A.P. Carvalho, Activated carbons for the adsorption of ibuprofen, Carbon 45 (2007) 1979-1988. https://doi.org/10.1016/j.carbon.2007.06.005

[54] V. Datsyuk, M. Kalyva, K. Papagelis, J. Parthenios, D. Tasis, A. Siokou, I. Kallitsis, C. Galiotis, Chemical oxidation of multiwalled carbon nanotubes, Carbon 46 (2008) 833-840. https://doi.org/10.1016/j.carbon.2008.02.012

[55] R. Förch, H. Schönherr, A. Tobias, A. Jenkins, Appendix C: Contact Angle Goniometry, in Surface Design: Applications in Bioscience and Nanotechnology, Wiley-VCH Verlag GmbH & Co. KGaA, 2009, pp. 471-473.

[56] C. Martínez, M. Canle L, M.I. Fernández, J.A. Santaballa, J. Faria, Kinetics and mechanism of aqueous degradation of carbamazepine by heterogeneous photocatalysis using nanocrystalline TiO_2, ZnO and multi-walled carbon nanotubes–anatase composites, Appl. Catal. B: Environ. 102 (2011) 563-571.

[57] H. Wang, H.-L. Wang, W.-F. Jiang, Solar photocatalytic degradation of 2,6-dinitro-p-cresol (DNPC) using multi-walled carbon nanotubes (MWCNTs)–TiO_2 composite photocatalysts, Chemosphere 75 (2009) 1105-1111. https://doi.org/10.1016/j.chemosphere.2009.01.014

[58] Y. Zhang, Z.-R. Tang, X. Fu, Y.-J. Xu, TiO_2−Graphene Nanocomposites for Gas-Phase Photocatalytic Degradation of Volatile Aromatic Pollutant: Is TiO_2−Graphene Truly Different from Other TiO_2−Carbon Composite Materials?, ACS Nano 4 (2010) 7303-7314.

[59] SRC, PhysProp Database, Ed. 2011.

[60] T.X. Bui, H. Choi, Influence of ionic strength, anions, cations, and natural organic matter on the adsorption of pharmaceuticals to silica, Chemosphere 80 (2010) 681-686. https://doi.org/10.1016/j.chemosphere.2010.05.046

Chapter 10

Removal of Malachite Green from Water and Wastewater by Low-Cost Adsorbents

Qasimullah[1*,] A. Mohammad[1], P.F. Rahman[2] and Mahfoozurrahaman Khan[3]

[1]Department of Chemistry, School of Sciences, Maulana Azad National Urdu University, Gachibowli, Hyderabad- 500032, India

[2] Department of Zoology, School of Sciences, Maulana Azad National Urdu University, Gachibowli, Hyderabad- 500032, India

[3]Department of Applied Chemistry, Zakir Hussain College of Eng. & Technology, Aligarh Muslim University Aligarh, 202002, India

*drqasimullah@gmail.com

Abstract

This chapter describes the use of various low-cost adsorbents for the removal of malachite green (MG) from water and wastewater. Malachite green is a cationic dye which is widely used worldwide in the aquaculture industries (silk, wool, cotton, leather, paper etc.) as a therapeutic agent to treat parasites, fungal, and bacterial infections in fish and fish eggs and as an antiseptic for external application on wounds and ulcers. Despite its extensive use, MG has several toxic properties, which are known to cause carcinogenesis, mutagenesis, teratogenesis, and respiratory toxicity. Its oral consumption is also hazardous and carcinogenic. Therefore, it is essential to remove MG from water and wastewater using different low-cost adsorbents. A literature survey reveals that several methods are available for the dyes removal from water and wastewater such as photocatalytic degradation, combined photo-Fenton and biological oxidation, advanced oxidation processes, aerobic degradation, nanofiltration membranes, ozonation, coagulation, fluid extraction, solid phase extraction, and adsorption. Among these, adsorption is a well-known equilibrium separation process and an effective method for water decontamination application. Adsorption has been found to be superior to other techniques for water re-use in terms of initial cost, flexibility, and simplicity of design, ease of operation, and insensitivity to toxic pollutants. Adsorption also does not result in the formation of harmful substances.

Keywords

Adsorption, Malachite Green, Activated Carbon, Bio-adsorbent, Agricultural Waste, Industrial Waste

Contents

1. Introduction

Water is the most important and essential component on the earth for vital activities of living beings. Unfortunately, the water quality of our water resources is deteriorating continuously due to the geometrical growth of population, industrialization, agricultural activities, and other geological as well as environmental changes. Dyes are widely used in textile, paper, plastic, food and cosmetic industries. The wastes coming from these industries affect our environment and create pollution. The level of pollutants even at a very low concentration is highly visible and it affects aquatic life. Hence, contaminations due to dyes pose not only a severe public health concern but also many serious environmental problems because of their persistence in nature and non-biodegradable characteristics. At present, more than 100,000 commercially available dyes exist and more than 7×10^5 tonnes /yr and mostly associated with water pollution [1,2]. The wastewater containing dye is difficult to treat by conventional treatment methods as most of the dyes are stable to light and oxidizing agents.

Malachite green (MG), a tri-phenylmethanedye is a cationic dye which is used worldwide as an antifungal, anti-bacterial and anti-parasitical therapeutic agent in aquacultures and animal husbandry. It is also widely used as a direct dye for silk, wool, jute, and leather. MG has detrimental effects on liver, gill, kidney, intestine, and gonads of aquatic organisms [3]. When it is inhaled or ingested by a human, it may cause irritation to the gastrointestinal tract and even cancer [4]. Contact of MG with skin causes irritation with redness and pain. Intermediate degradation products of MG have also been reported as being carcinogenic [3]. Therefore, the use of MG in aquaculture has been banned in several countries. However, MG in fish, animal milk, and other foodstuff is still detected due to illegal use of MG [5] which alarm the health hazards against a human being.

Various treatment technologies have been employed for the removal of textile dyes from wastewaters which include ion-exchange [6], electrochemical [7], membrane [8], ozonation [9], coagulation [10] and Fenton reagent [11]. Among all these techniques, adsorption is a well-known equilibrium separation process and an effective method for water decontamination application. Adsorption has been found to be superior to other techniques for water re-use in terms of initial cost, flexibility, and simplicity of design, ease of operation, and insensitivity to toxic pollutants. Adsorption generally does not result in the formation of harmful substances.

2.   Adsorbents

Adsorption using activated carbon has been found to be effective for the removal of dyes, but it was an expensive process as both regeneration and disposal of the used carbon have been very difficult. Therefore, it was necessary to identify and develop easily available, economically viable and highly effective adsorbents for efficient and facile removal of dye.

2.1   Bio- adsorbents

Biosorption is defined as the accumulation and concentration of pollutants from aqueous solutions by the use of biological materials. Biological materials, such as yeast, fungi or bacterial biomass and biopolymers have been used as adsorbents for the removal of MG from aqueous media. The adsorption capacities and other parameters of important bio-adsorbents in respect of removal of MG from water are presented in Table 1.

Luffa cylindrical an algae was used by Seki et al.[19] in MG removal from aqueous solution. As the pH was increased from 3 to 5, the amount of sorbed dye also increased. Langmuir, Freundlich and Dubinin-Radushkevich isotherms were investigated to explain equilibrium data, but Langmuir isotherm exhibited the best fit with the experimental data. Monolayer sorption capacity increased with the increase in temperature.

Table 1: Adsorption capacities and other parameters of bio-adsorbents for adsorption of MG.

Adsorbent	Adsorption capacity (mg/g)	Isotherm	pH	Equilibrium time (h)	Ref.
Pithophora sp. A fresh water algae	117.64	Redlich-peterson	5.0	8 .0	[12]
Sccharomyces cerevisiae	17.0	Freundlich	NA	NA	[13]
Aerobic granules	56.80	Langmuir	Alakline pH	1.0	[14]
Heat – treated anaerobic granular sludge	149.1	Langmuir	5.0	2 .0	[15]
Freshwater macrophyte alligator weed	185.54	Langmuir	6.0	NA	[16]
Brown marine algae turbinariaconoides	66.60	Langmuir	8.0	2.5	[17]
Chlorella- based biomass	18.4	NA	7.0	NA	[18]
Lufiacylindrica	29.40	Langmuir	5.0	5.0	[19]
Chitin hydrogel	33.6	Langmuir	7.0	2.5	[20]
Hydrillaverticilata biomass	91.97	Freundlich	8.0	3.5	[21]
Hydrillaverticilata biomass	69.88	Langmuir	8.0	2.5	[21]
Yeast S. CerevisiaeMTTC-174	NA	NA	6.88	1.0	[22]
Ulvalactuca and systoceira algae based activated carbon	400	Langmuir	4.0	1.5	[23]
Amylopectin and poly(acrylic acid) (Ap-g-PAA)	352.11	Langmuir	7.0	0.5	[24]
Dried biomass of ulvalactuca (L.)	200	Langmuir	6.0	1.0	[25]
Cellulose modified with maleic anhydride (CMA)	370	Langmuir	NA	NA	[26]
Bacillus cerus M^1 16	485	Langmuir and Redlich-Peterson	5.0	6.0	[27]

NA= Not Available

Tang et al. [20] used chitin hydrogel (biopolymer) as an adsorbent to remove MG from aqueous solution. They found that the maximum adsorption capacity of hydrogel for MG was 26.88 mg/g. Scanning electron microscope and fourier-transform infrared spectroscopy results indicated that the adsorption of MG on the hydrogel was through a physical process. Marine algae *Ulva Lactuca* (ULV-AC) and *Systoceirastricta* (SYS-AC) [23] based activated carbons have proven to be potential adsorbents for the removal of hazardous cationic dye including MG. The surfaces of these algae were treated with phosphoric acid and subsequently air activated at 600 °C for 3 h. Model equations such as Langmuir, Freundlich and Temkin isotherms were used to analyze the adsorption equilibrium data and the best fits to the experimental data were provided by the first two isotherm models and the maximum adsorption capacity for MG and safranine O were 400 and 526 mg/ g on the SYS-AC and ULV-AC respectively. *Ulva Lactuca (L.)* biomass has been used for the removal of methylene blue (MB) and MG dyes [25]. Correlation coefficient values were close to unity which suggested that adsorption data were in favor of Langmuir and Freundlich models. The pseudo-second-order kinetic model was found favorable to describe the adsorption behavior of both the dyes. The intraparticle diffusion was a prominent process right from the beginning of dye-solid interaction and the adsorption involved monolayer surface coverage and heterogeneous adsorption mechanism. Dry cells of B. cereus M1 16 [27] were found to be efficient biosorbent for the removal of MG from aqueous solution. Dry biosorbent was found to have a high affinity for charged species and the maximum dye uptake capacity of 485 mg/g was observed at pH 5.0. Langmuir and Redlich–Peterson isotherm models were used to describe the adsorption process.

2.2   Industrial solid wastes adsorbents

Industrial solid wastes such as coal fly ash, sugarcane bagasse and red mud etc. being low-cost materials and their local availability have been frequently used as adsorbents for dyes removal. However, according to a literature survey, few papers have been published on the use of industrial solid wastes as adsorbents for the removal of MG from water and wastewater samples.   Table 2 presents the adsorption capacities and other related parameters of such adsorbents.

Fly ash [31], an industrial by-product abundant in India, has been widely used as a low-cost adsorbent after modification with alkali for the removal of MG from aqueous solutions by batch adsorption method. The adsorption was found to be strongly dependent on pH of the medium and the adsorption capacity decreased with an increase in temperature. The Langmuir isotherm model showed a good fit to the equilibrium adsorption data at all temperatures. The mean free energy estimated from the Dubinin-

Radushkevich model indicated that the adsorption mechanism was chemical ion exchange. The kinetic data were found to follow the pseudo-second-order kinetic model.

Table 2: Adsorption capacities and parameters of industrial solid wastes bio-adsorbents for adsorption of MG

Adsorbent	Adsorption capacity (mg/g)	Isotherm	pH	Equilibrium time (h)	Ref.
Bagasse fly ash	170.33	Freundlich	7.0	4.0	[28]
Poly (methyacrylic acid) modified sugarcane bagasse	103.2	Langmuir	6.0	3.0	[29]
Ethylenediamminetetraac eticdianhydride (DTAD) modified sugarcane bagasse	157.20	Langmuir	6.0	3.0	[30]
Ca(OH)$_2$.treated fly ash	17.383	Langmuir	7.0	2.0	[31]
Flyash	40.65	Langmuir	>6.0	NA	[32]
Acid- activated sintering processs red mud (ASRM)	336.40	Langmuir	>3.2	3.0	[33]
Coal fly ash	2.23	Freundlich	8.0	2 .0	[34]

NA=Not Available

Fly ash was used as a waste material for the removal of cationic dyes (methylene blue and MG) from their aqueous solutions by Witek-Krowiak et al. [32]. Effects of initial dye concentration, contact time, pH, adsorbent dosage, solution temperature, surfactant addition and ionic strength on the fly ash sorption of dyes were studied. The isothermal data for sorption followed the Langmuir model. The maximum sorption capacities were reported for methylene blue and MG were 36.05 and 40.65 mg/g respectively. Kinetic studies indicated the pseudo-second-order for the sorption on fly ash.

Acid-activated sintering process red mud (ASRM) was used as an attractive adsorbent for the removal of cationic dyes, especially MG   from aqueous solution. The maximum adsorption (336.4 mg/g)was realized at pH higher than 3.2. Langmuir isotherm was best fitted to describe the adsorption process. The thermodynamic parameters (ΔH) indicated that the adsorption process of MG was endothermic. [33].Gopal et al. [34] used coal fly ash as an adsorbent for the removal of MG dye from aqueous solution. For maximum adsorption, the experimental conditions were pH 8, adsorbent dose 2 g, contact time 2 h and temperature 303K. The maximum adsorption capacity of coal fly ash was 2.23 mg/g and hence it can be considered as an attractive adsorbent for removing basic dyes from aqueous solutions.

2.3   Carbonaceous adsorbents

Adsorption is a surface phenomenon which results out of binding forces between atoms, molecules, and ions of adsorbate and the surface of the adsorbent. Adsorption also depends on the number of sites available, porosity, and specific surface area of adsorbent as well as various types of interactions. Generally, carbonaceous materials have been widely used adsorbents, as these are known, for a long time to be capable of adsorbing various organic compounds. Activated carbon due to its high surface area and porosity is very efficient to remove dyes from water and wastewater. A variety of activated carbon materials have been prepared from different sources for their use as adsorbents for the removal of MG from aqueous water systems. These materials have been listed in Table 3 with their adsorption capacities and other important parameters particularly surface area.

Table 3: Adsorption capacities and parameters of activated carbon for adsorption of MG

Adsorbent	Adsorption capacity (mg/g)	Isotherm study	pH	Equilibrium time (h)	Specific area. (m^2/g)	Ref.
Commercial grade activated carbon	14.8	Freundlich	NA	0.25	NA	[35]
Laboratory grade activated carbon	42.18	Redich-Peterson	7.0	4.0	NA	[35]
Waste *apricot*	116.27	Langmiur	NA	1.0	1060	[36]
Ground nut shell	222.22	Freundlich	NA	2.0	1200	[37]
Bamboo	263.58	Langmuir	5.0	2.8	1724	[38]
Arundo donax root	8.69	Langmuir	5.0	3.0	158	[39]
Rice husk	63.85	Langmuir and Freundlich	N.A.	0.66	N.A.	[40]
Degreased coffe bean (DCB)	55.3	Langmuir and Freundlich	4.0	2.0	120	[41]
Annona squamosa	25.91	Langmuir and Freundlich, Temkin	7.0	N.A.	N.A.	[42]

Borassus aethiopum flower	48.48	Langmuir	6.0-8.0	24	9.57	[43]
Polygonum orientale	480	Langmuir	10.0	2.5	1398	[44]
Epicarp of *Ricinuscommunis*	27.78	Langmuir	7.0	0.82	NA	[45]
Terminalia catappa	74.96	Freundlich	NA	066	NA	[46]
Rice husk	NA	Langmuir and Freundlich	NA	NA	180.5	[47]
Rambutan peel	329.49	Freundlich	8.0	24	988.24	[48]
Kapok hull	30.16	N.A.	N.A.	N.A.	158.228	[49]
Coca shell	37.03	Langmuir	NA	2.0	N.A.	[50]
Rubber seed coat	72.92	Langmuir	6.0	N.A.	N.A.	[51]
Coconut shell	214.63	Langmuir	8-11	2.0	945.74	[52]
Musa paradisiaca stalk	141.76	Langmuir	8.0	2.0	684.11	[53]
Oil palm fruit	356.27	Langmuir	6.0-9.0	0.75	1254	[54]
Spent tea leaves	256.4	Langmuir	4.0	N.A.	134	[55]
Lemon peel	66.67	Langmuir and Freundlich	7.0	0.41	1158	[56]
Durian seed	476.19	Freundlich	8.0	20-22 h	980.62	[57]
Almond shell	333.3	Langmuir	N.A.	N.A.	1492.3	[58]
Pomelo (Citrus grandis) peels	178.43	Langmuir	8.0	N.A.	N.A.	[59]

N.A.= Not Available

Ahmad et al. [53] used banana stalk activated (BSAC) carbon by activation with KOH at an impregnation ratio of 1:1 (KOH pellets: char). The specific surface area of the BSAC was calculated using the BET equation and was reported 684.11m2/g. The result showed that 97.3% of the dye removal was observed at pH 8.0. The equilibrium time was 90 and

120 min for maximum adsorption. Negative values of $\Delta G°$ indicated that the process of MG dye adsorption onto BSAC was spontaneous whereas the positive values of $\Delta H°$ and $\Delta S°$ suggested that the process of dye adsorption was endothermic.

Bello et al. [54] used oil palm fruit fibre activated carbon (OPFAC) by activation with 10 % KOH at an impregnation ratio of 1:1 (KOH pellets: char) to remove (MG) dye from its aqueous solution. The positive values of $\Delta H°$ and $\Delta S°$ indicated that the process is endothermic in nature and showed increased the randomness of the adsorbate molecules on the adsorbent. Cost analysis revealed that OPFAC is 20 times cheaper than commercially available activated carbon. Akar et al. [55] used spent tea leaves activated carbon (STAC) by activation with 4% NaOH solution for the removal MG. STAC was effective to remove 94% of MG at pH 4. Various isotherms were used for the adsorption study, but Langmuir isotherm exhibited the best fit with the experimental data. The negative value of free energy ($\Delta G°$) and positive value of enthalpy ($\Delta H°$) change indicated the spontaneous and endothermic nature of adsorption process.

Activated carbon prepared from durian seed was very effective for the removal of MG from aqueous solution. Results showed that 97% of dye removal was observed at pH 8. Various adsorption isotherms were used to describe the process, but the Freundlich model was fitted the best. The Freundlich constant(1/n) is less than 1, indicated that the adsorption process is favorable.Thermodynamic studies also revealed that adsorption process was spontaneous and endothermic in nature. The positive value of ΔS showed the increased randomness of the adsorbate molecules on the solid surfaces than in the solution [57].

Ahmad et al. [59] used activated carbon produced from pomelo peels (PPAC) for the removal of MG dye from aqueous solution. Dye removal was pH dependent through which 95.06% removal of MG was observed at pH 8.0. Various adsorption isotherms were used to describe the adsorption process, but Langmuir isotherm fit the adsorption data most with maximum monolayer adsorption capacity of 178.43 mg/g The kinetic data fitted the pseudo-second-order model with a correlation coefficient greater than 0.99. Thermodynamic parameters studies indicated that the adsorption process was spontaneous and endothermic in nature.

2.4   Natural materials as adsorbents

Natural clay minerals are well known to mankind from the earliest days of civilization. Because of their low cost, abundance in most continents of the world, high sorption properties, and potential for ion exchange, clay materials have been considered most useful adsorbents. These adsorbents possess layered structure and are considered as host materials for the adsorbates and counter ions. In recent years there has been an increasing interest in utilizing clay minerals such as cloisite, clinoptilolite, eluthrilithe, kerolite, faujasite, montmorillonite, and palygorskite to adsorb not only inorganic ions but also organic molecules. More recently, low-cost adsorbents, for example, organoclay complex adsorbents, have been investigated as an alternative to activated carbon. Table 4 lists the relative adsorption capacities of natural sorbents used for the removal of MG from the water.

Table 4: Adsorption capacities and parameters of natural materials for adsorption of MG

Adsorbent	Adsorption capacity (mg/g)	Isotherm study	pH	Equilibrium time (h)	Ref.
Bentonite clay	7.72	Langmuir, Freundlich and D-R	9.0	0.16	[60]
Zeolite	46.35	Langmuir	6.0	NA	[61]
Bentonite	178.60	Langmuir	5.0	1.0	[62]
Perlite	3.36	Freundlich	8.0-9.0	0.66	[63]
Montmorillonite	NA	Langmuir	NA	NA	[64]
Clayey soil of Indian origin	78.57	Langmuir	6.0	1.5	[65]
Halloysite nanotubes (NTs)	99.60	Langmuir	9.5	0.5	[66]
Aminopropyl functionalized magnesium phyllosilicate (AMP) clay	334.80	Langmuir	9.8	3.0	[67]
Natural rarasaponin surfactants modified kaolin	NA	Sips	NA	NA	[68]
Talc	20.96	Langmuir	8.0	NA	[69]
Organically modified clay (OC)	56.82	Langmuir	6.0	0.5	[70]
Activated sintering process red mud	336.4	Langmuir	>3.2	NA	[71]
Zeolite- iron oxide magnetic nano-composite	13.8	Langmuir	8.0	1.0	[72]

NA= Not Available

The clayey soil of Indian origin [65] has been proven as a potential adsorbent for the removal of MG from its aqueous solutions. Langmuir isotherm model was fitted the best. The activation energy of the adsorption process was calculated by using Arrhenius equation which indicated that the adsorption of MG onto clayey soil was chemisorptions in nature. Thermodynamic studies also revealed that the adsorption process was spontaneous and exothermic in nature. Halloysite nanotubes [66] (HNTs) were used as nano-adsorbent for the adsorption of MG from aqueous solutions. The maximum adsorption was observed after 30 min of contact time and at pH 9.5. Thermodynamic parameters studies indicated that the adsorption process was spontaneous and endothermic. Nano-sized aminopropyl functionalized magnesium phyllosilicate (AMP) clay [67] was used as an adsorbent for the treatment of MG by adsorption and precipitation methods. An AMP clay dosage of 0.1 mg/mL showed a maximum removal capacity of 334.80 mg/g. Kaolin modified with natural surfactants rarasaponin was used for the removal of MG from the aqueous solution by Ismadji et al.[68]. Various isotherms were used to study the adsorption process but Sips was best fitted among all the isotherms. Acid-activated sintering red mud (ASRM) was investigated [71] as an adsorbent for the removal of MG and crystal violet (CV) from aqueous solution. The pH higher than 3.2 was favorable for the adsorption of both the dyes on ASRM. From the Langmuir isotherm, the maximum adsorption capacities were found 336.4 and 60.5 mg/g for MG and CV respectively. Thermodynamic studies revealed that adsorption process was endothermic for MG whereas it was exothermic for CV.

Table 5: Adsorption capacities and parameters of agricultural solid waste for adsorption of MG.

Adsorbent	Adsorptio n capacity (mg/g)	Isotherm study	pH	Equilibrium time (h)	Ref.
De- oiled soya	20.44	Langmuir	5.0	NA	[73]
Citric acid modified rice straw	256.41	Langmuir	6.0	10h	[74]
Jute fibre carbon	136.58	Freundlich	8.0	3.6	[75]
Rubber wood saw dust	36.45	Langmuir	NA	3.0	[76]
Lemon peel	51.73	Langmuir and Redlich-Peterson	NA	24.0	[77]
Orange peel	483.63	Langmuir and Redlich-Peterson	NA	NA	[78]
Rattan saw dust	62.71	Langmuir	4.0 - 9.0	3.5	[79]

Oil palm trunck fibre (OPTF)	149.35	Langmuir	8.0	3,25	[80]
Chitosan bead	98.5	Langmuir	8.0	5.0	[81]
Dead pine needless	33.56	Freundlich	-	5.5	[82]
Eucalyptus bark	59.88	Langmuir	5.0	4.5	[83]
Treated ginger waste	188.6	Langmuir	9.0	2.5	[84]
Almond shell	29.0	Langmuir	NA	3.0	[85]
Sea shell powder	42.33	Langmuir	8.0	2.0	[86]
Mango seed husk	47.9	Freundlich	5.0	2.0	[87]
Rice husk	15.5	Freundlich	NA	NA	[88]
Pine tree root decayed by brown root fungi	42.63	Langmuir	4.0	24	[89]
Conch shell powder	92.254	Langmuir	8.0	2.0	[90]
Odina bark					[91]
Waste rapeseed press cake	17.85	Langmuir	6.5	3.0	[92]
Scots pine saw dust (Pinussylvestria)	71.67	Langmuir	5.0	NA	[93]
Sawdust a timber waste	13.87	Freundlich	NA	1.6	[94]
Tartaric acid modified Cinnamomum camphora sawdust	156.70	Langmuir	8.0	3.25	[95]
Citric acid modified Cinnamomum camphora sawdust	260.40	Langmuir	8.0	3.0	[95]
Citric acid treated pea shell	14.49	Freundlich	7.0	0.66	[96]
Walnut shell	90.80	Langmuir	5.0	1.5	[97]
Carrot leaves powder (CLP)	52.60	Freundlich	7.0	0.5	[98]
Carrot stem powder (CSP)	43.40	Freundlich	7.0	0.5	[98]
Coco- peat	276.80	Sips	7.0	1.0	[99]
Hydrothermally carbonized pine needles (HTC- APN)	97.08	Langmuir	7.0	NA	[100]
Hydrothermally carbonized pine needles (HTC- PN)	52.91	Langmuir	7.0	NA	[100]
Sawdust modified with triethyl amine	NA	Freundlich	5.08	6.0	[101]

NA= Not Available

2.5   Agricultural solid wastes adsorbents

There are various agricultural solid wastes, easily available due to their abundance in nature which are used as adsorbents for the removal of MG from aqueous solution. The adsorption capacities of these adsorbents are reported in Table 5.

The removal of MG from aqueous solutions by adsorption onto rubber wood sawdust showed that adsorption process reached to equilibrium after 2.5 h and at optimum pH of 5 and the adsorption process was exothermic in nature. Equilibrium data fitted well in Langmuir isotherm model ($R^2 = 0.7776$). The maximum adsorption capacity was found 27.4 mg/g at temperature 310 K. [76]. Sawdust (SD), a timber waste has been used as bio – sorbent for the removal of a cationic dye, MG from aqueous solutions by Uma et al. [94].The experimental results indicated that 5 g / L of SD was able to remove 93 % of dye from an initial concentration of 20 mg L^{-1} in 100 min of equilibrium time, The removal of dye decreased with the increase in temperature confirming adsorption process as is exothermic. Adsorption data were better fitted in Freundlich isotherm model and followed the pseudo-second-order kinetics. The maximum adsorption capacity of adsorbent was 13.87 mg/g at 298 K.   Wang et al [95] used natural adsorbent (*Cinnamomum camphora* sawdust) modified by organic acids (oxalic acid, citric acid, and tartaric acid) for the removal of MG dye in aqueous media through a batch process. The extent of MG adsorption onto modified sawdust increased with increasing organic acid concentration, pH, contact time, and temperature but decreased with increasing the adsorbent dosage and ionic strength. Kinetic studies indicated that the pseudo-second-order kinetic model could best describe the adsorption kinetics of MG. Equilibrium data were found to fit well with the Langmuir model, and the maximum adsorption capacity of the three kinds of organic acids-modified sawdust was 280.3, 222.8, and 157.5 mg/g, respectively. Thermodynamic parameters also suggested that the sorption of MG was through the endothermic process.

Yinghua Song et al. [101] synthesized   a new adsorbent from sawdust, a forest residue, in which methanol was used as a solvent and triethylamine as a modification agent under the following optimum conditions: 25 °C of reaction temperature, 1 : 8.75 of the ratio of sawdust to triethylamine (g:mL) and 1h of reaction time. The results indicated that the maximum adsorption capacity can be achieved at 5.08 of pH value and adsorption equilibrium can be reached in 6h. The Freundlich isotherm model provided a better description for the adsorption equilibrium when compared with the Langmuir equation. The adsorption process of MG on sawdust tended to be controlled simultaneously by film mass transfer and intra-particle diffusion. It showed that the sawdust modified with triethylamine had good performance for cationic dyes.

Rice husk - Das et al [88] used rice husk treated with NaOH, as a low-cost adsorbent for the removal of MG from aqueous solution in batch adsorption procedure. The adsorption was found to be strongly dependent on pH of the medium. The Freundlich isotherm model showed a good fit to the equilibrium adsorption data and the maximum adsorption capacity was found to be 15.5 mg/g. The mean free energy (E) estimated from the Dubinin–Radushkevich model indicated that the main mechanism governing the sorption process was chemical ion-exchange. The kinetics of adsorption followed the pseudo-second-order model and the rate constant increased with increase in temperature indicating endothermic nature of adsorption. Thermodynamic studies suggested the spontaneous and endothermic nature of adsorption of MG by treated rice husk. The isosteric heat of adsorption was also determined from the equilibrium information using the Clausius–Clapeyron equation.

Eucalyptus bark a forest waste was evaluated for its ability to remove MG from aqueous solutions [83]. The equilibrium sorption data of MG by eucalyptus bark were analyzed by Langmuir and Freundlich isotherm models. The results indicated that both the Langmuir and Freundlich equations provided good correlation of the experimental data, but the Langmuir expression fit the equilibrium data better and the maximum sorption capacity of eucalyptus bark was found to be 59.88 mg/g at 293K. Among the kinetic models studied, the pseudo-second-order was the best applicable model to describe the sorption of MG by eucalyptus bark. The overall rate of dye uptake was found to be controlled by external mass transfer at the beginning of sorption, while intraparticle diffusion controlled the overall rate of sorption at a later stage. The results indicated the potential of eucalyptus bark as biosorbent for the removal of basic dye from aqueous solution. Batch adsorption behavior of MG from aqueous solution by using OdinaWodier bark carbon (OWC) [91] was investigated as a function of parameters such as initial MG concentration, adsorbent dose, pH, contact time and temperature. Freundlich and Langmuir adsorption models were applied to describe the equilibrium. The thermodynamic parameters ($\Delta G°$, $\Delta H°$ and $\Delta S°$) indicated that the adsorption process was favorable. The pseudo-first-order, pseudo second-order, elovich and intraparticle diffusion models were used to describe the kinetic parameters.

Table 6: Adsorption capacities and parameters of miscellaneous for adsorption of MG

Adsorbent	Adsorption capacity (mg/g)	Isotherm	pH	Equilibrium time (h)	Ref.
Hen feathers	26.10	Langmuir	5.0	2.5	[102]
Melamine/maleic anhydride (S-Me/MA)	641.03	Langmuir	9.0	24.0	[103]
Cyclodextrin based adsorbent	91.90	Langmuir	8.0	2.0	[104]
Surfactant- modified alumina (SMA)	185.00	Langmuir	NA	1.0	[105]
Activated-carbon / $CoFe_2O_4$ composite	89.20	Langmuir	5.0	0.09	[106]
Egg shell	56.76	Langmuir	>3.	3.0	[107]
Fish scale	38.46	Langmuir	8.0	1.0	[108]
Vinyl- modified Mesoporouspoly Acrylicacid)/Sio_2(PAA/Si O_2)Composite Nanofibremembranes	220.49	Redlich-Peterson	NA	4.0	[109]
Functionalized multi walled carbon nanotubes	142.85	Langmuir	7.0	1.4	[110]
Novel superparamagnetic sodium alginate- coated Fe_3O_4 nano particles	47.84	Langmuir	9.0	0.67	[111]
amylopectin and poly(acrylic acid) (ap-g-ppa)	352.11	Langmuir	NA	0.5	[112]
cadmium hydroxide nanowires loaded on activated carbon	19.0	Langmuir	NA	NA	[113]
copper nano wires loaded on activated carbon	434.8	Langmuir	5.0	0.34	[114]
magnetic β-cyclodextrin- graphene oxide nano composites	990.10	Langmuir	7.0	2.0	[115]
zno-nanorod loaded activated carbon	59.17	Langmuir	NA	NA	[116]

NA= Not Available

2.6   Miscellaneous adsorbents

The adsorbents, not included under specified categories of adsorbents in this chapter, have been summarized as miscellaneous adsorbents in Table 6.

Activated carbon/CoFe$_2$O$_4$ composite (AC/CFO) [106] was synthesized by a simple one-step refluxing route and was used as an adsorbent for the removal of MG dye from water. The results indicated that CoFe$_2$O$_4$ particles deposited on the surface of activated carbon in the composite were uniform with the particle size in the range of 14–20 nm. The composite adsorbents exhibited a clearly hysteretic behavior under applied magnetic field, which allowed their magnetic separation from water. Batch experiments were carried out to investigate adsorption isotherms and kinetics of MG onto the composite. The experimental data fitted well with the Langmuir model with a monolayer adsorption capacity of 89.29 mg /g. The adsorption kinetics was found to follow pseudo-second-order kinetic model.

Vinyl-modified mesoporous poly(acrylic acid)/SiO$_2$ (PAA/SiO$_2$) composite [109] nanofiber membranes were prepared by a sol-gel electro-spinning process and the sorption behavior of MG on the resultant membranes was examined. Fourier transform infrared (FTIR) results demonstrated that vinyl groups were grafted onto the silica skeleton. Transmission electron microscopy (TEM) images confirmed the formation of mesopores on the electrospun nanofibers and the pore size was 3.8 nm based on the Barrett–Joyner–Halenda (BJH) model. According to Brunauer–Emmett–Teller (BET) method, the specific surface area of the membranes was 523.84 m^2/g. Three widely used isotherms, Freundlich, Langmuir, and Redlich–Peterson isotherms, were used to model the experimental data of MG adsorption on PAA/SiO$_2$nanofiber membranes. The best fit was found to be Redlich–Peterson isotherm and the equilibrium adsorption capacity was 220.49 mg/g. The adsorption kinetics followed a pseudo-second-order model. The removal of MG from the aqueous phase within 240 min was 98.8%. The membranes can be regenerated to reuse for multiple cycles, which is beneficial for practical application. Shirmadi et al. [110] studied the adsorption of MG as a cationic dye onto functionalized multi-walled carbon nanotubes. The results showed that by increasing of contact time, pH and adsorbent dose, the removal of dye increased, while by increasing initial dye concentration, the removal efficiency was decreased. The experimental data were correlated with the Langmuir isotherm with maximum adsorption capacity and regression coefficient of 142.85 mg/g and 0.997, respectively.

Sarkar et al.[112] investigated the application of a high-performance biodegradable adsorbent based on amylopectin and poly(acrylic acid) (AP-g-PAA) for removal of toxic

MG from aqueous solution. The adsorbent showed excellent potential with the removal of 99.05% of MG within 30 min from aqueous solution. It has been observed that point to zero charges (pzc) of graft copolymer played a significant role in adsorption efficacy. The adsorption kinetics and isotherm followed pseudo-second-order and Langmuir isotherm models, respectively. Thermodynamic parameters suggested that the process of dye uptake was spontaneous. Finally, the desorption study showed excellent regeneration efficiency of adsorbent. The Magnetic β-cyclodextrin- graphene oxide (Fe_3O_4- CD/GO) nanocomposites [115] have been utilized for the adsorptive removal of MG from aqueous solutions. The factors such as adsorption time, adsorption temperature, pH of the solution, adsorption kinetics and isotherms were investigated. The results indicated that the Fe_3O_4⁻ CD/GO nanocomposites had good adsorption ability with maximum adsorption capacity of 990.10 mg/ g at pH 7. The adsorption capacity reduced to 80% after five cycles. The adsorption process with MG was found fitted pseudo-second-order kinetics equations and the Langmuir adsorption model. Thermodynamic parameters were also calculated which indicated that the adsorption process was spontaneous and endothermic in nature.

3.   Conclusions and future perspectives

This chapter describes the use of various low-cost materials as adsorbents including agriculture solid waste. A literature survey revealed that amongst all the adsorbents, agricultural solid wastes and activated carbon, have been extensively used for the removal of MG from aqueous solutions as well as from industrial effluents. The Langmuir and Freundlich adsorption isotherm models have been frequently applied to evaluate the adsorption capacity of various adsorbents for MG. Thermodynamic studies for the adsorption of MG showed that most of the processes were spontaneous in nature. Since most of the reported studies had been performed using batch and small-scale column adsorption tests, further research is required for the development of more effective adsorbents, modeling of adsorption mechanism, regeneration of spent adsorbents and treatment of real industrial wastewater. More and more nanomaterials are expected to be used as effective adsorbents for the removal of organic dyes in future. According to the available results on the use of different materials as adsorbents during 2005-2015 for the removal of MG from aqueous media (Fig. 1), following trends is evident-: Agricultural solid waste > activated carbon > natural materials >biosorbents> industrial solid waste.

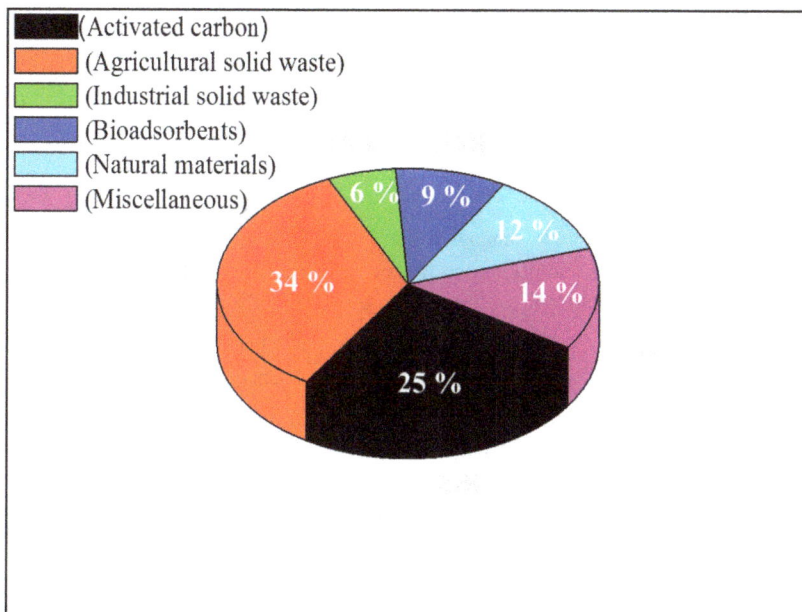

Figure 1: Preferential trend of using different adsorbents during 2005-2015 for the removal of malachite green dye from aqueous media.

References

[1]   A. Mittal, J. Mittal, V.K. Gupta, Adsorptive removal of hazardous anionic dye "Congo Red" from wastewater using waste materials and recovery by desorption, J. Colloid Interface Sci. 340 (2009) 16–26. https://doi.org/10.1016/j.jcis.2009.08.019

[2]   V.K. Gupta, R. Jain, A. Nayak, Removal of the hazardous dye-tartrazine by photodegradation on titanium dioxide surface, Mater. Sci. Eng. C 31 (2011) 1062–1067. https://doi.org/10.1016/j.msec.2011.03.006

[3]   S. Srivastava, R. Sinha, D. Roy, Review: Toxicological effects of malachite green, Aquat. Toxicol. 66 (2004) 319-329. https://doi.org/10.1016/j.aquatox.2003.09.008

[4]   V.K.Garg, R. Kumar, R. Gupta, Removal of malachite green dye from aqueous solution by adsorption using agro-industry waste: a case study of Prosopis

cineraria, Dyes and Pigments 62 (2004) 1-10.
https://doi.org/10.1016/j.dyepig.2003.10.016

[5]   S.J. Srivastava, N.D. Singh, A.K. Srivastava, R. Sinha, Acute toxicity of malachite green and its effects on certain blood parameters of a catfish, Heteropneustesfossilis, Aquat. Toxicol. 31 (1995) 241-247. https://doi.org/10.1016/0166-445X(94)00061-T

[6]   M. Wawizkiewicz, Anion exchange resins as effective sorbents for acidic dye removal from aqueous solutions and wastewaters, Solvent Extraction and Ion exchange 30 (2012) 507–523. https://doi.org/10.1080/07366299.2011.639253

[7]   M. Muthukumar, M.T. Karuppiah, G.B. Raju. Electrochemical removal of CI acid orange 10 from aqueous solutions, Sep. and Purif. Technol. 55 (2007) 198–205. https://doi.org/10.1016/j.seppur.2006.11.014

[8]   X. Zhu, Y. Zheng, Z. Chen, Q. Chen, B. Gao, S. Yu, Removal of reactive dye from textile effluent through submerged filtration using hollow fiber composite nanofiltration membrane, Desal. Water Treat. 51 (2013) 6101–6109. https://doi.org/10.1080/19443994.2013.770225

[9]   S. Wijannaronga, S. Aroonsrimorakota, P. Thavipokea, A. Kumsopaa, S. Sangjanb, Removal of reactive dyes from textile dyeing industrial effluent by ozonation process, APCBEE Proced. 5 (2013) 279 –282. https://doi.org/10.1016/j.apcbee.2013.05.048

[10]   V.P. Kasperchik, A.L. Yaskevich, A.V. Bil'dyukevich, Wastewater treatment for removal of dyes by coagulation and membrane processes, Petr. Chem. 52 (2012) 545–556. https://doi.org/10.1134/S0965544112070079

[11]   M. El Haddad, A.B. Regti, M.R. Laamari, R. Mamouni, N. Saffaj, Use of Fenton reagent as advanced oxidative process for removing textile dyes from aqueous solutions, J. Mater. Environ. Sci. 5 (2014) 667-674.

[12]   K.V. Kumar, S. Sivanesan, V. Ramamurthi, Adsorption of malachite green onto Pithophora sp., a fresh water algae: equilibrium and kinetic modelling, Process Biochem. 40 (2005) 2865–2872. https://doi.org/10.1016/j.procbio.2005.01.007

[13]   P.T. Godbole, A.D. Sawant, Removal of malachite green from aqueous solutions using immobilised Saccharomyces cerevisiae, J. Sci. Ind. Res. 65 (2006) 440–443.

[14]   X.F. Sun, S.G .Wang, X.W. Liu, Biosorption of malachite green from queous solutions onto aerobic granules: Kinetic and equilibrium studies, Bioresour. Technol. 99 (2008) 3475–3483. https://doi.org/10.1016/j.biortech.2007.07.055

[15] W. Cheng, S.G .Wang, L. Lu, Removal of malachite green (MG) from aqueous solutions by native and heat-treated anaerobic granular sludge, Biochem. Eng. J. 39 (2008) 538–546. https://doi.org/10.1016/j.bej.2007.10.016

[16] X.S. Wang, Invasive freshwater macrophyte alligator weed: novel adsorbent for removal of malachite green from aqueous solution, Water Air Soil Pollut. 206 (2009) 215–223. https://doi.org/10.1007/s11270-009-0097-6

[17] R.R. Kannan, M. Rajasimman, N. Rajamohan, B. Sivaprakash, Brown marine algae Turbinariaconoides as biosorbent for malachite green removal: equilibrium and kinetic modelling, Front Sci. Eng. China 4 (2009) 116–122. https://doi.org/10.1007/s11783-010-0006-7

[18] W.T. Tsai, H.R. Chen, Removal of malachite green from aqueous solution using low-cost chlorella-based biomass, J. Hazard. Mater. 175 (2010) 844–849. https://doi.org/10.1016/j.jhazmat.2009.10.087

[19] A. Altinisik A, E. Gur, Y. Seki, A natural sorbent, Luffacylindrica for the removal of a model basic dye, J. Hazard. Mater. 179 (2010) 658–664. https://doi.org/10.1016/j.jhazmat.2010.03.053

[20] H. Tang, W. Zhou, L. Zhang, Adsorption isotherms and kinetics studies of malachite green on chitin hydrogels, J. Hazard. Mater. 209 (2012) 218–225. https://doi.org/10.1016/j.jhazmat.2012.01.010

[21] R. Rajeshkannan, M. Rajasimman, N. Rajamohan, Removal of malachite green from aqueous solution using hydrillaverticillata-optimization, equilibrium and kinetic studies, Int. J. Civ. Environ. Eng. 2 (2010) 222–229.

[22] A. Singh, S .M. Rani, N.R. Bishnoi, Malachite green dye decolorization on immobilized dead yeast cells employing sequential design of experiments, Ecol. Eng. 47 (2012) 291–296. https://doi.org/10.1016/j.ecoleng.2012.07.001

[23] A. Salima, B. Benaouda, B. Noureddine, L. Duclaux, Application of Ulvalactuca and Systoceirastricta algae-based activated carbons to hazardous cationic dyes removal from industrial effluents, Water Res. 47 (2013) 3375–3388. https://doi.org/10.1016/j.watres.2013.03.038

[24] A.K. Sarkar, A. Pal. S. Ghorai, Efficient removal of malachite green dye using biodegradable graft copolymer derived from amylopectin and poly (acrylic acid), Carbohydr. Polym. 111 (2014) 108–115. https://doi.org/10.1016/j.carbpol.2014.04.042

[25] R. Deokar, A. Sabale, Biosorption of methylene blue and malachite green from binary solution onto Ulvalactuca, Int. J. Curr. Microbiol. App. Sci. 3 (2014) 295-304.

[26] Y. Zhou, Y. Min, H. Qiao, Improved removal of malachite green from aqueous solution using chemically modified cellulose by anhydride, Int. J. Biol. Macromol. 74 (2015) 271–277. https://doi.org/10.1016/j.ijbiomac.2014.12.020

[27] J. Nath, L. Ray, Biosorption of malachite green from aqueous solution by dry cells of Bacillus cereus MMTCC, J. Environ. Chem. Eng. 3 (2015) 386-394. https://doi.org/10.1016/j.jece.2014.12.022

[28] I.D. Mall, V.C. Srivastava, N.K. Agarwal, I.M. Mishra, Adsorptive removal of malachite green dye from aqueous solution by bagasse fly ash and activated carbon-kinetic study and equilibrium isotherm analyses, Colloids Surf. Physico-Chem. Eng. Asp. 264 (2005) 17–28. https://doi.org/10.1016/j.colsurfa.2005.03.027

[29] Y. Xing, G. Wang, Poly(methacrylic acid)-modified sugarcane bagasse for enhanced adsorption of cationic dye, Environ. Technol. 30 (2009) 611–619. https://doi.org/10.1080/09593330902838098

[30] Y. Xing, D. Deng, Enhanced adsorption of malachite green by EDTAD-modified sugarcane bagasse, Sep. Sci. Technol. 44 (2009) 2117–2131. https://doi.org/10.1080/01496390902775588

[31] S. Chowdhury, P. Saha, Adsorption thermodynamics and kinetics of malachite green onto Ca(OH)$_2$-treated fly ash, J. Environ. Eng. 137 (2011) 388–397. https://doi.org/10.1061/(ASCE)EE.1943-7870.0000334

[32] A. Witek-Krowiak, R.G. Szafran, S. Modelski, A. Dawiec, Removal of cationic dyes from aqueous solutions using microspherical particles of fly ash, Water Environ. Res. Res. Publ. Water Environ. Fed. 84 (2012) 162–169. https://doi.org/10.2175/106143011X13233670703657

[33] L. Zhang, H. Zhang, W. Guo, Y. Tian, Removal of malachite green and crystal violet cationic dyes from aqueous solution using activated sintering process red mud, Appl. Clay Sci. 93–94 (2014) 85–93. https://doi.org/10.1016/j.clay.2014.03.004

[34] G. Gehlot, S. Verma, S. Sharma, N. Mehta, Adsorption isotherm studies in the removal of malachite green dye from aqueous solution by using coal fly ash, Inter. J. Chem. Studies 3 (2015) 42-44.

[35]   I.D. Mall, V.C. Srivastava, N.K. Agarwal, I.M. Mischra, Adsorptive removal of malachite green dye from aqueous solution by bagasse fly ash and activated carbon-kinetic study and equilibrium isotherm analyses, Colloids and Surfaces A 264 (2005) 17-28. https://doi.org/10.1016/j.colsurfa.2005.03.027

[36]   C.A. Basar, Applicability of the various adsorption models of three dyes adsorption onto activated carbon prepared waste apricot, J. of Hazard. Mater. 135 (2006) 232-241. https://doi.org/10.1016/j.jhazmat.2005.11.055

[37]   R. Malik, D.S Ramteke, S.R Wate, Adsorption of malachite green on groundnut shell waste based powdered activated carbon, Waste management 27 (2007) 1129-1138. https://doi.org/10.1016/j.wasman.2006.06.009

[38]   B.H. Hameed, M.I. El Khaiary, Equilibrium kinetics and mechanism of malachite green adsorption on activated carbon prepared from bamboo by K_2CO_3 activation and subsequent gasification with CO_2, J. Hazard. Mater. 157 (2008) 344-351. https://doi.org/10.1016/j.jhazmat.2007.12.105

[39]   J. Zhang, Y. Li, C. Zhang, Y. Jing, Adsorption of malachite green from aqueous solution onto carbon prepared from Arundodonax root, J. Hazard. Mater. 50 (2008) 774-782. https://doi.org/10.1016/j.jhazmat.2007.05.036

[40]   Y.C. Sharma, B. Singh, Uma, Fast removal of malachite green by adsorption on rice husk activated carbon, The Open Environ. Pollution & Toxic. J. 1 (2009) 74-78.

[41]   M.H. Baek, C.O. Ijagbemi, S. Jin O, D.S. Kim, Removal of malachite green from aqueous solution using degreased coffee bean, J. Hazard. Mater. 176 (2010) 820-828. https://doi.org/10.1016/j.jhazmat.2009.11.110

[42]   T. Santhi, S. Manonmani, T. Smitha, Kinetics and isotherm studies on cationic dyes adsorption onto Annonasquamosa seed activated carbon, Int. J. Eng. Sci. Technol. 2 (2010) 287-295.

[43]   S. Nethaji, A. Sivasamy, G. Thennarasu, S. Saravanan, Adsorption of malachite green dye onto activated carbon from Borassusaethiopum flower biomass, J. Hazard. Mater. 181 (2010) 271 -280. https://doi.org/10.1016/j.jhazmat.2010.05.008

[44]   L. Wang, J. Zhang, R. Zhao, C. Li, Y. Li, C. Zhang, Adsorption of basic dyes on activated carbon prepared from Polygonumorientale Linn: equilibrium, kinetics and thermodynamic studies, Desal. 254 (2010) 68-74. https://doi.org/10.1016/j.desal.2009.12.012

[45] T. Santhi, S. Manonmani, T. Smitha, Removal of malachite green from aqueous solution by activated carbon prepared from the Epicarpof Ricinuscommunis by adsorption, J. Hazard. Mater. 179 (2010) 178-186. https://doi.org/10.1016/j.jhazmat.2010.02.076

[46] B.R. Venkatraman, K. Hema, V. Nandhakumar, S. Arivoli, Adsorption thermodynamics of malachite green dye onto acid activated low lost carbon, J. Chem. Pharm. Res. 3 (2011) 637-649.

[47] Y.C. Sharma, Adsorptive characteristics of a low cost activated carbon for the reclamation of coloured effluents containing malachite green, J. Chem. Eng. Data 56 (2011) 478–484. https://doi.org/10.1021/je1008922

[48] M.A. Ahmad, R. Alrozi, Removal of malachite green dye from aqueous solution using Rambutanpeel-based activated carbon: Equilibrium, kinetic and thermodynamic studies, Chem. Eng. J. 171 (2011) 510–516. https://doi.org/10.1016/j.cej.2011.04.018

[49] S.P.S Syed, Study of the removal of malachite green from aqueous solution by using solid agricultural waste, Res. J. Chem. Sci. 1 (2011) 88-104.

[50] C. Theivarasu, S.Mylsamy, Removal of malachite green from aqueous solution by activated carbon developed from cocoa (Theobroma Cacao) shell - A kinetic and equilibrium studies. E. J. of Chem. 8 (2011) S363-S371. https://doi.org/10.1155/2011/714808

[51] M.N. Idrisa, Z.A. Ahmada, M.A. Ahmad, Adsorption equilibrium of malachite green dye onto rubber seed coat based activate carbon, Inter. J. Basic App. Sci. 11 (2011) 38-43.

[52] O.S. Bello, M.A. Ahmad, Coconut (Cocosnucifera) shell based activated carbon for the removal of malachite green dye from aqueous solutions, Sep. Sci. Technol. 47 (2012) 903–912. https://doi.org/10.1080/01496395.2011.630335

[53] O.S. Bello, M.A. Ahmad, N. Ahmad, Adsorptive features of banana (Musa paradisiaca) stalk-based activated carbon for malachite green dye removal, Chem. Ecol. 28 (2012) 153–167. https://doi.org/10.1080/02757540.2011.628318

[54] O.S. Bello, Adsorptive removal of malachite green with activated carbon prepared from oil palm fruit fibre by KOH activation and CO_2 gasification, S. Afr. J. Chem. 66 (2013) 32–41.

[55]   A. Emine, A. Aylin, S. Yoldas, Using of activated carbon produced from spent tea leaves for the removal of malachite green from aqueous solution, Ecol. Eng. 52 (2013) 19–27. https://doi.org/10.1016/j.ecoleng.2012.12.032

[56]   S.Z. Mohammadi, M.A. Karimi, S.N. Yazdy, T. Shamspu, H. Hamidian, Removal of Pb(II) ions and malachite green dye from wastewater by activated carbon produced from lemon peel, Quim. Nova 37 (2014) 804-809. https://doi.org/10.5935/0100-4042.20140129

[57]   M.A. Ahmad, N. Ahmad, O.S. Bello, Adsorptive removal of malachite green dye using durian seed-based activated carbon, Water Air Soil Pollut. 225 (2014) 2057-2075. https://doi.org/10.1007/s11270-014-2057-z

[58]   S. Takute, S. Singh, M.K.N. Yenkie, Removal of malachite green from aqueous solution by activated carbon prepared almond shell, Int. J. Chem. Sci. 12 (2014) 663-671.

[59]   O.S. Bello, M.A. Ahmad, B. Semire, Scavenging malachite green dye from aqueous solutions using pomelo (Citrus grandis) peels: kinetic, equilibrium and thermodynamic studies, Desal. Water Treat. 56 (2015) 521-535. https://doi.org/10.1080/19443994.2014.940387

[60]   S.S Tahir, N. Rauf, Removal of a cationic dye from aqueous solutions by adsorption onto bentonite clay, Chemosphere 63 (2006) 1842–1848. https://doi.org/10.1016/j.chemosphere.2005.10.033

[61]   S. Wang, E. Ariyanto, Competitive adsorption of malachite green and Pb ions on natural zeolite, J. Colloid Interface Sci. 314 (2007) 25–31. https://doi.org/10.1016/j.jcis.2007.05.032

[62]   E. Bulut, M.O. zacar, I.A.S. Engil, Equilibrium and kinetic data and process design for adsorption of Congo Red onto bentonite, J. Hazard Mater. 154 (2008) 613–622. https://doi.org/10.1016/j.jhazmat.2007.10.071

[63]   V.Govindasamy, R. Sahadevan, S. Subramanian, D.K. Mahendradas, Removal of malachite green from aqueous solutions by perlite, Int. J. Chem. React. Eng. 7 (2009) 1-22. https://doi.org/10.2202/1542-6580.1889

[64]   H. Tahir, U. Hammed, M. Sultan, Q. Jahanzeb, Batch adsorption technique for the removal of malachite green and fast green dyes by using montmorillonite clay as adsorbent, Afr. J. Biotechnol. 9 (2010) 8206–8214. https://doi.org/10.5897/AJB10.911

[65] P.Saha, S. Chowdhury, S. Gupta, I. Kumar, Insight into adsorption equilibrium, kinetics and thermodynamics of malachite green onto clayey soil of Indian origin, Chem. Eng. J. 165 (2010) 874–882. https://doi.org/10.1016/j.cej.2010.10.048

[66] G. Kiani, M. Dostali, A. Rostami, A.R. Khataee, Adsorption studies on the removal of malachite green from aqueous solutions onto halloysite nanotubes, Appl. Clay Sci. 54 (2011) 34–39. https://doi.org/10.1016/j.clay.2011.07.008

[67] Y.C. Lee, E.J. Kim, J.W. Yang, H.J. Shin Removal of malachite green by adsorption and precipitation using aminopropyl functionalized magnesium phyllosilicate, J. Hazard. Mater. 192 (2011) 62–70. https://doi.org/10.1016/j.jhazmat.2011.04.094

[68] A.C. Suwandi, N. Indraswati, S. Ismadji, Adsorption of N-methylated diaminotriphenyl methane dye (malachite green) on natural rarasaponin modified kaolin, Des. Water Treat. 41 (2012) 342–355. https://doi.org/10.1080/19443994.2012.664738

[69] Y.C. Lee, J.Y. Kim, H.J. Shin, Removal of malachite green (MG) from aqueous solutions by adsorption, precipitation, and alkaline fading using talc, Sep. Sci. Technol. 48 (2013) 1093–1101. https://doi.org/10.1080/01496395.2012.723100

[70] S.A. Cardenas, S.L. Cortez, M.C. Mazon J.C.M. Gutierrez, Study of malachite green adsorption by organically modified clay using a batch method, Appl. Surf. Sci. 280 (2013) 74–78. https://doi.org/10.1016/j.apsusc.2013.04.097

[71] L. Zhang, H. Zhang, W. Guo, Y. Tian, Removal of malachite green and crystal violet cationic dyes from aqueous solution using activated sintering process red mud, Appl. Clay Sci. 93 (2014) 85–93. https://doi.org/10.1016/j.clay.2014.03.004

[72] N. Jain, M.K. Dwivedi, R. Agarwal, P. Sharma, Removal of malachite green from aqueous solution by zeolite iron oxide magnetic nanocomposite, J. Envi. Sci. Toxicol. Food Technol. 9 (2015) 42-50.

[73] A. Mittal, L. Krishnan, V.K. Gupta, Removal and recovery of malachite green from wastewater using an agricultural waste material, de-oiled soya, Sep. Purif. Technol. 43 (2005) 125–133. https://doi.org/10.1016/j.seppur.2004.10.010

[74] R. Gong, Y. Jin, F. Chen, Enhanced malachite green removal from aqueous solution by citric acid modified rice straw, J. Hazard. Mater. 137 (2006) 865–870. https://doi.org/10.1016/j.jhazmat.2006.03.010

[75] K. Porkodi, K.V. Kumar Equilibrium, kinetics and mechanism modeling and simulation of basic and acid dyes sorption onto jute fiber carbon: Eosin yellow,

malachite green and crystal violet single component systems, J. Hazard. Mater. 143 (2007) 311–327. https://doi.org/10.1016/j.jhazmat.2006.09.029

[76]   K.V. Kumar, S. Sivanesan, Isotherms for malachite green onto rubber wood (Heveabrasiliensis) sawdust: comparison of linear and non-linear methods, Dyes Pigm. 72 (2007) 124-129. https://doi.org/10.1016/j.dyepig.2005.07.021

[77]   K.V. Kumar, Optimum sorption isotherm by linear and nonlinear methods for malachite green onto lemon peel, Dyes Pigm. 74 (2007) 595–597. https://doi.org/10.1016/j.dyepig.2006.03.026

[78]   K.V. Kumar, K. Porkodi, Batch adsorber design for different solution volume/adsorbent mass ratios using the experimental equilibrium data with fixed solution volume/adsorbent mass ratio of malachite green onto orange peel, Dyes Pigm. 74 (2007) 590–594. https://doi.org/10.1016/j.dyepig.2006.03.024

[79]   B.H. Hameed, M.I. El. Khaiary, Malachite green adsorption by rattan sawdust: isotherm, kinetic and mechanism modelling, J. Hazard. Mater. 159 (2008) 574–579. https://doi.org/10.1016/j.jhazmat.2008.02.054

[80]   B.H. Hameed, M.I. El-Khaiary, Batch removal of malachite green from aqueous solutions by adsorption on oil palm trunk fibre: equilibrium isotherms and kinetic studies, J. Hazard. Mater. 154 (2008) 237–244. https://doi.org/10.1016/j.jhazmat.2007.10.017

[81]   Z.O. Bekc, C. Zveri, Y. Seki, K. Yurdakoc, Sorption of malachite green on chitosan bead, J. Hazard. Mater. 154 (2008) 254–261. https://doi.org/10.1016/j.jhazmat.2007.10.021

[82]   O. Hamdaoui, M. Chiha, E. Naffrechoux, Ultrasound-assisted removal of malachite green from aqueous solution by dead pine needles, Ultrason Sonochem. 15 (2008) 799–807. https://doi.org/10.1016/j.ultsonch.2008.01.003

[83]   S. Boutemedjet, O. Hamdaoui, Sorption of malachite green by eucalyptus bark as a non-conventional low-cost biosorbent, Desalin. Water Treat. 8 (2009) 201–210. https://doi.org/10.5004/dwt.2009.684

[84]   R. Ahmad, R. Kumar, Adsorption studies of hazardous malachite green onto treated ginger waste, J. Environ. Manag. 91 (2010) 1032–1038. https://doi.org/10.1016/j.jenvman.2009.12.016

[85]   D. Ozdes, A. Gundogdu, C. Duran, H.B. Senturk, Evaluation of adsorption characteristics of malachite green onto Almond Shell (Prunusdulcis), Sep. Sci. Technol. 45 (2010) 2076–2085. https://doi.org/10.1080/01496395.2010.504479

[86]   S. Chowdhury, P. Saha, Sea shell powder as a new adsorbent to remove Basic Green 4 (malachite green) from aqueous solutions: Equilibrium, kinetic and thermodynamic studies, J. Chem. Eng. 164 (2010) 168–177. https://doi.org/10.1016/j.cej.2010.08.050

[87]   A.S. Franca, L.S. Oliveira, S.A. Saldanha, Malachite green adsorption by mango (Mangiferaindica L.) seed husks: Kinetic, equilibrium and thermodynamic studies. Desalin. Water Treat. 19 (2010) 241–248. https://doi.org/10.5004/dwt.2010.1105

[88]   S. Chowdhury, R. Mishra, P. Saha, P. Khushwaha, Adsorption thermodynamics, kinetics and isosteric heat of adsorption of malachite green onto chemically modified rice husk, Desalina. 265 (2011) 159-168. https://doi.org/10.1016/j.desal.2010.07.047

[89]   H. Zhang, Y. Tang, X. Liu, Improved adsorptive capacity of pine wood decayed by fungi Poriacocos for removal of malachite green from aqueous solutions. Desalination 274 (2011) 97–104. https://doi.org/10.1016/j.desal.2011.01.077

[90]   S. Chowdhury, P.D. Saha, Mechanistic, Kinetic and thermodynamic evaluation of adsorption of hazardous malachite green onto conch shell powder, Sep. Sci. Technol. 46 (2011) 1966–1976. https://doi.org/10.1080/01496395.2011.584930

[91]   V. Vijayakumaran, S. Arivoli, Equilibrium and kinetic modeling on the removalof malachite green from aqueous solution using odinawodier bark carbon, J. Mater. Environ. Sci. 3 (2012) 525-536.

[92]   A. Jasinska, P. Bernat, K. Paraszkiewicz, Malachite green removal from aqueous solution using the system rapeseed press cake and fungus Myrotheciumroridum, Desalin. Water Treat. 51 (2013) 7663–7671. https://doi.org/10.1080/19443994.2013.779939

[93]   A. Witek-Krowiak, Biosorption of malachite green from aqueous solutions by pine sawdust: equilibrium, kinetics and the effect of process parameters, Desal. Water Treat. 51 (2013) 3284–3294. https://doi.org/10.1080/19443994.2012.749053

[94]   Uma, Y.C. Sharma, Removal of malachite green from aqueous solutions by adsorption on to timber waste, Inter. J. Envir. Eng. Manag. 4 (2013) 631-638.

[95]   H. Wang, X. Yuan, G. Zeng Removal of malachite green dye from wastewater by different organic acid-modified natural adsorbent: kinetics, equilibriums, mechanisms, practical application, and disposal of dye-loaded adsorbent, Environ. Sci. Pollut. Res. Int. 21 (2014) 11552–11564. https://doi.org/10.1007/s11356-014-3025-2

[96] T.A. Khan, R. Rahman, I. Ali Removal of malachite green from aqueous solution using waste pea shells as low-cost adsorbent—adsorption isotherms and dynamics. Toxicol. Environ. Chem. 96 (2014) 569–578. https://doi.org/10.1080/02772248.2014.969268

[97] M.K. Dahri, M.R.R. Kooh, L.B.L.Lim , Water remediation using low cost adsorbent walnut shell for removal of malachite green: equilibrium, kinetics, thermodynamic and regeneration studies, J. Environ. Chem. Eng. 2 (2014) 1434–1444. https://doi.org/10.1016/j.jece.2014.07.008

[98] A.K. Kushwaha, N. Gupta, M.C. Chattopadhyaya Removal of cationic methylene blue and malachite green dyes from aqueous solution by waste materials of Daucuscarota. J. Saudi Chem. Soc. 18 (2014) 200–207. https://doi.org/10.1016/j.jscs.2011.06.011

[99] K. Vijayaraghavan, Y. Premkumar, J. Jegan, Malachite green and crystal violet biosorption onto coco-peat: characterization and removal studies, Desal. Water Treat. 57 (2015) 6423–6431. https://doi.org/10.1080/19443994.2015.1011709

[100] H.H. Hammud, A. Shmait, N. Hourani, Removal of malachite green from water using hydrothermally carbonized pine needles, RSC Adv. 5 (2015) 7909–7920. https://doi.org/10.1039/C4RA15505J

[101] Y. Song, S. Ding, S. Chen, H. Xue, Y. Mei, J. Ren, Removal of malachite green in aqueous solution by adsorption on sawdust, Korean J. Chem. Eng. 32 (2015) 2443–2448. https://doi.org/10.1007/s11814-015-0103-1

[102] A. Mittal Adsorption kinetics of removal of a toxic dye, malachite green, from wastewater by using hen feathers, J. Hazard. Mater. 133 (2006) 196–202. https://doi.org/10.1016/j.jhazmat.2005.10.017

[103] X. Rong, F. Qiu, J. Q. Jiao, Y. H. Zhao, D. Yang. Removal of malachite green from the contaminated water using a water-soluble melamine/maleic anhydride sorbent, J. Ind. Eng. Chem. 20 (2007) 3808- 3814. https://doi.org/10.1016/j.jiec.2013.12.083

[104] G. Crini, N.H. Peindy, F. Gimbert, C. Robert, Removal of C.I. Basic Green 4 (malachite green) from aqueous solutions by adsorption using cyclodextrin-based adsorbent: kinetic and equilibrium studies, Sep. Purif. Technol. 53 (2007) 97–110. https://doi.org/10.1016/j.seppur.2006.06.018

[105] A. Das, A. Pal, S. Saha, S.K. Maji, Behaviour of fixed-bed column for the adsorption of malachite green on surfactant modified alumina, J. Environ. Sci.

Health Part A Tox. Hazard. Subst. Environ. Eng. 44 (2009) 265–272.
https://doi.org/10.1080/10934520802597929

[106] L. Ai, H. Huang, Z. Chen, Activated carbon/$CoFe_2O_4$ composites: facile synthesis, magnetic performance and their potential application for the removal of malachite green from water, Chem. Eng. J. 156 (2010) 243–249.
https://doi.org/10.1016/j.cej.2009.08.028

[107] S. Chowdhury, P.D. Saha, Fixed-bed adsorption of malachite green onto binary solid mixture of adsorbents: seashells and eggshells, Toxicol. Environ. Chem. 94 (2012) 1272–1282. https://doi.org/10.1080/02772248.2012.703205

[108] S. Chowdhury, P.D. Saha, U. Ghosh, Fish (Labeorohita) scales as potential low-cost biosorbent for removal of malachite green from aqueous solutions, Bioremediat. J. 16 (2012) 235–242.
https://doi.org/10.1080/10889868.2012.731444

[109] R. Xu, M. Jia, Y. Zhang, F. Li, Sorption of malachite green on vinyl-modified mesoporous poly(acrylic acid)/SiO_2 composite nanofiber membranes, Microporous Mesoporous Mater. 149 (2012) 111–118.
https://doi.org/10.1016/j.micromeso.2011.08.024

[110] M. Shirmardi, A.H. Mahvi, B. Hashemzadeh, The adsorption of malachite green (MG) as a cationic dye onto functionalized multi walled carbon nanotubes, Korean J. Chem. Eng. 30 (2013) 1603–1608. https://doi.org/10.1007/s11814-013-0080-1

[111] A. Mohammadi, H. Daemi, M. Barikani, Fast removal of malachite green dye using novel superparamagnetic sodium alginate-coated Fe_3O_4 nanoparticles, Int. J. Biol. Macromol. 69 (2014) 447–455.
https://doi.org/10.1016/j.ijbiomac.2014.05.042

[112] A.K. Sarkar, A. Pal, S. Ghorai, N.R. Mandre, S. Pal, Efficioent removal of malachite green dye using biodegradable graft copolymer derived from amylopectin and poly(acrylic acid), Carbohydr. Polym. 111 (2014) 108-115.
https://doi.org/10.1016/j.carbpol.2014.04.042

[113] M. Ghaedi, N. Mosallanejad, Study of competitive adsorption of malachite green and sunset yellow dyes on cadmium hydroxide nanowires loaded on activated carbon, J. Ind. Eng. Chem. 20 (2014) 1085–1096.
https://doi.org/10.1016/j.jiec.2013.06.046

[114] M. Ghaedi, E. Shojaeipour, A.M. Ghaedi, R. Sahraei, Isotherm and kinetics study of malachite green adsorption onto copper nanowires loaded on activated carbon:

artificial neural network modeling and genetic algorithm optimization, Spectrochim. Acta A, Mol. Biomol. Spectrosc. 142 (2015) 135–149. https://doi.org/10.1016/j.saa.2015.01.086

[115] D. Wang, L. Liu, X. Jiang, Adsorption and removal of malachite green from aqueous solution using magnetic β-cyclodextrin-graphene oxide nanocomposites as adsorbents. Colloids Surf. Physicochem. Eng. Asp. 466 (2015) 166–173. https://doi.org/10.1016/j.colsurfa.2014.11.021

[116] F.N. Azad, M. Ghaedi, K. Dashtian, S. Hajati, A. Goudarzi, M. Jamshidi, Enhanced simultaneous removal of malachite green and safraninby ZnO nanorod-loaded activated carbon: modeling, optimization and adsorption isotherms, New J. Chem. 39 (2015) 7998-8005. https://doi.org/10.1039/C5NJ01281C

Keyword Index

About the Editors

Dr. Inamuddin is currently working as Assistant Professor in the Chemistry Department, Faculty of Science, King Abdulaziz University, Jeddah, Saudi Arabia. He is a permanent faculty member (Assistant Professor) at the Department of Applied Chemistry, Aligarh Muslim University, Aligarh, India. He obtained Master of Science degree in Organic Chemistry from Chaudhary Charan Singh (CCS) University, Meerut, India, in 2002. He received his Master of Philosophy and Doctor of Philosophy degrees in Applied Chemistry from Aligarh Muslim University (AMU) in 2004 and 2007, respectively. He has extensive research experience in multidisciplinary fields of Analytical Chemistry, Materials Chemistry, and Electrochemistry and, more specifically, Renewable Energy and Environment. He has worked on different research projects as project fellow and senior research fellow funded by University Grants Commission (UGC), Government of India, and Council of Scientific and Industrial Research (CSIR), Government of India. He has received Fast Track Young Scientist Award from the Department of Science and Technology, India, to work in the area of bending actuators and artificial muscles. He has completed four major research projects sanctioned by University Grant Commission, Department of Science and Technology, Council of Scientific and Industrial Research, and Council of Science and Technology, India. He has published 100 research articles in international journals of repute and seventeen book chapters in knowledge-based book editions published by renowned international publishers. He has published twenty two edited books with Springer, United Kingdom, Elsevier, Nova Science Publishers, Inc. U.S.A., CRC Press Taylor & Francis Asia Pacific, Trans Tech Publications Ltd., Switzerland and Materials Research Forum LLC, U.S.A. He is the member of various editorial boards and also serves as associate editor for journals such as Environmental Chemistry Letter, Applied Water Science, Springer-Nature and editorial board member for Scientific Reports-Nature and editor, Eurasian Journal of Analytical Chemistry. He has attended as well as chaired sessions in various international and national conferences. He has worked as a Postdoctoral Fellow, leading a research team at the Creative Research Initiative Center for Bio-Artificial Muscle, Hanyang University, South Korea, in the field of renewable energy, especially biofuel cells. He has also worked as a Postdoctoral Fellow at the Center of Research Excellence in Renewable Energy, King Fahd University of Petroleum and Minerals, Saudi Arabia, in the field of polymer electrolyte membrane fuel cells and computational fluid dynamics of polymer electrolyte membrane fuel cells. He is a life member of the Journal of the Indian Chemical Society. His research interest includes ion exchange materials, a sensor for heavy metal ions, biofuel cells, supercapacitors and bending actuators.

Mohd Imran Ahamed is a Research Scholar at Department of Chemistry, Aligarh Muslim University, Aligarh, India. He is working towards his Ph.D. thesis entitled Synthesis and characterization of inorganic-organic composite heavy metals selective cation-exchangers and their analytical applications. He has published several research and review articles in the journals and book editions of international repute. He has completed his Bachelor of Science (Chemistry) from Aligarh Muslim University, Aligarh, India, and Masters in Chemistry (Organic Chemistry) from Baba Bheem Rao Ambedkar University, Agra, India. His research work includes ion exchange chromatography, wastewater treatment, and analysis, bending actuator and electrospinning.

Dr. Shadi W. Hasan holds BSc, MASc, and PhD in Chemical, Mechanical and Environmental Engineering, respectively. Dr. Hasan's field of expertise includes water purification, wastewater treatment and reuse, water desalination, and nanotechnology. Dr. Hasan has worked on several projects in water desalination using forward osmosis, reverse osmosis, and membrane distillation technologies in Abu Dhabi; particularly on process design and scale up, and modeling and simulation. He was also the principal investigator, in collaboration with MIT, on the design of hybrid wastewater treatment system. Dr. Hasan has also established research collaboration with well-known international universities and institutes such as MIT, Harvard and Yale (USA), Manchester, Oxford, Brunel London, and Bradford (UK), Concordia (Canada), UTS (Australia), Korea (Korea) and Salerno (Italy), among others. Throughout his academic career, Dr. Hasan has been recognized as an excellent designer (first to design the novel submerged membrane electro-bioreactor for municipal wastewater treatment at a pilot scale - Montreal, Canada) and has obtained a number of prestigious awards including the FQRNT. Dr. Hasan has authored and co-authored over 85 articles in peer reviewed journals and conference proceedings; and generated significant research interest abroad in the USA and overseas. Dr. Hasan's research has significant impacts and led to several funding opportunities both internally and externally. Dr. Hasan is an active reviewer for more than 30 scientific journals, and has supervised more than 25 MSc and PhD students, and research fellows. The "Water Purification Lab" established by Dr. Hasan at KU is well-equipped and has several analytical instruments for water quality analysis, and membrane fouling monitoring and characterization.

www.ingramcontent.com/pod-product-compliance
Lightning Source LLC
Chambersburg PA
CBHW071330210326
41597CB00015B/1403